SpringerWienNewYork

CISM COURSES AND LECTURES

The series presents lecture notes, monographs, edited works and proceedings in the field of Mechanics, Engineering, Computer Science and Applied Mathematics.
Purpose of the series is to make known in the international scientific and technical community results obtained in some of the activities organized by CISM, the International Centre for Mechanical Sciences.

INTERNATIONAL CENTRE FOR MECHANICAL SCIENCES

COURSES AND LECTURES - No. 453

MOVING INTERFACES IN CRYSTALLINE SOLIDS

EDITED BY

FRANZ DIETER FISCHER
MONTANUNIVERSITÄT LEOBEN, AUSTRIA
AND
ERICH SCHMID INSTITUTE FOR MATERIALS SCIENCE,
AUSTRIAN ACADEMY OF SCIENCES, LEOBEN, AUSTRIA

SpringerWienNewYork

The publication of this volume was co-sponsored and co-financed by the UNESCO Venice Office - Regional Bureau for Science in Europe (ROSTE) and its content corresponds to a CISM Advanced Course supported by the same UNESCO Regional Bureau.

This volume contains 126 illustrations

SPIN 11355274

In order to make this volume available as economically and as rapidly as possible the authors' typescripts have been reproduced in their original forms. This method unfortunately has its typographical limitations but it is hoped that they in no way distract the reader.

ISBN 3-211-23899-9 SpringerWienNewYork

PREFACE

Materials with a changing microstructure are a matter of fact in many fields of materials research as well as in the application of materials in diverse fields of technical practice. Such a change may consist in one or more phase transformations in the material, the growth or shrinking of grains, or the appearance of the material in a different geometrical configuration like twinning. Both physical and thermomechanical activities are necessary to drive or stop or, better, to control such a process. The goal of the course "Moving Discontinuities in Crystalline Solids", on which this booklet is based, was to bring together experts from materials physics and materials mechanics to explain the fundamental phenomena of moving interfaces accompanied by the change of the material on both sides of the interface. The application of both classical and modern concepts of materials physics was demonstrated. Furthermore, the findings of continuum mechanics and numerical methods were discussed ranging from Eshelby's derivation of the thermodynamical force on an interface to numerical concepts under development for multiparticle, multicomponent systems. The authors of the various chapters, however, have tried hard not to present simply diverse concepts but to bridge the gaps between the numerous physical and mechanical approaches to this wide field of knowledge. Such a task is not easy and rather new. For many years researchers from mechanics and physics have tended to take different roads which met only by chance. Within the last fifteen years both groups felt that they had to come nearer, and finally, the authors of this booklet have the impression that they have come together. The inherent interdisciplinarity of the subject has been one of the strongest motivations to perform such a course and to write the booklet at hand. Computational methods, like the Monte Carlo Method, Ab-initio Modelling and "Enriched" Finite Elements have contributed a lot to a better understanding of what is behind the change of the microstructure. One may recognize this in the ever increasing number of courses, seminars, conferences and corresponding papers on "Modelling and Simulation". The authors also feel that the common view on the controlling mechanisms for the microstructure will have an increasing impact on industrial application. More effective materials that are better adapted to their respective functions and require shorter development times are more or less a must in the world of today's technology.

Some comments are also necessary with respect to the layout of this booklet. Since a group of researchers got together from different fields, very often their symbols and notations differ. A full unification would take too much time although it is a demanding task for the future. Therefore, the authors decided to write each chapter in a self-contained and self-explaining way with a list of notations at the beginning of the chapter. So each chapter has a textbook character starting from the basics and stating carefully the assumptions and limitations of the application of the theoretical framework. The authors try to present examples that are easy to understand. The

authors have also performed a mutual reading of their chapters with the goal to bring their own contribution in line with the related chapters.
The authors hope that the booklet will provide a sufficient basis for the understanding of this interdisciplinary field of materials mechanics and materials physics and will further mutual understanding. Both researchers and industrial developers should profit from this rather unique presentation of the motion of interfaces in solids.

The authors express their thanks to the Director of CISM, Prof. M.G. Velarde, for supporting the course with the staff of CISM and for the strong encouragement they received to write this booklet. Finally, the authors are grateful to Prof. C. Tasso for accepting this booklet for publication and his help in the editing.

Franz Dieter Fischer

CONTENTS

Application of Configurational Mechanics to Elastic Solids with Defects and Cracks

R. Kienzler[1] and G. Herrmann[2]

[1]University of Bremen, Bremen, Germany
[2]Stanford University, Stanford, CA, USA

Abstract. Classical mechanics, i. e., in *physical* space is concerned with forces, stresses and strains and attempts to describe the motion and/or deformation of bodies with mass. In this context, the notions of tractions, trajectories, balance and conservation laws, stability of equilibrium etc. are well established. Mechanics in material space (or configurational mechanics) describes the behaviour of defects (e. g., voids, dislocations, cracks) as they move relatively to the material, in which they find themselves. Concerning this change of configuration, similar notions, as given above, are introduced in material space. After providing the elements of configurational mechanics the method is applied to elastic solids with defects and cracks. In particular, the hole-dislocation interaction problem is discussed and the use of path-independent integrals and local failure criteria in fracture mechanics are demonstrated.

List of Notations

a	abbreviation
a	crack length
a_i	material properties (not further specified)
A	area, mostly used as differential area element dA
A	cross-sectional area mostly in connection with compressional stiffness EA
A	action integral
$\boldsymbol{b}, b_i, b$	Burgers vector, its components and its magnitude
b_{ij}	components of Eshelby tensor, material momentum
B	material force in one-dimensional problem
B	body, domain of integration
c_{ijkl}	components of tensor of elasticity
C	compliance
d	thickness of bar
E	Young's modulus
E^*	modified Young's modulus
EA	compressional or axial stiffness
$\boldsymbol{E}_j$	Euler operator

f_i	characteristics (Neutral Action method)
$\boldsymbol{f}, f_i$	vector of body forces and its components
$f_{ij}^{I}, f_{ij}^{II}, f_{ij}^{III}$	dimensionless geometry functions correlated with crack-opening modes *I*, *II* and *III*
$\boldsymbol{F}, F_i, F$	(physical) force vector, its components and its magnitude
g	gravitational acceleration
$g()$	function
G	shear modulus
$\underset{n}{\boldsymbol{G}}, \underset{n}{G}_i$	material traction vector acting across a surface with normal vector $\boldsymbol{n}$ and its components
$\mathcal{G}$	energy release rate
$\overline{\mathcal{G}}$	Energy release rate of strength-of-material theories
h	height of bar
${}_{\sigma}I_1$	first invariant of stress tensor
J	material force (J integral)
$\boldsymbol{J}, J_i$	vector of material forces and its components (components of J integral)
K_I, K_{II}, K_{III}	stress-intensity factors correlated with crack-opening modes, *I*, *II* and *III*
ℓ	length of bar
L	Lagrangian
$\widetilde{\boldsymbol{L}}$	null Lagrangian
$\boldsymbol{L}, L_n, L$	vector of material momentums ($\boldsymbol{L}$ integral) its components and $L = L_3$ in plane problems
m	mass
M	material virial (M integral)
$\boldsymbol{n}, n_i$	unit outward normal vector and its components
N	axial force in one-dimensional problem
$0(\delta^n)$	terms of (vanishing) order n and higher
$\boldsymbol{P}, P_i, P$	i -component vector (current), its components and current in one-dimensional problem
$\boldsymbol{pr}^{(n)}$	n-th prolongation
Q_α	characteristics (Noether's formalism)
r	radial coordinate in polar coordinate system
r_o	radius of hole
s	arc length, mostly used as differential arc-length element ds
s	abbreviation
S	surface, domain of integration
S_t	surface of body B with prescribed tractions
S_u	surface of body B with prescribed displacements

$\underset{n}{\boldsymbol{t}}, \underset{n}{t}_j$	(physical) traction vector acting across a surface with normal vector $\boldsymbol{n}$ and its components
$\boldsymbol{u}, u_i, u$	displacement vector, its components and longitudinal displacement in one-dimensional problem
$u_{i,j}$	displacement gradient
V	volume, mostly used as differential volume element dV
V	potential of external forces (density)
$\boldsymbol{w}$	infinitesimal generator
W	strain-energy density
$\overline{W}$	strain energy per unit of length of bar
$\boldsymbol{x}, x_i, x$	vector of independent variables, its components and independent variable in one-dimensional problem
γ	dimensionless distance
Γ	(closed) line or path
$\delta()$	variation of ()
δ_{ij}	Kronecker tensor of unity
ε	infinitesimal group parameter
ε_{ij} , ε	components of (linearized) strain tensor, axial strain in one-dimensional problem
ε_{ijk}	permutation tensor, Levi-Cività tensor
κ	material constant
λ_k, λ	rigid (material) translation
ν	Poisson's ratio
ξ, ξ_i	position
ξ_i	transformation functions of independent variables
Π	total (potential) energy
Π^a, Π^i	external potential and internal energy
ρ	dimensionless distance
σ_{ij}	components of the stress tensor, physical momentum
φ	circumferential coordinate in a polar coordinate system
ϕ_a	transformation functions of dependent variables
ψ	angle of transformation
$\boldsymbol{\omega}, \omega_i, \omega$	vector of rotation, its components and $\omega = \omega_3$ in plane problems
$\partial() / \partial() x_i$	partial differential operator
$()_{,j}$	total derivative
$()^{/}$	total derivative in a one-dimensional problem

$()^*$	transformed quantity involving transformation of dependent and independent variables
$[()]$	jump of ()

1 Introduction

Every real material contains, at least on some scale, defects, such as vacancies, voids, dislocations, inclusions, cracks, etc. Due to external loading, to changes of temperature or to other circumstances these defects may move relatively to the ambient material. The so-called driving force, or material force, or configurational force that causes this movement results from the fact that the energy of the system (or body) under consideration is changed by the change of the configuration, e. g., energy is released.

Energy-release rates per unit of defect advance are strongly related to material conservation laws and path- or domain-independent integrals.

It is the aim of the present contribution to introduce the notion of material forces and tractions, material conservation and balance laws, trajectories and stability of defect configurations and so forth in an as simple as possible setting, i. e., in the framework of the linear theory of elasticity, and to elucidate the physical background by some illustrative examples.

Therefore, we start with the governing equations of the theory (Chapter 2) and discuss in Chapter 3 energy-release rates by means of a "thought experiment" suggested by Eshelby (1975 a) where the energy change of a system is calculated due to the movement of a defect. We further introduce various path-independent integrals involving the Eshelby tensor (some times called energy-momentum tensor of material momentum tensor).

The usual mechanics in physical space (Newtonian mechanics) and the not so common mechanics in material space (Eshelbian mechanics) are juxtaposed in Chapter 4 to emphasise their correspondence or duality.

In Chapter 5 the mathematical background is sketched concerning the derivation of conservation laws in a systematic manner by Noether's formalism and by the Neutral-Action method.

The path-independent integrals are applied first to hole-dislocation-interaction problems (Chapter 6) and then to plane problems of fracture mechanics (Chapter 7).

Since the Eshelby tensor is the most prominent ingredient of the integrals, its physical significance is studied in Chapter 8 and fracture criteria are proposed based on its local properties.

Finally, configurational mechanics is applied in Chapter 9 to the theory of strength-of-materials. Remarkably simple formulae are derived to calculate stress-intensity factors for bars with cracks.

2 Elements of Linear Elastostatics

In these lecture notes, we are merely concerned with the linear theory of elasticity. Therefore, it seams to be appropriate to recollect briefly the governing equations and to introduce the notation used. In Cartesian coordinates $(x_i, i = 1,2,3)$ we have, cf., e. g., Timoshenko and Goodier (1970)

$$\sigma_{ji,j} + f_i = 0\,, \tag{2.1}$$

$$\varepsilon_{kl} = \frac{1}{2}(u_{k,l} + u_{l,k}), \tag{2.2}$$

$$\sigma_{ij} = c_{ijkl}\varepsilon_{kl}\,. \tag{2.3}$$

The equations of equilibrium (2.1) connect the partial derivatives of the stress tensor σ_{ji} with the vector of the body forces f_i; by the kinematic equations (2.2), the components of the strain tensor ε_{kl} are determined from the components of the symmetric part of the displacement gradient $u_{k,l}$; Hooke's law (2.3) relates stresses to strains by the 4th-rank tensor c_{ijkl} of elastic constants. Commas followed by an index denote partial differentiation with respect to the indicated coordinate; the summation convention is implied for repeated indices; indices are of the range 1,2,3 if not indicated otherwise.

Due to the balance of angular momentum, the stress tensor is symmetric

$$\varepsilon_{kij}\sigma_{ij} = 0\,, \tag{2.4}$$

involving the completely skew-symmetric 3rd-rank permutation tensor ε_{kij} .The strain tensor is symmetric by definition.

The strain-energy density W per unit of volume is defined as

$$W = \frac{1}{2}c_{ijkl}\varepsilon_{ij}\varepsilon_{kl} \tag{2.5}$$

with the constitutive relations

$$\sigma_{ji} = \frac{\partial W}{\partial \varepsilon_{ij}} = \frac{\partial W}{\partial u_{i,j}}\,. \tag{2.6}$$

Thus, the strain energy my be interpreted as

$$W = W(x_k, u_{i,j}) . \tag{2.7}$$

W might depend on x_k explicitly, as indicated above, if the material constants are functions of x_k i. e., the components c_{ijkl} vary from position to position (material inhomogeneity).

Similarly to W (2.6), a potential of external forces V per unit of volume might be introduced, if the force field is conservative, as

$$f_i = -\frac{\partial V}{\partial u_i} \tag{2.8}$$

If the body foces result from dead loads, the potential of external forces is given by

$$V = f_i u_i \tag{2.9}$$

Also $V = V(x_k, u_i)$ might be an explicit function of x_k if the forces per unit of volume f_i are functions of x_k (physical nonhomogeneity).

With (2.7) and (2.9), the Lagrangian function is

$$L = -(W + V) = L(x_k, u_i, u_{i,j}) . \tag{2.10}$$

If the material is homogeneous, both physically and materially, L does not depend explicitely on x_k leading to

$$L = L(u_i, u_{i,j}) . \tag{2.11}$$

In the absence of body forces $(f_i = 0)$, L depends merely on the deformation gradient

$$L = L(u_{i,j}) . \tag{2.12}$$

The so-called action integral is defined as the integral of the Lagrangian over a body B with volume V

$$A = \int_B L(x_k, u_i, u_{i,j}) dV . \tag{2.13}$$

In the following, we will use also the internal energy Π^i

$$\Pi^i = \int_B W dV , \tag{2.14}$$

the external energy Π^a

$$\Pi^a = \int_B V dV - \int_{St} \overset{*}{\underset{S}{t_i}} u_i dA , \tag{2.15}$$

the total energy Π

$$\Pi = \Pi^i + \Pi^a , \tag{2.16}$$

and Clapeyron's theorem, (e. g., Fung (1968))

$$\Pi^a = -2\Pi^i . \tag{2.17}$$

The three sets of field equations (2.1) - (2.3) form a set of 15 equations for the 15 unknowns u_i, ε_{ij} and σ_{ij}. The aim of the theory of elasticity is to solve these field equations for a given set of appropriate boundary conditions and body forces (direct problem).

Consider a body B with surface S and unit outward normal vector $\boldsymbol{n}$. The surface S might be divided into two parts, a surface S_t at which the traction vector $\overset{*}{\underset{S}{\boldsymbol{t}}}$ is prescribed and a surface S_n on which the displacement vector $\overset{*}{\underset{S}{\boldsymbol{u}}}$ is given

$$\overset{*}{\underset{S}{t_i}} = (\sigma_{ji} n_j)_{S_t} \quad \text{at} \quad S_t . \tag{2.18}$$

$$\underset{S}{u_i} = (u_i)_{S_u} \quad \text{at} \quad S_u . \tag{2.19}$$

It is assumed that every point of the surface belongs either to S_t or S_u.

3 Energy-Release Rates, Conservation Laws and Path-Independent Integrals

Since we are concerned with moving discontinuities in solids let us start with a thought-experiment suggested by Eshelby (1975 a). The picture on the left of Fig. 1 shows an elastic body containing a singularity indicated by a black spot. We call this body the original system. The picture on the right represents an exact replica of the original body. We use a stencil to mark an arbitrary surface S on to the *undeformed* original body enclosing the singularity. Using the same stencil we mark a surface S' on to the replica. Again, the surface S' encloses the singularity but has suffered a vector shift, i. e., a rigid translation, by $-\delta\lambda_k$,

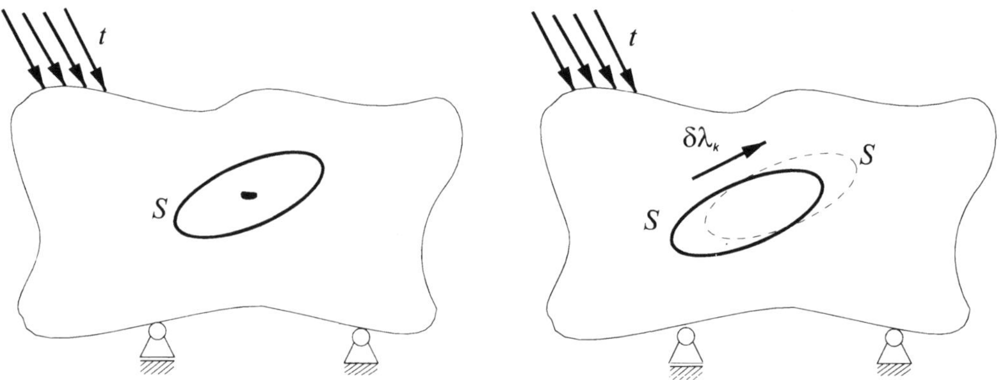

Fig. 1: Original body (left) and replica (right) with defect

say, in comparison to S. Now both bodies are loaded by the same tractions $\boldsymbol{t}$ (no body forces f_i for simplicity). It is our object to calculate the change of the total energy due to the displacement of the singularity by $+\delta\lambda_k$ inside of the *deformed* body. In order to do so we shall carry out the following imaginary procedures: After both bodies come to an equilibrium state under the mechanical loading, we cut out the material enclosed by S in the original system and throw it away. then we apply equivalent surface tractions on the resulting hole to prevent the rest of the original system from relaxing. We then cut out the materials inside S' from the replica and maintain, now, the tractions on the surface of the cut-out piece to prevent relaxation. The internal energy $\Pi^i_{S'}$ (2.14) inside S' differs from the energy Π^i_S inside S of the original system by the addition of energy in the crescent-shaped region 1 and the subtraction of energy in the crescent-shaped region 2. This difference can be calculated by expanding the energy about Π^i_S,

$$\Pi^i_{S'} = \Pi^i_S - \frac{d\Pi^i_S}{dx_k}\delta\lambda_k + 0(\delta\lambda_k)^2$$

which leads to

$$\delta\Pi^i = \Pi^i_{S'} - \Pi^i_S = -\delta\lambda_k \int_B \frac{dW}{dx_k} dV + 0(\delta\lambda_k)^2$$

$$= -\delta\lambda_k \int_S W n_k dA + 0(\delta\lambda_k)^2$$

$$= -\delta\lambda_k \int_S W\delta_{jk} u_j dA + 0(\delta\lambda_k)^2 \tag{3.1}$$

with the Kronecker tensor of unity δ_{jk}.

Here the divergence theorem is used only in the material region 1 and 2 where there is no mathematical singularity. The strain energy density W or its derivative may be infinite in the region occupied by the defects, but this region is subtracted out and thus does not contribute to the energy change.

So far no energy has been changed in the remaining material or the loading mechanism of the original system. We shall now try to fit S' into the hole S of the original system. the boundaries of S and S' would coincide in the *natural* state because one is a mere vector shift of another. But once the deformation has taken place the displacements on S' will differ from those on S by

$$\delta u_i = -\delta\lambda_k u_{i,k} + 0(\delta\lambda_k)^2. \tag{3.2}$$

In order to fit S' into the hole, we have to supply a displacement of an amount given by expression (3.2) on the surface of the hole. This requires work, which will increase the external energy (2.15) of the original system by the amount

$$\delta\Pi^a = -\int_S \delta u_i \sigma_{ji} n_j dA + 0(\delta\lambda_k)^2$$

$$= +\delta\lambda_k \int_S \sigma_{ji} u_{i,k} n_j dA + 0(\delta\lambda_k)^2. \tag{3.3}$$

During this process, the surface tractions change according to

$$\sigma_{ji} n_j \to \sigma_{ji} n_j + 0(\delta\lambda_k), \tag{3.4}$$

whose extra terms contribute an amount of order $(\delta\lambda_k)^2$ to the energy change.

We can now fit S' into S and weld across the interface. Though the displacements are now continuous across the interface, the stresses are not. They differ by an order of $\delta\lambda_k$. As we relax these quantities, the displacements would be of order $\delta\lambda_k$ and an amount of energy of order $(\delta\lambda_k)^2$ is extracted, which can be ignored in comparison with (3.1) and (3.3).

We have now finished the task of moving the defect in the original body by $+\delta\lambda_k$. The total energy change $\delta\Pi = \delta\Pi^i + \delta\Pi^a$ during this procedure is

$$\delta\Pi = \delta\lambda_k \int_S (\sigma_{ji} u_{i,k} - W\delta_{jk}) n_j dA + 0(\delta\lambda_k{}^2) .$$

Let us define the material force on the defect as the negative gradient of the total energy of the body B with respect to the change in position of the defect within the body, this force can be expressed as

$$J_k = - \lim_{\delta\lambda_k \to 0} \frac{\delta\Pi}{\delta\lambda_k} = \int_S (W\delta_{jk} - \sigma_{ji} u_{i,k}) n_j dA , \tag{3.5}$$

and the corresponding operation is denoted as material translation.

As we apply the displacements δu_i (3.2) to S, the tractions across S change according to (3.4). the extra term will depend on the way the forces exerted by the loading mechanism vary as their points of application move. However, as the extra term in (3.4) only alters (3.3) by a quantity $(\delta\lambda_k)^2$, the external forces do not enter the final result explicitely.

In a similar way, it may be shown that the change of energy of the system due to an infinitesimal rotation ω_n of the defect with respect to the origin of the coordinate system is given as

$$L_n^0 = -\frac{\partial\Pi}{\partial\omega_n} = \int_S \varepsilon_{nlk} \left[x_l (W\delta_{jk} - \sigma_{ji} u_{i,k}) + u_l \sigma_{jk} \right] n_j dA , \tag{3.6}$$

and $L_n{}^0$ may be referred to as material momentum on the defect.

Finally we calculate the energy-release rate due to a self-similar expansion

$$x_i \to \alpha\, x_i$$

of the defect, such as a cavity or crack. The result is

$$M^0 = -\frac{\partial\Pi}{\partial\alpha} = \int_S \left[x_k (W\delta_{jk} - \sigma_{ji} u_{i,k}) - \frac{1}{2} u_k \sigma_{jk} \right] n_j dA . \tag{3.7}$$

It may be mentioned that in two-dimensional problems the last term in (3.7) has to be dropped. M is called the material virial. Change of reference point for the L_n - and M - integrals leads to, cf. Bakker (1984),

$$\begin{aligned} M &= M^0 - x_i{}^0 J_i , \\ L_n &= L_n{}^0 - \varepsilon_{nij} x_i{}^0 J_j , \end{aligned} \tag{3.8}$$

where x_i^0 are the coordinates of the new, arbitrarly fixed point of reference. Analogous equations for changing the reference point exist for a system of physical forces as known in elementary statics, cf. Schweins (1849).

In all three expressions (3.5), (3.6) and (3.7) one and the same term in round brackets appears. It is the by-now well-known energy-momentum or, material-momentum or, as we would like to call it, Eshelby tensor

$$b_{jk} = \mathrm{W}\,\delta_{jk} - \sigma_{ji} u_{i,k} \,. \tag{3.9}$$

The interals J_k , L_n and M are extremely useful in fracture mechanics and have been introduced by Günther (1962) and independently by Rice (1968) and Budiansky and Rice (1973). We will come back to these integrals later.

In the absence of any singularities or defects inside S , Gauss' theorem may be applied leading to

$$\begin{aligned} \int_S b_{jk} n_j dA &= \int_B b_{jk,j} dV \\ &= \int_B (W_{,k} - \sigma_{ji,j} u_{i,k} - \sigma_{ji} u_{i,kj}) dV \,. \end{aligned} \tag{3.10}$$

If the material is homogeneous, i. e., the strain energy W does not depend explicitely on x_k , we have with (2.6)

$$W_{,k} = \frac{\partial W}{\partial u_{i,j}} u_{i,jk} = \sigma_{ji} u_{i,kj} \,. \tag{3.11}$$

If, additionally, body forces would be absent $(f_i - 0)$, the integral (3.10) vanishes and due to (2.1) and (3.11) we are left with a conservation law is in global form

$$\int_S b_{jk} n_j dA = 0 \,, \tag{3.12}$$

or, in local form

$$b_{jk,j} = 0 \,. \tag{3.13}$$

Conservation laws play an important role in defect, fracture and damage mechanics. In this Chapter they are introduced in a more - or - less ad-hoc fashion. Systematic procedures to derive conservation laws will be dealt with in Chapter 5.

Once established, conservation laws are helpful in establishing path-independence of the corresponding integrals in the presence of defects. Consider a body containing a defect or a singularity that is again represented in Fig. 2 by a black spot.

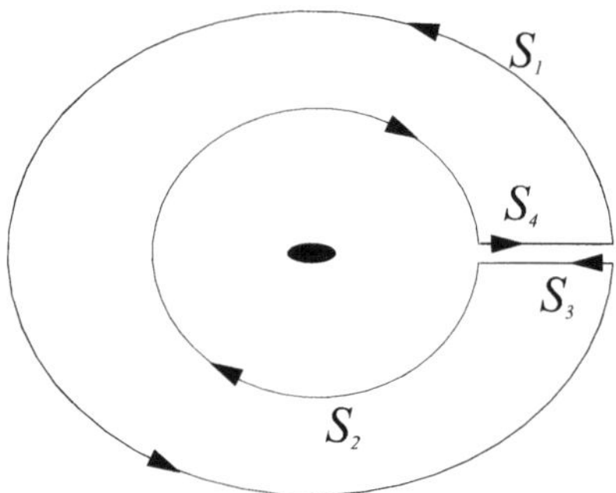

Fig. 2: Contour for path-independent integrals in the presence of a defect.

For the surface shown in Fig. 2 the homogeneous relations $J_i = 0$, $L_n = 0$ and $M = 0$ still hold, because there is no defect inside the indicated surface

$$\int_{S_1+S_2+S_3+S_4} (\ ) dA = 0 . \tag{3.14}$$

However, the contributions along the parallel parts of the surface cancel each other, because they are calculated far from the singularity, and the integrands are smooth, regular functions. By reversing the path of integration along S_2 we can conclude that the quantities J_k, L_n and M evaluated along surfaces close to the defect (S_2) will be equal to those evaluated along arbitrary surfaces far from the defect (S_1), as long as the surfaces enclose the same defect and no other defects are included.

Having established path-independence of the integrals $\boldsymbol{J}, \boldsymbol{L}$ and M, any convenient path surrounding the defect can be chosen in evaluating them.

4 Mechanics in Material and Physical Space

So far, we have introduced notions like material translation, rotation and self-similar expansion as well as material forces, moments and virials. It may be mentioned, however, that, e. g., the material or Eshelbian force is radically different from the physical or Newtonian concept of force. The Eshelby-type force is always understood as a relative change of the total energy of a given system, with respect to some quantity which alters the configuration of that system. The latter quantity might be the displacement of a foreign or missing atom in a lattice, the change in location of a dislocation, the change in size or shape of a crack, cavity or inclusion or the progress of a phase-transition front in a material. All such changes of configuration of certain objects occur within the material in which they find themselves, by contrast to changes in the configuration deformation of a structure under some Newtonian loadings, which occur in what might be called physical space of our surroundings, in which the structure finds itself and in which Newton's laws are valid.

Thus the term space, whether physical or material, is used here not in a strictly mathematical sense, as possessing a certain metric and possibly other properties, but in essentially a descriptive meaning.

In both physical space and material space the concepts of force are introduced as negative gradients of the total energy. Further quantities and concepts which can be introduced both in physical and material spaces concern tractions, momentum (or stress), application of conservation laws, both in local and integral form, as well as with and without source terms. Free body diagrams can be sketched and such typical notions of mechanics as stability of equilibrium and trajectories of moving objects can be discussed both in physical as well a in material space.

To emphasize the parallelism (or correspondence) between the (usual) mechanics in physical space and the (not common) mechanics in material space the notions like force, traction, trajectories, stability of equilibrium are juxtaposed in the following.

Mechanics in

Physical Space	**Material Space**

Forces on discrete objects

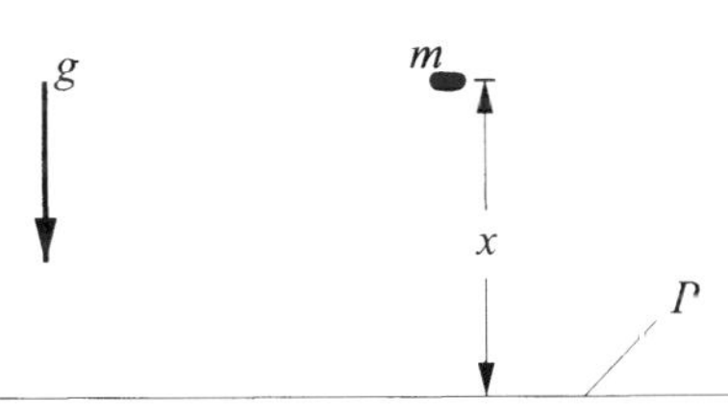

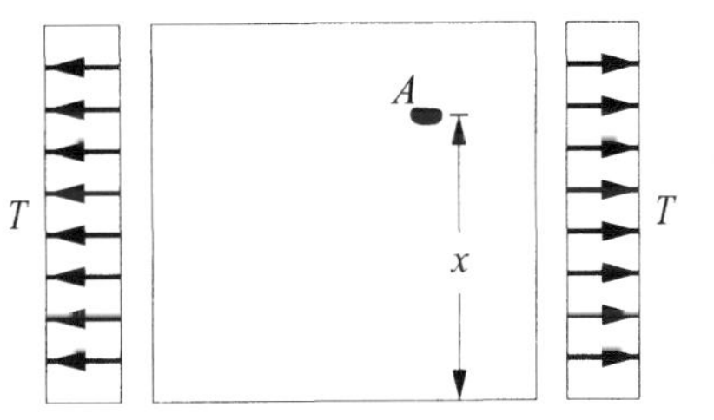

Particle of mass m in gravity field (acceleration g) above a reference plane P at distance x	Elastic plate of given dimensions and material properties a_i under tension T with defect A at distance x
Total energy Π	Total energy
$\Pi = mgx$	$\Pi = \Pi(a_i, T, x)$

Physic. force F acting on mass m

$$F = -\frac{\partial \Pi}{\partial x} = -mg$$

Mater. force J acting on defect A

$$J = -\frac{\partial \Pi}{\partial x}$$

Lagrangian for continua

$$L = L(u_i(x_k), u_{i,j}(x_k), x_k)$$

Cauchy stress σ_{ij} (or physical momentum) introduced as derivative of Lagrangian associated with translation in physical space

$$\sigma_{ij} = -\frac{\partial L}{\partial u_{j,i}}$$

Eshelby stress b_{ij} (or material momentum) introduced as total derivative of Lagrangian associated with translation in material space

$$b_{ij} = u_{k,j} \frac{\partial L}{\partial u_{k,i}} - L\delta_{ij}$$

Tractions in Continua

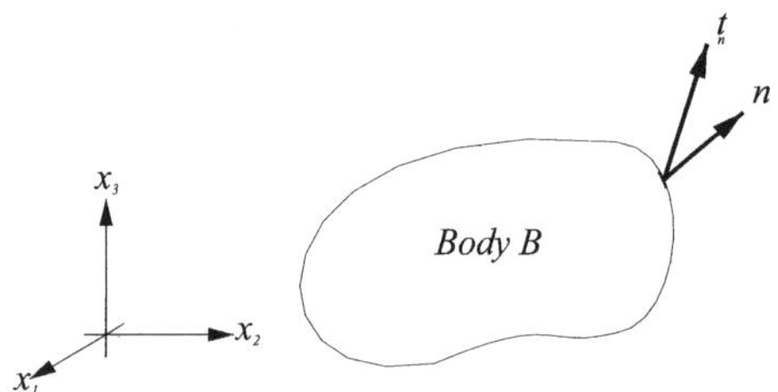

Physical tractions $\underset{n}{\boldsymbol{t}}$

Physical force acting across a unit area with normal n_i, i. e., the virtual work if the unit area undergoes a virtual displacement by a unit in x_i-direction

Stress σ_{ij} (or physical momentum) introduced through the Cauchy relation via physical traction $\underset{n}{t}_j$

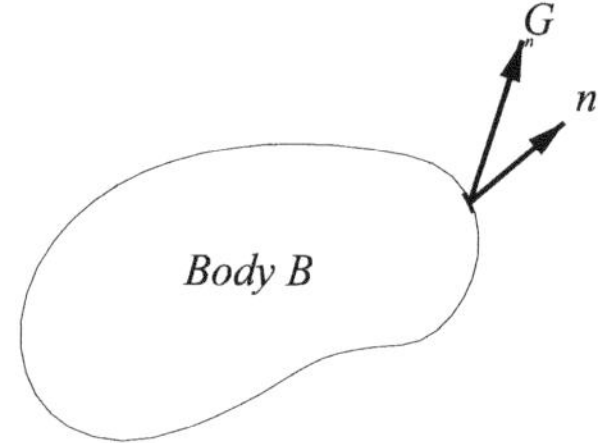

Material tranctions $\underset{n}{\boldsymbol{G}}$

Material force "acting" across a unit area with normal n_i, i. e., the negative change of total energy if the unit area undergoes a material displacement by a unit in x_j-direction

Material traction $\underset{n}{G}_j$ introduced via material stress b_{ij} (or material momentum), i.e., the Eshelby tensor

$\underset{n}{t}_j = \sigma_{ij} n_i$ $\underset{n}{G}_j = b_{ij} n_i$

Trajectories of motion

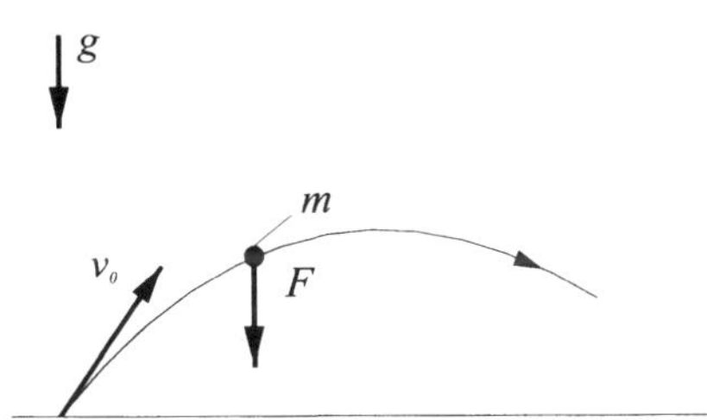

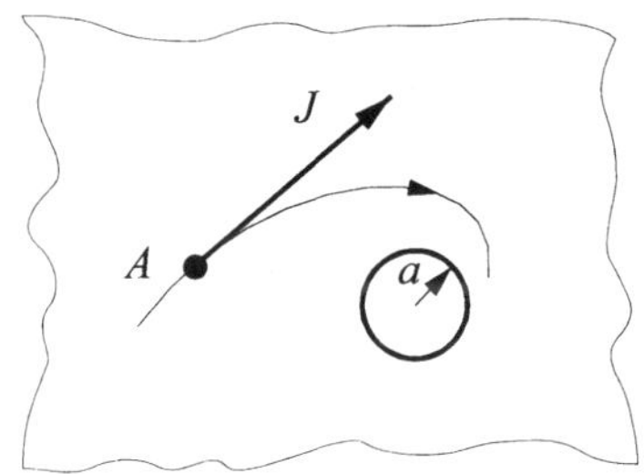

Plane trajectory of a mass particle m in field of gravity g with assigned initial velocity v_0 and determined on the basis of Newton's equations of motion. (Physical force $\boldsymbol{F}$ not tangential to trajectory, but rather in the direction of the acceleration)

Plane trajectory of an inclusion A near a cavity of radius a in inhomogeneous stress field produced by remote physical tractions. In the absence of a "true" equation of motion, material force $\boldsymbol{J}$ is assumed to be tangential to the trajectory

Stability of Equilibrium of Discrete Objects

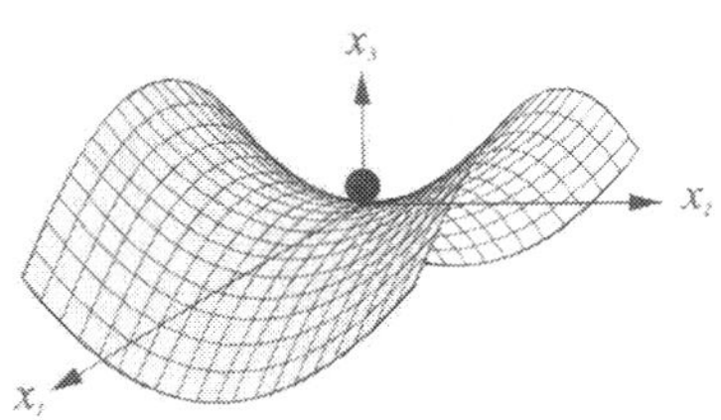

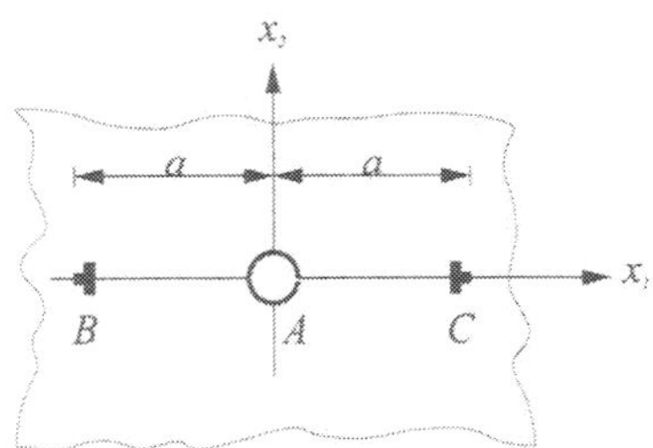

Mass particle m on a saddle-shaped surface in a field of gravity g (acting in negative x_3-direction)	Hole A in an elastic material in a plane stress field produced by dislocations B and C
Total energy $\Pi = mgx_3(x_1, x_2)$; at $x_1 = x_2 = 0$ it is	Material forces acting on A
$$F_1 = -\frac{\partial \Pi}{\partial x_1} = 0 ,$$ $$F_2 = -\frac{\partial \Pi}{\partial x_2} = 0$$	$$J_1 = -\frac{\partial \Pi}{\partial x_1} = 0 ,$$ $$J_2 = -\frac{\partial \Pi}{\partial x_2} = 0$$
because of equilibrium	because of symmetry
Stability of equilibrium	Calculation shows that
unstable $\frac{\partial^2 \Pi}{\partial x_1^2} < 0$,	unstable $\frac{\partial^2 \Pi}{\partial x_1^2} < 0$,
stable $\frac{\partial^2 \Pi}{\partial x_2^2} > 0$.	stable $\frac{\partial^2 \Pi}{\partial x_2^2} > 0$.
Equilibrium at a saddle point	Equilibrium at a saddle point

5 Noether's Formalism and the Neutral Action Method

As has already been mentioned in Chapter 3, conservation laws are essential tools to assess defects in solids. The object of this Chapter is to only sketch the systematic procedures available to derive conservation laws. A more comprehensive presentation of the subject may be found in Kienzler and Herrmann (2000).

Noether (1918) employed a formalism for systems governed by a Lagrangian. Adopting (2.10) we have

$$L = L(x_i, u_j, u_{j,k}) , \tag{5.1}$$

and the action integral A is given by

$$A = \int_B L(x_i, u_j, u_{j,k}) dV \, . \tag{5.2}$$

Without knowledge of the specific form of L, i. e., its specific dependence on $x_i, u_j, u_{j,k}$, we can calculate the variation δA of A by varying u_i under the integral sign

$$u_j \rightarrow u_j^* = u_j + \delta u_j \, , \tag{5.3}$$

but doing nothing to x_i.

The usual procedure, cf., e. g., Gelfand and Fomin (1963) leads to the Euler-Lagrange equations, i. e., the equations of motion.

$$\boldsymbol{E}_j(L) = \frac{\partial L}{\partial u_j} - \left(\frac{\partial L}{\partial u_{j,n}} \right)_{,k} = 0 \tag{5.4}$$

In extension of the usual variation (5.3), Noether applied a one-parametric group of infinitesimal (point) transformations (ε is the infinitesimal parameter)

$$\begin{aligned} x_i &\rightarrow x_i^* = x_i + \varepsilon \xi_i(x_j, u_k) \, , \\ u_i &\rightarrow u_i^* = u_i + \varepsilon \phi_i(x_j, u_k) \, , \end{aligned} \tag{5.5}$$

i. e., a transformation of the dependent *and* independent variables u_i and x_i, respectively. The change of the action integral turns out to be

$$\begin{aligned} \delta A &= \varepsilon \int_B (\boldsymbol{pr}^{(1)} \boldsymbol{w} + \xi_{i,i}) L dV \\ &= \varepsilon \int_B \left(P_{j,j} + Q_j \boldsymbol{E}_j(L) \right) dV \, . \end{aligned} \tag{5.6}$$

The infinitesimal generator $\boldsymbol{w}$ and its first prolongation $\boldsymbol{pr}^{(1)} \boldsymbol{w}$ are, respectively, given by

$$\begin{aligned} \boldsymbol{w} &= \xi_i \frac{\partial}{\partial x_i} + \phi_j \frac{\partial}{\partial u_j} \, , \\ \boldsymbol{pr}^{(1)} \boldsymbol{w} &= w + (\phi_{j,i} - u_{j,k} \xi_{k,i}) \frac{\partial}{\partial u_{j,i}} \, . \end{aligned} \tag{5.7}$$

The characteristics Q_j and the currents P_i are defined as

$$Q_j = \phi_j - \xi_i u_{j,i} ,$$
$$P_j = \phi_i \frac{\partial L}{\partial u_{i,j}} + \xi_k \left(L\delta_{jk} - \frac{\partial L}{\partial u_{i,j}} u_{i,k} \right) . \tag{5.8}$$

Noether's theorem, in essence, states that if the action integral A is invariant with respect to the action of a symmetry group (5.5) then there exists a conservation law. In fact, along solutions of the field equations, the right term in (5.6 b) vanishes and from $\delta A = 0$ the conservation law

$$P_{i,i} = 0 \tag{5.9}$$

follows. The unknown transformation functions ξ_i and ϕ_i can be calculated from the invariance condition (5.6 a) once a specific Larangian is given. In turn, the characteristics Q_i (5.8 a) are determined by the transformation chosen, and the conserved currents P_i can be constructed immediately from equation (5.8 b).

The conservation law corresponding to a material translation is given by

$$\xi_k = k_k = const ,$$
$$\phi_i = 0 . \tag{5.10}$$

The invariance condition (5.6 a) is satisfied, if

$$\left. \frac{\partial L}{\partial x_k} \right|_{\exp l.} = 0 . \tag{5.11}$$

i. e., if the material is homogeneous (2.11), both physically (constant body forces) and materially (space independent material constants). In the absence of body forces, the Lagrangian L reduces to minus the strain-energy density W (2.12) and with (2.6) the conservation law (5.9) reduces for arbitrarily chosen constants k to

$$(W\delta_{jk} - \sigma_{ji} u_{i,k})_{,j} = 0 , \tag{5.12}$$

i. e., with (3.9) to the statement that the Eshelby tensor is divergence-free

$$b_{jk,j} = 0 . \tag{5.13}$$

Without striving for completeness, it may be mentioned that the transformation leading to the

L_n integral (3.6) is given by

$$\xi_j = l_n \varepsilon_{nij} x_i ,$$
$$\phi_j = l_n \varepsilon_{nij} u_i , \qquad l_n = const . \tag{5.14}$$

provided the material is homogeneous and isotropic, and body forces are absent. The transformation

$$\xi_j = m x_j ,$$
$$\phi_j = m u_j , \qquad m = const . \tag{5.15}$$

finally leads to the M integral (3.7) for a homogeneous material in the absence of body forces.

For a system without a Lagrangian (e. g., for systems with dissipation), the governing equations are no longer the Euler-Lagrange equations of a variational problem, but rather a set of partial differential equations which has been obtained on the basis of some balance laws (e. g., balance of linear momentum) and some postulated constitutive relations.

Until recently, no systematic procedure existed to derive conservation laws for such systems, but a few years ago, a novel methodology was advanced by Honein et al. (1991).

In order to develop this new procedure, the concept of a null Lagrangian has to be introduced.

Let $g_i = g_i(x_k, v_j, v_{j,k}, ...)$ be some arbitrary function. It can be shown, cf., e. g., Olver (1993), that the following relation is valid

$$\widetilde{L} = g_{i,i} \Leftrightarrow E_i(\widetilde{L}) = 0 , \tag{5.16}$$

i. e., if the Lagrangian is a divergence than the Euler-Lagrange equations are satisfied identically and if the Euler-Lagrange are satisfied identically than $\widetilde{L}$ is a divergence. Such a Lagrangian is called a "null Lagrangian".

The basic idea of the Neutral-Action method is as follows. As we have seen above, a conservation law $P_{i,i} = 0$ is expressable in terms of the Euler-Lagrange equations $E_\alpha(L)$ as (5.6 b)

$$P_{i,i} = -Q_i \boldsymbol{E}_i(L) , \tag{5.17}$$

where the elements of the set Q_i are the characteristics of the conservation law which are fixed by the transformation function ξ_i and ϕ_i (5.8 a).

Now, instead of the Euler-Lagrange equations $E_i(L) = 0$, we have a set of differential equations which we can abbreviate as

$$\Delta_i(x_j, u_k, u_{k,l}) = 0 . \tag{5.18}$$

The possibility is not excluded that they may be Euler-Lagrange equations.

Now, instead of the characteristics Q_i being specified by the transformation functions ξ_i and ϕ_i we determine Q_i employing the symbol - f_α instead (in order to avoid confusion), such that

$$f_i \Delta_i = P_{i,i}, \tag{5.19}$$

where the function f_i are not considered to be pre-determined as Q_i were.

Thus, it follows from (5.19) that a requirement for the existence of conservation laws is that

$$E_j(f_i \Delta_i) = 0, \tag{5.20}$$

since $E_i(P_{i,i}) = 0$ as shown above. The sum $f_i \Delta_i$ implies that it is formally a null Lagrangian whose action integral

$$A = \int_B \tilde{L} dV = \int_B f_i \Delta_i dV = \int_S P_j n_j dA \tag{5.21}$$

has vanishing variation for any dependent variable u_j, i. e., $\delta A = 0$. In other words, in order to construct conservation laws for any system (whether or not it possesses a Lagrangian), governed by a set of differential equations $\Delta_i = 0$, we try to construct a product of $f_i \Delta_i$ whose action A does not change variationally, i. e., the action A behaves neutrally under its variation. Thus the name "Neutral Action" (NA) method was given to this procedure, Chien (1992).

The governing differential equations of linear elasticity are the Navier-Lamé equations. They are obtained by elimination of σ_{ij} and ε_{ij} from (2.1) - (2.3). For a homogeneous and isotropic material they are given as, e. g., Timoshenko and Goodier (1970)

$$\Delta_i = \mu u_{i,jj} + (\lambda + \mu) u_{j,ji} = 0, \tag{5.22}$$

with the Lamé constants λ and μ. We assume that

$$f_i = f_i(x_j, u_m, u_{m,n}), \tag{5.23}$$

and the unknown functions f_i have to be determined from (5.20). It turns out, Kienzler and Herrmann (2000), that

$$f_i = k_k u_{i,k},$$

$$f_i = l_n \varepsilon_{nmk}(x_m u_{i,k} + \delta_{mi} u_k), \tag{5.24}$$

$$f_i = m(x_k u_{i,k} + u_k)\,,$$

lead to the conservation laws (3.5), (3.6) and (3.7), respectively.

As it turns out, Kienzler and Herrmann (2000), even if a Lagrangian function exists and thus $\boldsymbol{E}_i(L)$ is available, the Neutral Action method leads to conservation laws in a more straight-forward manner than application of Noether's theorem.

6 Hole-Dislocation Interaction

As an example of the application of path-independent integrals we consider the interaction of a circular hole and an edge dislocation in plane elastostatics depicted in Fig. 3. It may be mentioned that the rim of the hole is traction free and that the cavity does not carry an eigenstress field with itself as it is sometimes assumed in physics. Let the edge dislocation lie at the origin of a Cartesian coordinate system (x_1, x_2) and have Burgers vector components b_1 and b_2. The eigen-stress field produced by this dislocation, Volterra (1907), is given as

$$\sigma_{11}^{(0)} = \frac{2G}{r^4 \pi(\kappa+1)}\left[-b_1 x_2 (3x_1^2 + x_2^2) + b_2 x_1 (x_1^2 - x_2^2)\right],$$

$$\sigma_{22}^{(0)} = \frac{2G}{r^4 \pi(\kappa+1)}\left[+b_1 x_1 (x_1^2 - x_2^2) + b_2 x_1 (x_1^2 + 3x_2^2)\right], \tag{6.1}$$

$$\sigma_{12}^{(0)} = \frac{2G}{r^4 \pi(\kappa+1)}\left[+b_1 x_1 (x_1^2 - x_2^2) + b_2 x_2 (x_1^2 + x_2^2)\right],$$

$$r = \sqrt{x_1^2 + x_2^2}$$

(shear modulus $G = 1/2E/(1-\nu), \kappa = 3-4\nu$ for plane strain, $\kappa = (3-\nu)/(1+\nu)$ for plane stress, Young's modulus E, Poisson's ratio ν).

This state of stress is altered due to the presence of a circular hole (radius r_0) centered at a distance $d + r_0$ from the origin on the x_1-axis (Fig 3). Kienzler and Duan (1987) derived a simple formula to obtain the hoop stresses at the boundary of a stress-free circular hole based on the stress distribution there would exist along the boundary of the hole in its absence.

Let $\sigma_{rr}^{(0)}(r_0, \varphi)$ and $\sigma_{\varphi\varphi}^{(0)}(r_0, \varphi)$ be the stresses calculated from (6.1) along the curve coinciding with the boundary of the hole (r, φ polar coordinates with respect to the center of the hole) and ${}_{\sigma}I_1^{(0)}$ the value of the first stress invariant at the center or the prospective hole

(here ${}_{\sigma}I_1^0 = (\sigma_{11}^{(0)} + \sigma_{22}^{(0)})\Big|_{\substack{x_1 = d + r_0 \\ x_2 = 0}}$).

The hoop stress $\sigma_{\varphi\varphi}(\varphi)$ at the boundary of the hole due to the applied load (in this case the eigen-stress field of the dislocation) follows from

$$\sigma_{\varphi\varphi}(\varphi)={}_{\sigma}I_1^{(0)}+2\left(\sigma_{\varphi\varphi}^{(0)}(r_0,\varphi)-\sigma_{rr}^{(0)}(r_0,\varphi)\right). \tag{6.2}$$

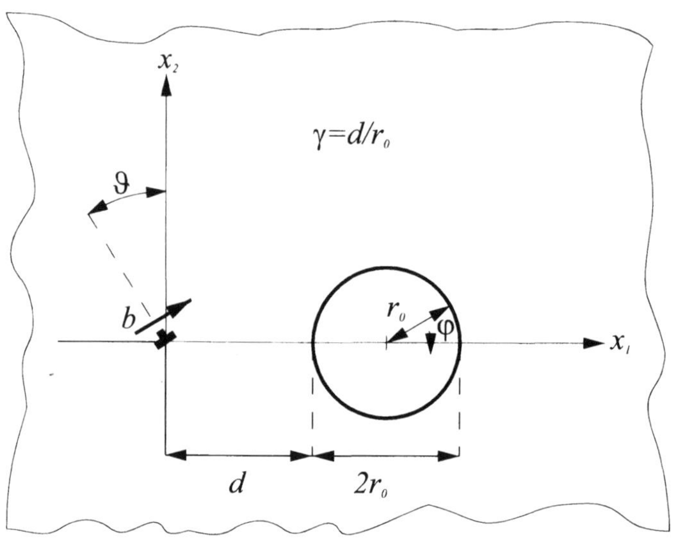

Fig. 3 Interaction between circular hole and dislocation.

Introducing the dimensionless distance

$$\gamma = d / r_0 \tag{6.3}$$

it follows from (6.2) with (6.1) and (6.3) after standard coordinate transformations and suitable rearrangements Kienzler and Kordisch (1990)

$$\sigma_{\varphi\varphi}(\varphi)=\frac{4G}{r_0\pi(\kappa+1)}\left\{b_1\sin\varphi\left[\left(\frac{\gamma(2+\gamma)}{2(1+\gamma)(1+\cos\varphi)+\gamma^2}\right)^2-1\right]+\right.$$

$$\left.+b_2\left[\frac{1}{1+\gamma}-4(1+\gamma)\frac{\sin^2\varphi\left[1+(1+\gamma)\cos\varphi\right]}{\left[2(1+\gamma)(1+\cos\varphi)+\gamma^2\right]^2}\right]\right\}. \tag{6.4}$$

We wish to calculate the material force exerted by the dislocation on the cavity or, vice versa, the force exerted by the cavity on the dislocation. It appears much simpler to evaluate the latter force by means of a path-independent integral choosing the contour along the rim of the hole. The only nonvanishing stress component is $\sigma_{\varphi\varphi}$ and the strain energy W is given by

$$W = \frac{\kappa + 1}{16G} \sigma_{\varphi\varphi}^2(\varphi) = \frac{1}{2E^*} \sigma_{\varphi\varphi}^2(\varphi) . \tag{6.5}$$

The constant E^* is defined as

$$E^* = \frac{8G}{1+\kappa} = \begin{cases} \frac{E}{1-\nu^2} & \text{for plane strain} \\ E & \text{for plane stress.} \end{cases} \tag{6.6}$$

The integrals (3.5) - (3.7) reduces to

$$\begin{aligned} J_1 &= \frac{r_0}{2E^*} \int_B^{2\pi} \sigma_{\varphi\varphi}^2(\varphi) \cos\varphi \, d\varphi , \\ J_2 &= \frac{r_0}{2E^*} \int_0^{2\pi} \sigma_{\varphi\varphi}^2(\varphi) \sin\varphi \, d\varphi , \\ M^0 &= \frac{r_0^2}{2E^*} \int_0^{2\pi} \sigma_{\varphi\varphi}^2(\varphi) \, d\varphi , \\ L^\perp &= \frac{r_0^2}{2E^*} (1+\varepsilon) \int_0^{2\pi} \sigma_{\varphi\varphi}^2(\varphi) \sin\varphi \, d\varphi . \end{aligned} \tag{6.7}$$

Thus, from the knowledge of $\sigma_{\varphi\varphi}(\varphi)$ (6.4), the integrals can readily be evaluated.

The reference point for M^0 has been chosen to be the center of the hole, while the reference point for $L_3 = L^\perp$ is the origin of the coordinate system. L and M for an arbitrary point of reference may be calculated by the transformation (3.8)

$$\begin{aligned} M &= M^0 - x_i^0 J_i , \\ L &= L^\perp - \varepsilon_{3ij} x_i^\perp J_j , \end{aligned} \tag{6.8}$$

with x_i^0 and $x_i^\perp$ the distance from the respective reference point. The evaluation of the integrals (6.7) is cumbersome but straightforward. The results are

$$J_1 = -\frac{E^*}{4r_0\pi(1+\lambda)}\left[b_1^2\left(\frac{1}{\gamma(2+\gamma)}+\frac{1}{(1+\gamma)^2}\right)+b_2^2\frac{1}{\gamma(2+\gamma)}\right],$$

$$J_2 = +\frac{E^*}{4r_0\pi(1+\gamma)}\frac{b_1b_2}{(1+\gamma)^2},$$

$$M^0 = -r_0(1+\gamma)J_1, \tag{6.9}$$

$$L^{\perp} = +r_0(1+\gamma)J_2 .$$

The equivalence of the relations between M^0 and J_1 (6.9 c) on one hand and between $L^{\perp}$ and J_2 (6.9 d) on the other is remarkable. By choosing $x_1^{\perp} = d + r_0$ and $x_2^{\perp} = 0$ in (6.8 b), it turns out that L^0 referred to the centre of the circular hole vanishes, i. e., the lines of action of the material forces J_1 and J_2 intersect each other at the centre of the hole.

In order to discuss the integrals in terms of energy-release rates, with respect to the change of the position of the size of the hole relative to the dislocation, it is stated that the energy of our system depends - geometrically - only on two parameters, γ and ϑ. With

$$b_1 = b\cos\vartheta ,$$
$$b_2 = b\sin\vartheta , \tag{6.10}$$

$b = \sqrt{b_1^2 + b_2^2}$ is the intensity of the inclined edge dislocation and ϑ is the angle of inclination (cf. Figure 3). Consider now an infinitesimal transformation of the hole with respect to the coordinate system consisting of four parts: a translation δ_1 in the x_1-direction, a translation δ_2 in the x_2-direction, a rotation ω about the origin and a self-similar "expansion" $r_0 \to \alpha r_0$. The geometrical parameters change to

$$\gamma \to \gamma + \delta_1 / r_0 - \alpha(1+\gamma),$$
$$\vartheta \to \vartheta + \delta_2 /[r_0(1+\gamma)] - \omega . \tag{6.11}$$

The path-independent integrals are related to energy release rates via (3.5) - (3.7) as

$$J_1 = -\frac{\partial \Pi^{tot}}{\partial \delta_1} = -\frac{\partial \Pi^{tot}}{\partial \gamma}\frac{\partial \gamma}{\partial \delta_1} - \frac{\partial \Pi^{tot}}{\partial \vartheta}\frac{\partial \vartheta}{\partial \delta_1} = -\frac{1}{r_0}\frac{\partial \Pi^{tot}}{\partial \gamma},$$

$$J_2 = -\frac{\partial \Pi^{tot}}{\partial \delta_2} = -\frac{\partial \Pi^{tot}}{\partial \gamma}\frac{\partial \gamma}{\partial \delta_2} - \frac{\partial \Pi^{tot}}{\partial \vartheta}\frac{\partial \vartheta}{\partial \delta_2} = +\frac{1}{r_0(1+\gamma)}\frac{\partial \Pi^{tot}}{\partial \gamma},$$

(6.12)

$$M = -\frac{\partial \Pi^{tot}}{\partial \alpha} = +(1+\gamma)\frac{\partial \Pi^{tot}}{\partial \gamma},$$

$$L^{\perp} = -\frac{\partial \Pi^{tot}}{\partial \omega} = +\frac{\partial \Pi^{tot}}{\partial \vartheta}.$$

From (6.12 a) and (6.12 c) the relation (6.9 c) is rederived and (6.12 b) with (6.12 d) confirm Eqn. (6.9 d). Integration of (6.12 a) and (6.12 d) leads to the total energy Π^{tot} of our system up to an insignificant constant of integration

$$\Pi^{tot} = \frac{Gb^2}{\pi(\kappa+1)}\left\{ln\left[\frac{\gamma(2+\gamma)}{(1+\gamma)^2}\right] - \cos^2\vartheta\frac{1}{(1+\gamma)^2}\right\} + C. \tag{6.13}$$

Equation (6.13) is in agreement with a formula given by Dundurs and Mura (1964), who studied the interaction between an edge dislocation and a circular *inclusion*. They showed that under certain circumstances (stiffness ration of inclusion and matrix material, distance between inclusion and dislocation, strength of the dislocation) there exists a position of equilibrium $(J_1 = J_2 = 0)$ between inclusion and dislocation. From equation (6.9 a) it is obvious that such an equilibrium position will never exist (except when γ tends to infinity) between a *hole* and a dislocation. The force between the dislocation and the hole is always a force of attraction. Thus, if they would move, they would always move towards each other.

We would like to consider next the shape of trajectories envisaging the possibility that the cavity, by means of some mechanism (e. g., diffusion, cf. Stark (1976), could move towards the dislocation. Lacking equations of motion (of the type of Newton's second law for mass points, where, in general, the force is not tangential to the trajectory), it is usually assumed that the material force is tangential to the trajectory, i. e., path of motion.

We assume that the dislocation is fixed at the origin of the coordinate system and the angle ϑ is (without loss of generality) set $\vartheta = \pi/2$. The centre of a circular cavity of radius r_0 is placed in an arbitrary point x_1, x_2, thus at distance $r = \sqrt{x_1^2 + x_2^2}$ from the dislocation. By means of equation (6.9 a, b), the material forces J_1 and J_2 can be calculated. In terms of a polar coordinate system r, φ, see Figure 4, and of the dimensionless distance ρ between dislocation and center of the cavity

$$\rho = \frac{r}{r_0} = (1+\gamma), \rho > 1, \tag{6.14}$$

the material forces in radial and tangential directions are given by

$$J_r = -\frac{E^* b^2}{4\pi r_0 \rho}\left[\frac{1}{\rho^2 - 1} + \frac{\sin^2\varphi}{\rho^2}\right], \tag{6.15}$$

$$J_{\varphi} = -\frac{E^{*}b^{2}}{8\pi\, r_{0}\rho}\frac{\sin^{2}\varphi}{\rho^{2}}.$$

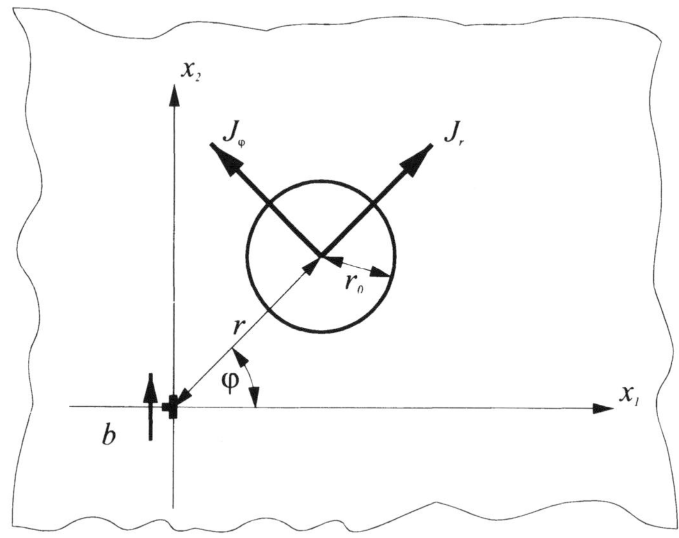

Fig. 4 Interaction between circular hole and dislocation, polar coodinates

The trajectories of possible motion of the cavity have to be determined from the differential equation

$$\frac{dx_2}{dx_1} = \frac{J_2}{J_1}, \tag{6.16}$$

or, in polar coordinates

$$\frac{d\rho}{d\varphi} = \rho\frac{J_r}{J_\varphi}. \tag{6.17}$$

If the distance between inclusion and hole is large, i. e., $\rho >> 1$, the differential equation can readily be integrated, leading to

$$\rho = \rho_0 \frac{\sin\varphi_0}{\cos^2\varphi_0}\frac{\cos^2\varphi}{\sin\varphi}. \tag{6.18}$$

Each trajectory is specified by the choice of φ_0 and ρ_0. Some of the trajectories, with the restriction $\rho >> 1$, are sketched in Figure 5.

Concluding this Chapter, we consider the stability of material equilibrium of a circular hole in the eigen-stress field of two symmetric dislocations as depicted in Figure 6.

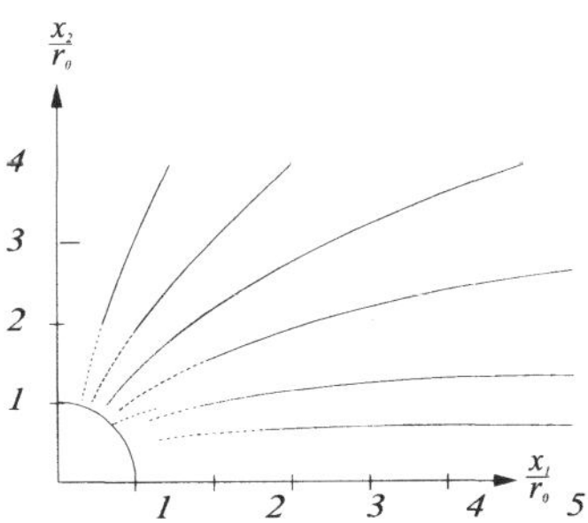

Fig. 5 Trajectories of motion of a cavity in an eigenstress field due to a dislocation

Since the theory is linear, the stress fields of the two dislocations may be superimposed, the results for J_1, J_2 or J_r, J_φ, however, do not follow from superposition, because they are quadratic forms in $\sigma_{\varphi\varphi}$ (6.7) and an interaction term will occur. Due to the symmetry of the problem it can be concluded, however, that $\xi_1 = \xi_2 = 0$ corresponds to an equilibrium position. The material force in the x_2-direction is zero and the material forces towards the dislocations in the x_1-direction are equal but opposite. If the hole is shifted by a small amount along the x_1-axis to the right, say, the attracting force of the right dislocation becomes larger than that of the left. The hole will thus move further to the right. Therefore, the equilibrium position is *unstable* with respect to x_1.

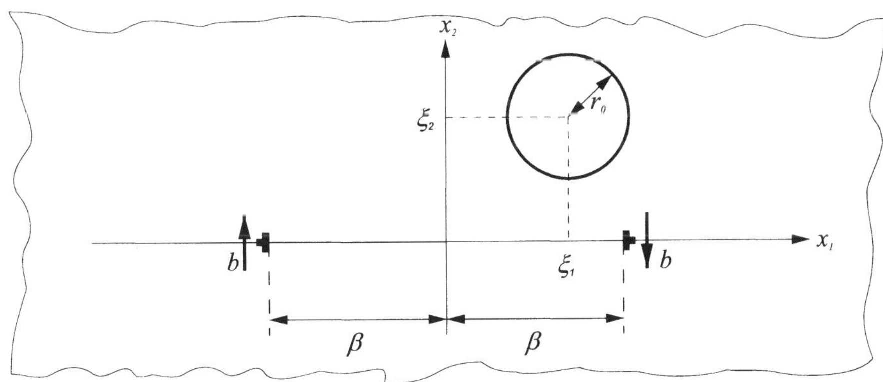

Fig. 6 Interaction of a circular hole with two dislocations

If the hole is shifted, on the other hand, by a small amount in x_2-direction, the x_1-components of the force exerted by the two dislocations are still equal and opposite. The forces, however, are now inclined with respect to the x_1-axis, both components in x_2-direction add and drive the hole back into the original position. Thus, the equilibrium position is *stable* with respect to x_2. The potential energy associated with this problem, therefore, possesses a saddle point at $\xi_1 = \xi_2 = 0$ and this equilibrium position is overall unstable. This and a corresponding situation in physical space are sketched in the Chapter 4.

7 Path-Independent Integrals of Fracture Mechanics

The broadest application of path-independent integrals is to be found in fracture mechanics, specifically it is Rice's J integral taken around a crack tip which serves as an important tool in ascertaining structural integrity of various components. It is the purpose of this Chapter to briefly summerize some of the most important concepts of fracture mechanics within the context of the linear theory of elasticity and its relation to the mechanics of material space outlined so far. Readers interested in comprehensive treatment of fracture mechanics should consult one of the numerous texts available, e. g., Kanninen and Popelar (1975), Freund (1993). A detailed listing of the classical literature on fracture mechanics may be found in Paris and Sih (1965).

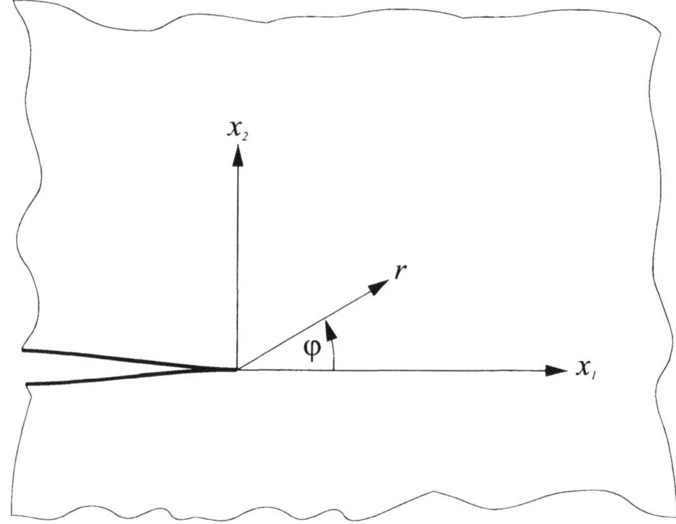

Fig. 7 Coordinates for the crack near-tip fields

Linear elasticity leads to stress singularities at a sharp crack tip and Williams (1957) has shown that the asymptotic behavior of stresses, strains and displacements in the vicinity of a crack tip is the same for every crack problem. In particular, the stress distribution is given universally by

$$\sigma_{ij} = \frac{1}{\sqrt{2\pi r}}\left(K_I f_{ij}^I(\varphi) + K_{II} f_{ij}^{II}(\varphi) + K_{III} f_{ij}^{III}(\varphi)\right) \tag{7.1}$$

with plane polar coordinates r, φ centered at the crack tip (Figure 7). The universal dimensionless function f_{ij} depend only on the angular coordinate φ and will be given explicitely further below. Different crack problems are distinguished by the values of K_I, K_{II} and K_{III}.

At the crack tip, the stresses are governed by a $1/\sqrt{r}$ singularity. The constant K_i, K_{II} and K_{III} are measures for the intensity of the increase of stresses near the crack tip. They depend on the geometry of the body under consideration, on the crack configuration and linearly on the external load. They are called *stress-intensity factors*.

Irwin (1958) defined three basic kinds of crack tip deformation - called *modes* - which are distinguished by the relative displacements of the crack edges (see Figure 8)

mode I: opening mode (K_I),

mode II: inplane sliding mode (K_{II})

mode III: antiplane sliding mode (K_{III}).

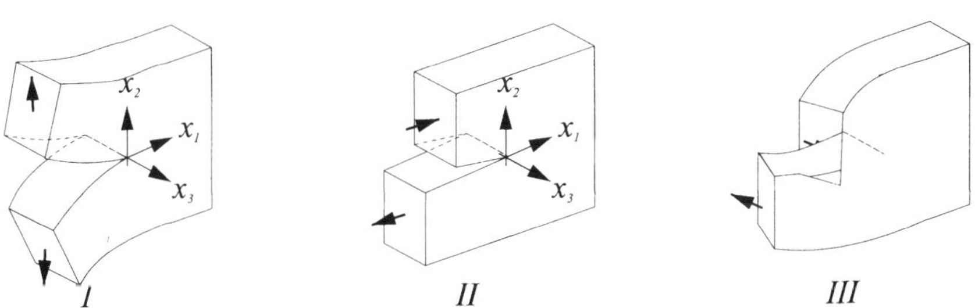

Fig. 8 Crack-opening modes

Generally, the modes of crack extension occur in combined form. This is referred to as mixed-mode condition. Stress-intensity factors are tabulated for several conditions of geometry and loading, e. g., Sih (1973), Tada et al. (1973).

The total energy Π of the system decreases as the crack length increases. This decrease of Π per unit crack tip extension is called the *energy-release rate* (or crack-extension force). For brittle materials, Irwin (1958) established the relation between the energy-release rate $\mathcal{G}$ (already discussed in Chapter 3 in the context of material force) and stress-intensity factors

$$\mathcal{G} = -\frac{\partial \Pi}{\partial a} = \frac{K_I^2 + K_{II}^2}{E^*} + \frac{K_{III}^2}{2G} . \tag{7.2}$$

As already introduced in (5.6), E^* is Young's modulus E for plane stress and $E/(1-\nu^2)$ for plane strain, $G = \frac{1}{2} E/(1+\nu)$ is the shear modulus and ν is Poisson's ratio. In the remainder of this Chapter we restrict ourselves to plane problems $(K_{III} = 0)$.

Having recognized that the total energy Π^{tot} of the system decreases by $\Delta\Pi = -\mathcal{G}$ as the crack length increases, the question arises as to where this energy might be transfered. The answer to this question is contained in the proposition of Griffith (1920), who assumed that the energy released by crack advance is absorbed by the newly created crack surfaces, thus allowing the formulation of an energy balance equation (first law of thermodynamics)

$$\Delta\Pi + \Delta\Gamma = 0 \qquad \text{or} \qquad \mathcal{G} = \Delta\Gamma ,$$

where Γ is the specific surface energy per unit crack advance.

The stresses and displacements near the tip of a crack are given by (cf., e. g., Kanninen and Popelar (1985))

$$\sigma_{11} = \frac{K_I}{\sqrt{2\pi\, r}}\cos\frac{\varphi}{2}(1-\sin\frac{\varphi}{2}\sin\frac{3\varphi}{2}) + \frac{K_{II}}{\sqrt{2\pi\, r}}\sin\frac{\varphi}{2}(-2-\cos\frac{\varphi}{2}\cos\frac{3\varphi}{2}) ,$$

$$\sigma_{22} = \frac{K_I}{\sqrt{2\pi\, r}}\cos\frac{\varphi}{2}(1+\sin\frac{\varphi}{2}\sin\frac{3\varphi}{2}) + \frac{K_{II}}{\sqrt{2\pi\, r}}\sin\frac{\varphi}{2}\cos\frac{\varphi}{2}\cos\frac{3\varphi}{2} ,$$

$$\sigma_{12} = \frac{K_I}{\sqrt{2\pi\, r}}\sin\frac{\varphi}{2}\cos\frac{\varphi}{2}\cos\frac{3\varphi}{2} + \frac{K_{II}}{\sqrt{2\pi\, r}}\cos\frac{\varphi}{2}(1-\sin\frac{\varphi}{2}\sin\frac{3\varphi}{2}) , \tag{7.3}$$

$$4Gu_1 = \sqrt{\frac{r}{2\pi}}\left[K_I\left((2\kappa-1)\cos\frac{\varphi}{2} - \cos\frac{3\varphi}{2}\right) + K_{II}\left((2\kappa+3)\sin\frac{\varphi}{2} + \sin\frac{3\varphi}{2}\right)\right] ,$$

$$4Gu_2 = \sqrt{\frac{r}{2\pi}}\left[K_I\left((2\kappa+1)\sin\frac{\varphi}{2} - \sin\frac{3\varphi}{2}\right) + K_{II}\left(-(2\kappa-3)\cos\frac{\varphi}{2} - \cos\frac{3\varphi}{2}\right)\right] .$$

The symbol κ equals $(3-\nu)/(1+\nu)$ for plane stress and $3-4\nu$ for plane strain.

For pure mode I, K_I may be used directly as a failure parameter. It turns out that a crack extends when a critical material-dependent value of K_I is reached, called fracture toughness K_{Ic}. The assessment of a crack under mixed mode conditions is more involved. A variety of failure criteria exist. Different hypotheses are compiled in, e. g., Kienzler (1993).

Let us consider a plane crack of length 2a in an infinite body referred to an (x_1, x_2) system of coodinates and subjected to some arbitrary loading as shown in Figure 9. The crack edges are free of surface tractions.

Rice (1968) defined the J-integral as

$$J = \int_{\Gamma}(W dx_2 - t_i u_{i,1} ds) \tag{7.4}$$

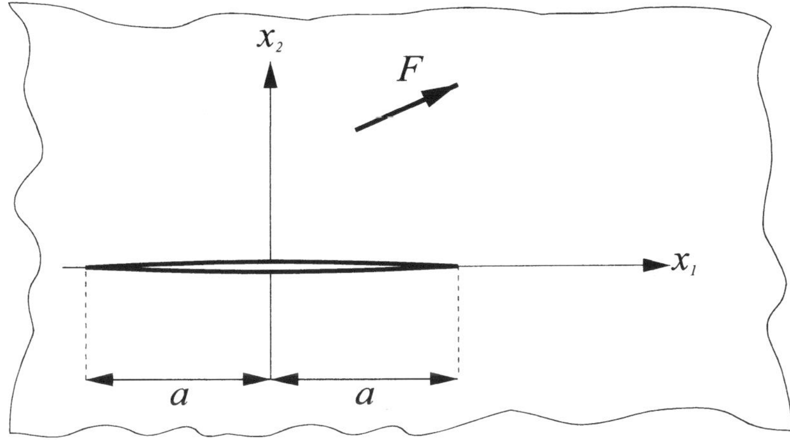

Fig. 9 Crack in the (x_1, x_2) plane

with the components of traction $t_i = \sigma_{ji} n_j$ introduced in equation (2.18). Comparing (3.5) with (7.4) it becomes obvious that J is identical to the component J_1

$$J = J_1 \tag{7.5}$$

of the vector quantity

$$J_i = \int (W\delta_{ji} - \sigma_{jk} u_{k,i}) n_j ds \, . \tag{7.6}$$

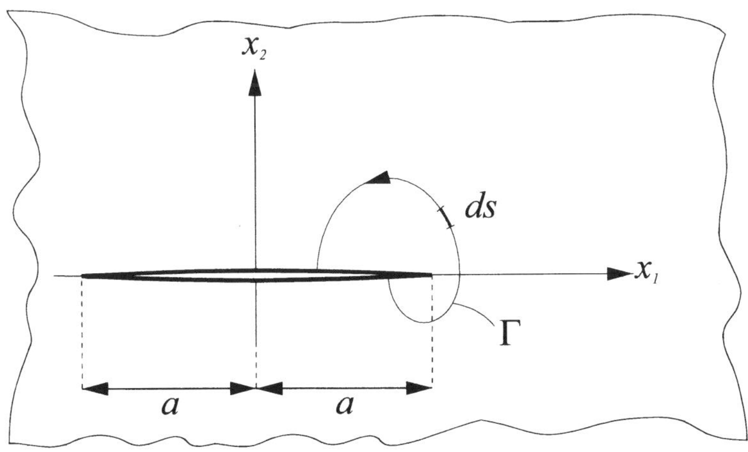

Fig. 10 Path of integration of the J integral around a crack tip

We want now to evaluate the J integral given by (7.4). In choosing the path Γ of integration (arc length s), it is sufficient to start at any point of the lower edge of the crack and procede counter-clockwise around the crack tip to any point of the upper edge, as sketched in Figure 10.

Two features are special to Rice's J integral. The first is that the contour Γ is not necessarily completely closed, the reason being that the integrand of (7.4) vanishes along the crack faces. Indeed, along the crack edges both dx_2 and t_i vanish. The second feature is that the defect, namely the crack, is not enclosed completely, but only a crack tip. The reason for the J-integral maintaining its path-independence is that in the x_1-direction there is only one concentrated material force (crack-extension force) at each crack tip and no material forces along the crack edges. This will be discussed further below.

The integrals J_i given by (7.6) around the crack tip can be evaluated by considering the singular stress and displacement fields given by (7.3) following a circular contour with infinitesimally small radius r, see Figure 11

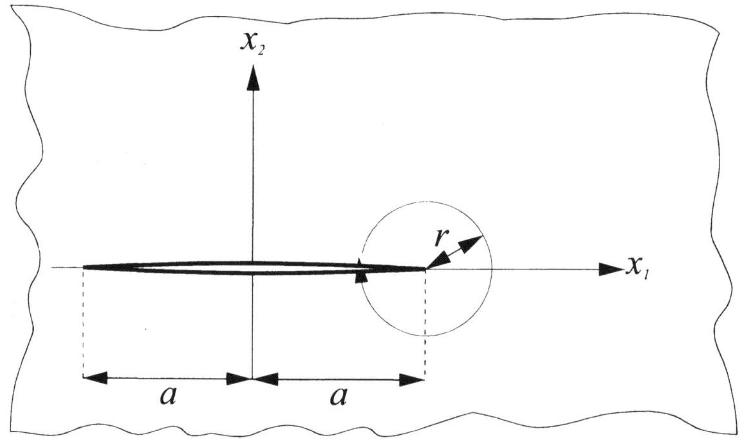

Fig. 11 Circular path around a crack tip with infinitesimal radius r

The results are

$$J_1 = \frac{K_I^2 + K_{II}^2}{E^*}, \tag{7.7}$$

$$J_2 = -2\frac{K_I \, K_{II}}{E^*}. \tag{7.8}$$

It is seen that J_1 is precisely the energy release rate $\mathcal{G}$ per unit crack-tip advance in its own plane as given b equation (7.2). By contrast, the expression for J_2 has no immediate physical interpretation and will also be discussed below. The discussion will further elucidate, why the component J_1 is path independent (such that the evaluation of J_1 in (7.6) along any arbitrary path will deliver (7.7)) and J_2 is path dependent (such that J_2 in (7.6) and (7.8) coincide only if the path is taken along a contour with infinitesimally small radius).

We proceed next to calculate the path-independent integrals $\boldsymbol{J}$, $L_3 = L$ and M based on a contour around the complete crack as indicated in Figure 12.

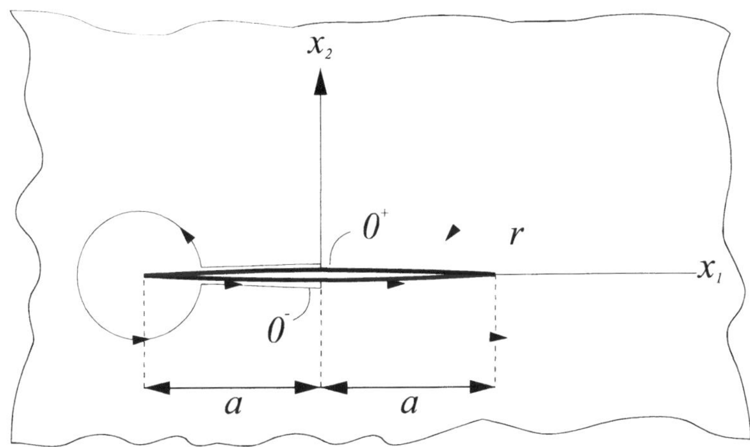

Fig. 12 Path enclosing the crack completely

We introduce the notation J_i^r for the value of J_i around the right crack tip and J_i^l for the value of J_i around the left crack tip, both taken in the counter-clockwise sense. The value of J_i around the complete crack will be denoted by J_i^{tot}. The following result is obtained

$$J_i^{tot} = J_i^r + J_i^l + \int_{-a}^{+a} [j_i]\, dx_l , \tag{7.9}$$

$$L = a(J_2^r - J_2^l) + \int_{-a}^{+a} x_1 [j_2]\, dx_l , \tag{7.10}$$

$$M = a(J_l^r - J_l^l) + \int_{-a}^{+a} x_1 [j_1]\, dx_l , \tag{7.11}$$

where $[j_i]$ is the jump in the material traction of G_i (cf. Chapter 4) across the crack.

$$[j_i] = \left[(W\delta_{ij} - \sigma_{jk} u_{k,i}) n_j \right]\Big|_{x_1,0^+} - \left[(W\delta_{ij} - \sigma_{jk} u_{k,i}) n_j \right]\Big|_{x_1,0^-} = G_i(x_1,0^+) - G_i(x,0^-) . \tag{7.12}$$

Since the crack faces are assumed to be free of applied tractions, (7.12) reduces to

$$[j_1]=0\,,$$
$$[j_2]=W(x_1,0^+)-W(x_1,0^-). \tag{7.13}$$

With the abbreviations

$$[W]:=\int_{-a}^{+a}\left(W(x_1,0^+)-W(x_1,0^-)\right)dx_1\,, \tag{7.14}$$

$$[H]:=\int_{-a}^{+a}\left(W(x_1,0^+)-W(x_1,0^-)\right)x_1dx_1 \tag{7.15}$$

(7.9) - (7.11), can be written as

$$J_1^{tot}=J_1^r+J_1^l\,, \tag{7.16}$$

$$J_2^{tot}=J_2^r+J_2^l+[W]\,, \tag{7.17}$$

$$L=a(J_2^r-J_2^l)+[H]\,, \tag{7.18}$$

$$M=a(J_1^r-J_1^l)\,. \tag{7.19}$$

If the objective of a stress analysis consists in determining stress intensity factors K_I, K_{II} in terms of, essentially, far-field quantities, evaluated numerically for a complicated boundary value problem, then the procedure might be as follows. We evaluate J_1^{tot}, J_2^{tot}, M and L in any convenient fashion using far-field quantities. Further, we evaluate $[W]$ and $[H]$ along the crack edges. It appears that this last calculation is not very sensitive with regard to mesh size in, e. g., a finite element analysis (cf. Kienzler and Kordisch (1990)). With (7.16) - (7.19) J_2^r and J_2^r at the right crack tip can then be given as

$$J_1^r=1/2(J_1^{tot}+M/a)\,, \tag{7.20}$$

$$J_2^r=1/2(J_1^{tot}-[W])+(L-[H]/a)\,, \tag{7.21}$$

and from these K_I^r and K_{II}^r at the right crack tip are found to be

$$K_I^r=\frac{1}{2}\left(\sqrt{E^*J_1^r-E^*J_2^r}+\sqrt{E^*J_1^r+E^*J_2^r}\right), \tag{7.22}$$

$$K_{II}^{r} = \frac{1}{2}\left(\sqrt{E^{*}J_{1}^{r} - E^{*}J_{2}^{r}} - \sqrt{E^{*}J_{1}^{r} + E^{*}J_{2}^{r}}\right). \tag{7.23}$$

Similarly, for the left crack tip we have

$$J_{1}^{r} = 1/2(J_{1}^{tot} - M/a), \tag{7.24}$$

$$J_{2}^{l} = 1/2\left[(J_{2}^{tot} - [W]) - (L - [H])/a\right] \tag{7.25}$$

leading to

$$K_{I}^{1} = \frac{1}{2}\left(\sqrt{-E^{*}J_{1}^{l} + E^{*}J_{2}^{l}} + \sqrt{-E^{*}J_{1}^{l} - E^{*}J_{2}^{l}}\right), \tag{7.26}$$

$$K_{II}^{1} = \frac{1}{2}\left(\sqrt{-E^{*}J_{1}^{l} + E^{*}J_{2}^{l}} - \sqrt{-E^{*}J_{1}^{l} - E^{*}J_{2}^{l}}\right). \tag{7.27}$$

As a specific example we consider a uniform far-field applied stress σ_{ij}^{∞} (homogeneous state of stress), cf. Golebiewska-Herrmann and Herrmann (1981). We will calculate the stress-intensity factors K_I and K_{II} from path-independent integrals. In our example, the complete analytic solution is known and the stress-intensity factors, obviously, would be calculated directly based on this solution, without recourse to any integrals. In a complicated boundary value problem, which would have to be treated numerically, the procedure to obtain the K-values using path-independent integrals is more efficient. Employing the analytical solution, the following expressions are obtained

$$[j_1] = 0, \tag{7.28}$$

$$[j_2] = -\frac{4}{E^{*}}\sigma_{12}^{\infty}(\sigma_{11}^{\infty} - \sigma_{22}^{\infty})\frac{x_1}{\sqrt{a^2 - x_1^2}}. \tag{7.29}$$

Integration leads to

$$[W] = 0, \tag{7.30}$$

$$[H] = -\frac{2\pi a^2}{E^{*}}\sigma_{12}^{\infty}(\sigma_{11}^{\infty} - \sigma_{22}^{\infty}), \tag{7.31}$$

and further

$$J_{1}^{tot} = J_{2}^{tot} = 0, \tag{7.32}$$

$$L = -2\frac{\pi a^2}{E^*}\sigma_{12}^{\infty}(\sigma_{11}^{\infty 2} + \sigma_{22}^{\infty 2}),$$

$$M = -2\frac{\pi a^2}{E^*}(\sigma_{22}^{\infty} - \sigma_{12}^{\infty}).$$

With (6.20 and (7.21) we find

$$J_1^r = \frac{\pi a^2}{E^*}(\sigma_{22}^{\infty 2} + \sigma_{12}^{\infty})^2 = J_1, \tag{7.33}$$

$$J_2^r = -2\frac{\pi a}{E^*}\sigma_{22}^{\infty}\sigma_{12}^{\infty} = J_2, \tag{7.34}$$

and by means of (7.24) and (7.25) we have

$$J_1^l = -J_1^r = -J_1, \tag{7.35}$$

$$J_2^l = -J_2^r = -J_2. \tag{7.36}$$

Substitution into (7.22), (7.23), (7.26) and (7.27) results finally in

$$K_I^l = K_I^r = \sigma_{22}^{\infty}\sqrt{\pi a}, \tag{7.37}$$

$$K_{II}^l = K_{II}^r = \sigma_{12}^{\infty}\sqrt{\pi a}. \tag{7.38}$$

As is frequently done in statics and strength of materials it might be helpful to introduce the concept of the free-body diagram in material space. We consider the crack to be our object and sketch all the material forces acting on it, see Figure 13.

Forces in the x_1-direction are present only at the two crack tips. They are equal in magnitude and opposite in direction. It is now graphically apparent why Rice's J-integral around a crack tip is path-independent and that the path does not have to be closed as already discussed. As regards material forces in the x_2-direction, it is seen that their distribution becomes singular at the crack tip and that their sum vanishes, i. e., $J_2^{tot} = 0$, as already found. The path dependence of J_2 taken around a crack tip is also obvious, because this resultant depends on the length of the crack segment enclosed. The expression $J_2 = -2K_I K_{II} / E^*$ merely represents the strength of the singularity at the crack tip in the x_2-direction. The L-integral is the total vector moment of the force distribution in the x_2-direction, taken with respect to the crack center. The M integral finally, represents the scalar moment (or virial) of the two material forces J_1^r and J_2^l with respect to the crack center.

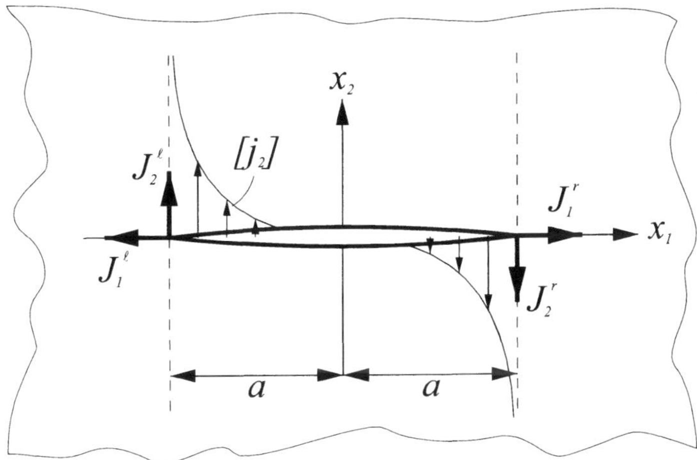

Fig. 13: Free-body diagram in material space of a crack in a remote stress field

If the remote loading is such that

$$\sigma_{11}^{\infty} = \sigma_{22}^{\infty} \tag{7.39}$$

then, as is seen from Equation (7.29), the jump $[j_2]$ vanishes

$$[j_2] = 0 \tag{7.40}$$

and the free-body diagram is reduced, as shown in Figure 14.

Only under this special circumstances does J_2 also become path-independent and equation (7.18) reduces with (7.34) and (7.36) to

$$L = 2aJ_2 \tag{7.41}$$

analogous to the relation, which follows from (7.19) with (7.33) and (7.35) as

$$M = 2aJ_1 \,. \tag{7.42}$$

The latter relation is also applicable without the restriction (7.39) and has been discussed by Freund (1972).

If the remote loading is such that

$$\sigma_{11}^{\infty} = -\sigma_{22}^{\infty} \,, \tag{7.43}$$

we see from equation (7.32 b) that the material vector moment L vanishes, because the vector moment [H] due to the material traction $[j_2]$, given in equation (7.31),

$$[H] = -+\frac{4\pi a^2}{E^*}\sigma_{22}^{\infty}\sigma_{12}^{\infty} \tag{7.44}$$

is just balanced by the moment of the material forces J_2^r and J_2^l, see equation (7.18), (7.34) and (7.36).

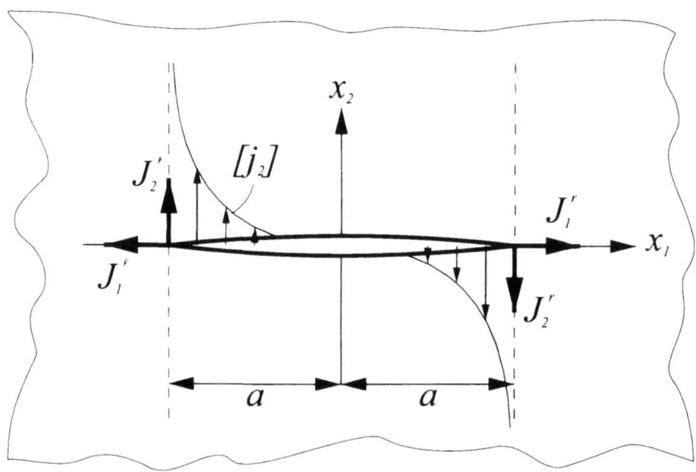

Fig. 14: Free-body diagram of a crack in a remote stress field with $\sigma_{11}^{\infty} = \sigma_{22}^{\infty}$

So far we discussed the situation of an infinite plane elastic domain subjected to constant remote stresses inducing a homogeneous state of stress in the absence of a crack. If the crack is inserted into this stress field, we found that the total material forces J_i^{tot} vanish. Since the material forces describe the change of energy of the system due to a unit translation of the whole crack in x_1-or x_2-direction, this result is obvious, because the energy of the system is the same wherever the position of the crack is. If the state of stress is not homogeneous, either due to non-uniform loadings or due to the presence of a boundary or interface, the situation is changed. Generally J_i^r will not be equal to - J_i^l causing a resultant material force $J_i^{tot} \neq 0$, i. e., the energy of the system is changed, if the position (and orientation) of the crack is varied.

It is tempting to apply the concepts of equilibrium of elementary statics to the free-body diagram in material space. In elementary statics, free-body diagrams are used to determine reaction forces from equilibrium conditions. If our free-body diagram (cf. Figure 13) would be one in physical space, the object (e. g., a beam of length 2a) would be translating and rotating in accordance with the principle of linear and angular momentum. To be held in equilibrium, a vector force and a vector moment equal but opposite to J_i^{tot} and L, respectively, would have to be applied additionally. By contrast, in material space our object, the crack, will not change its configuration, e. g., extend or move in some direction, unless the material forces are sufficiently large and beyond some critical value.

Although Rice's J integral is defined as an energy-release rate per unit crack extension, a finite value of J does not imply a change in configuration of the crack. A material transformation (translation, rotation or self-similar expansion) has to be understood in terms of *virtual* change in configuration. In physical space, it is common practice to apply the virtual-displacement theorem. Consider, e. g., a rigid body subjected to some arbitrary loading as depicted in Figure 15. The body is supported in such a manner, that any change in configuration is excluded.

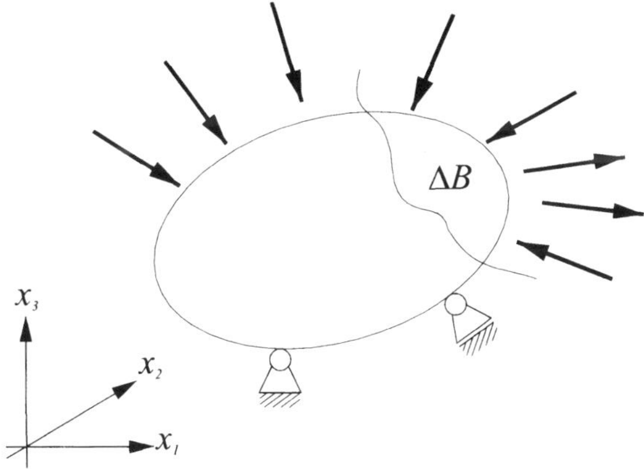

Fig. 15: Rigid body, rigidly supported

By means of the virtual-displacement theorem it is possible to calculate the resultant force exerted by the external loads across a part ΔB of the body in arbitrary, say x_1-direction. We apply a virtual translation δ_1 in the x_1-direction to ΔB, as shown in Figure 16, and calculate the virtual work ΔA of the forces in the translation $\delta_1 \boldsymbol{e}_1$

$$\Delta A = \sum_i F^i \bullet \delta_1 \mathbf{e}_1$$

leading to the component of the resultant force in the x_1-direction acting across ΔB

$$F_1^{res} = \frac{\Delta A}{\delta_1}.$$

Although no displacements u_i occur, it is possible to calculate the rate of energy change, here external work, due to a virtual displacement. Similarly, it is possible to calculate the contribution to the change of energy, if a part of crack is virtually displaced. Consider Figure 14 and, in evaluating J_2^r, say by the line integral (7.6) we determine the change of energy of the system due to a *virtual* translation of a part of the crack in the x_2-direction as shown in Figure 17.

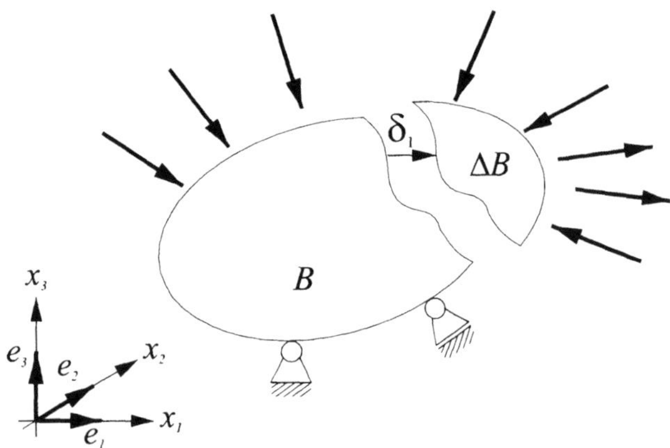

Fig. 16: Virtual displacement applied to the part ΔB of the body B

This does not imply that the crack becomes kinked or that the crack tip is displaced in some directions for any value of the change of energy. Similarly, in calculating the J_1 integral there is, in general, no crack extension in the x_1-direction involved. It is only if J_1 exceeds a critical, material dependent value J_{1c} (the so-called fracture toughness, to be determined experimentally) that crack extension will occur (see Figure 18, right part).

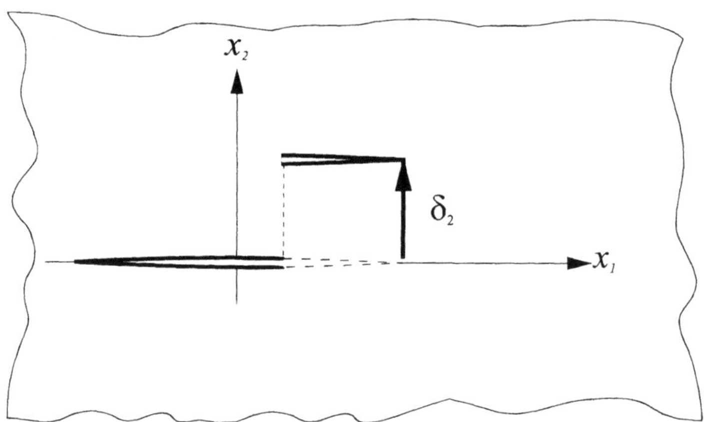

Fig. 17: Virtual translation of a part of the crack

As an analogy one might consider a solid block on a rough base subjected to tensile loading F, as depicted in Figure 18, left part. As long as the force F is smaller than the limiting friction force between the block and the base, the block will remain at rest. Only if the force F reaches the limiting friction force, or exceeds it, will the block lose its position of equilibrium and start moving.

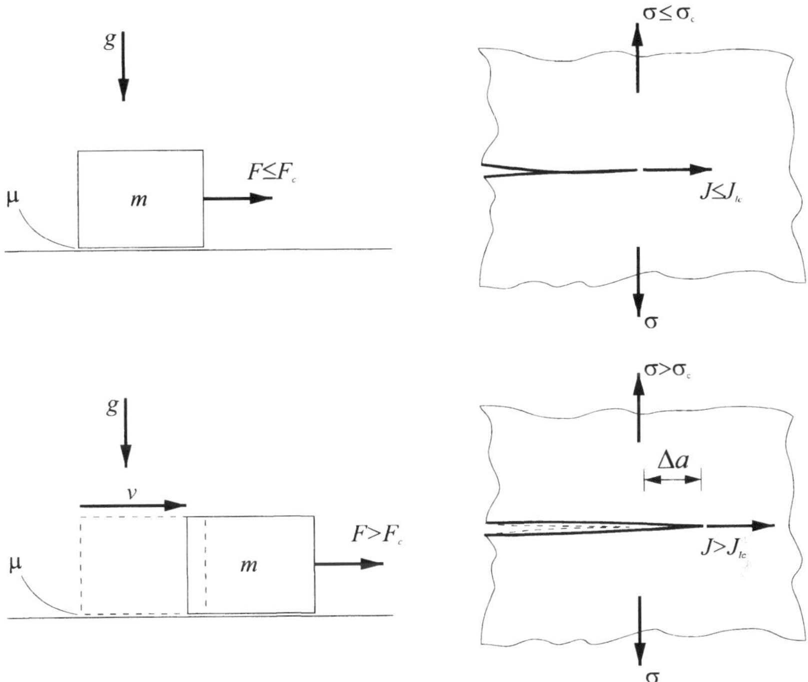

Fig. 18: Block on a rough surface (left) and crack under tensional loading (right)

8 Fracture Criteria Based on the Local Properties of the Eshelby Tensor

Although the $\boldsymbol{J}$ integral has been used ever since Rice published his well-known paper in 1968, the physical significance and the peculiar properties of the Eshelby tensor have been discussed only quite recently by Kienzler and Herrmann (1997).

In this Chapter, local fracture criteria based on components of the Eshelby tensor are proposed, guided by the following line of thinking. Consider a body B of volume V bounded by a surface S of area A with unit outward normal vector $\boldsymbol{n}$. As depicted in Fig. 18, the body is subjected to: (a) a concentrated force $\boldsymbol{F}$, or (b) contains a defect, e. g., a dislocation with intensity $\boldsymbol{B}$.

The physical resultant force action on B can be determined by integration of the associated tractions over S as

$$F_j = \int_S \sigma_{ij} n_i dA , \tag{8.1}$$

whereas the material force on the defect follows from

$$J_j = \int_S b_{ij} n_i dA . \tag{8.2}$$

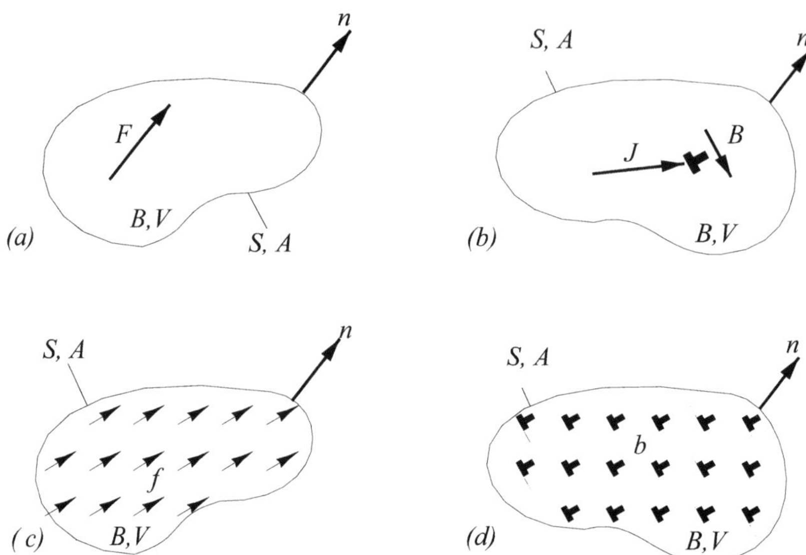

Fig. 19: Body B subjected to: (a) a concentrated force $\boldsymbol{F}$, (b) a "concentrated" dislocation $\boldsymbol{B}$, (c) a smoothly distributed body-force density $\boldsymbol{f}$ and (d) a smoothly distributed dislocation density $\boldsymbol{b}$

One and the same resultant force acting on B might be due to a concentrated force (Fig. 18 (a)) or due to a smoothly distributed body-force density $\boldsymbol{f}$ (Fig. 18 (c)). For the structure under consideration, the concentrated force is much more "dangerous" since it causes locally higher stresses than the smooth load distribution. Consequently, the assessment of structural integrity in the presence of physical forces is usually carried out on the basis of local considerations (e. g., some form of effective stress) and not on the basis of resultant forces as given by (8.1). For concentrated defects, such as a single crack, an integral criterion exemplified by the J integral (8.2) has proved to be reliable, but for distributed defects such as microcracks, voids, dislocations (all collected under the term damage), only a local failure criterion appears to be appropriate. Thus a criterion involving the components of the Eshelby tensor should be a suitable basis for this purpose. But even for a single crack, as shows below, an Eshebly-based criterion leads to useful results.

Since displacements and stresses in the near field of a crack tip are asymptotically in a state of plane strain cf. Riedel (1984), it is sufficient to investigate the components b_{ij} of the Eshelby tensor in two dimensions. Assuming a plane Cartesian coordinate system (x_1, x_2), b_{ij} is given as (3.9)

$$b_{ij} = W\delta_{ij} - \sigma_{ik}u_{k,j} \, . \tag{8.3}$$

The tensor is non-symmetric and, in the absence of defects, divergence-free

$$b_{12} \neq b_{21}\,, \tag{8.4}$$

$$b_{ij,i} = 0\,. \tag{8.5}$$

Replacing W by $\frac{1}{2}\sigma_{ij}u_{j,i}$ we find

$$b_{11} = -b_{22}\,. \tag{8.6}$$

The components b_{nn} and b_{nt} at a cross-section inclined at an angle ψ with respect to the x_1-axis are depicted in Fig. 19 and follow from the transformation equation

$$b_{nn} = \frac{1}{2}(b_{11}+b_{22}) + \frac{1}{2}(b_{11}-b_{22})\cos 2\psi + \frac{1}{2}(b_{12}+b_{21})\sin 2\psi\,,$$

$$b_{nt} = \frac{1}{2}(b_{12}-b_{21}) + \frac{1}{2}(b_{12}-b_{21})\cos 2\psi - \frac{1}{2}(b_{11}+b_{22})\sin 2\psi\,. \tag{8.7}$$

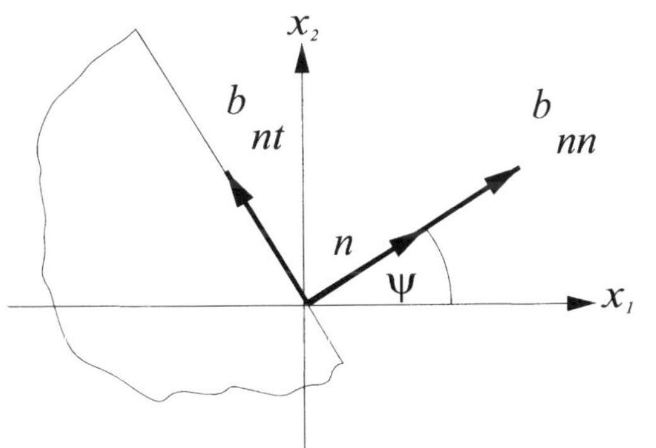

Fig. 20: Cross-section under an arbitrary angle ψ

Conveniently, the properties of the Eshelby tensor can be discussed in terms analogous to Mohr's circle (cf., e. g., Timoshenko and Goodier, 1970).

Elimination of the transformation angle ψ from (8.7) leads to

$$b_{nn}^2 + (b_{nt}+a)^2 = s^2,$$

$$a = \frac{1}{2}(b_{21}-b_{12}) \gtreqless 0\,, \tag{8.8}$$

$$s^2 = b_{11}^2 + \frac{1}{4}(b_{12}-b_{21})^2\,.$$

Mohr's circle for the Eshelby tensor is sketched in Fig. 21.

Any pair (b_{nn}, b_{nt}) that can be calculated by (8.7) from a given "state" b_{ij} lies on a circle with radius s and centre at $-a$ on the b_{nt}-axis. Principal directions, i. e., cross-sections with a vanishing b_{nt} component, exist only, if $|a| \leq s$. In general, they are not perpendicular to each other. Principal values $b_I = -b_{II}$ are real, if $|a| \leq s$ or conjugate complex otherwise. A rather detailed discussion is given in Kienzler and Herrmann (1997). For the following, it is essential that the extremal values of b_{ij} occur at the b_{nt}-axis and are given as

$$b_{extr.} = a \pm s \tag{8.9}$$

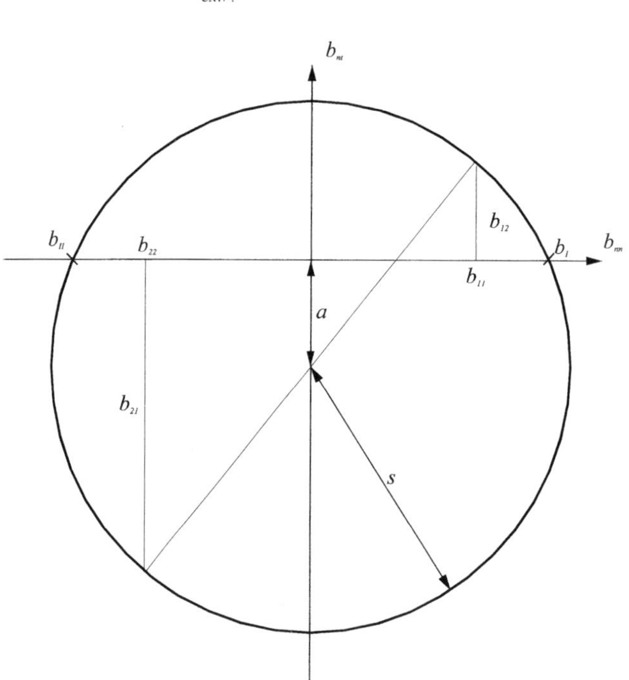

Fig. 21: Mohr's circle for the Eshelby tensor

By contraction of b_{jk} with the completely skew-symmetric permutation tensor ε_{lki}, Eshelby (1975 b) introduced a modified tensor $b_{ij}^{/}$ as

$$b_{ij}^{/} = \varepsilon_{3ki} b_{jk} = \begin{bmatrix} -b_{12} & +b_{11} \\ -b_{22} & +b_{21} \end{bmatrix}, \tag{8.10}$$

which reveals interesting features. The tensor is symmetric, divergence-free and its trace is a harmonic function

$$b'_{ij} = b'_{ji},$$
$$b'_{ij,i} = 0, \tag{8.11}$$
$$b'_{ii,jj} = 0.$$

Thus, b'_{ij} could describe a state of stress and we can calculate its principal values and a v.Mises-type of equivalent value

$$\begin{aligned} b'_{I,II} &= \frac{1}{2}(b'_{11} + b'_{22}) \pm \sqrt{\left(\frac{b'_{11} - b'_{22}}{2}\right)^2 + b'^2_{12}} \\ &= \frac{1}{2}(b_{21} - b_{12}) \pm \sqrt{\left(\frac{b_{21} + b_{12}}{2}\right)^2 + b^2_{11}} \\ &= \quad a \quad \pm \quad s \\ &= b_{extr.}, \end{aligned} \tag{8.12}$$

$$b'_v = \sqrt{b'^2_{11} + b'^2_{22} - b'_{11}b'_{22} + 3b'^2_{12}}. \tag{8.13}$$

The principal values $b'_{I,II}$ of b'_{ij} coincide with the extremal values $b_{extr.}$ of b_{ij}. Although no specific physical interpretation will be attached to b'_v, both $b'_I(|b'_I| \geq |b'_{II}|)$ and b'_v will used to establish fracture criteria.

Consider a crack in the x_1, x_2-plane. The origins of a (x_1, x_2) Cartesian coordinate system and a (r, φ) polar coordinate system coincide with the crack tip as shown in Fig. 7.

The near-tip crack fields for stresses and displacements are already given in (7.3). The calculation of b_{ij} from (8.3) is straight-forward (Kienzler, 1993). With the abbreviations

$$c = \cos\varphi, \qquad s = \sin\varphi, \tag{8.14}$$

we have

$$b_{ij} = \frac{1}{E^*\pi r}\begin{bmatrix} c(K_1 s + K_{II}c)^2 & -(K_I c - K_{II}s)(K_I sc + K_{II}(1 + c^2)) \\ s(K_1 s + K_{II}c)^2 & -c(K_I s + K_{II}c)^2 \end{bmatrix}. \tag{8.15}$$

With (8.10), (8.12),(8.13) and (8.15) we calculate

$$b'_{I,II} = \frac{1}{2E'\pi r}\left\{K_I{}^2 s + 2K_I K_{II} c - K_{II}{}^2 s \pm \right.$$

$$\left. \pm\sqrt{(K_I{}^2 s + 2K_I K_{II} c + K_{II}{}^2 s)^2 + (2K_{II}{}^2 c)^2}\right\} \tag{8.16}$$

$$b'_V = \frac{1}{E'\pi r}\sqrt{K_I{}^4 s^2 + 4K_I{}^3 K_{II} sc + K_I{}^2 K_{II}{}^2 (s^2 + 4c^2) + 2K_I K_{II}{}^3 sc + K_{II}{}^4 (s^2 + 3c^2)}. \tag{8.17}$$

As an example, we consider the infinite plane with an inclined crack of length $2a$ subjected to an applied far-field stress σ_0 in x_2-direction (see Fig. 22).

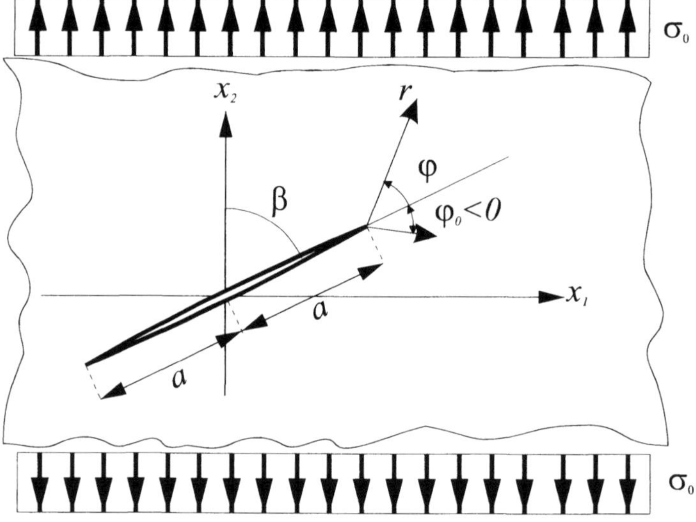

Fig. 22: Plane with inclined crack under far-field tension, crack-kinking angle φ_0

The stress intensity factors are (Sih, 1973)

$$K_I = \sigma_0\sqrt{\pi a}\,\sin^2\beta \ ,$$
$$K_{II} = \sigma_0\sqrt{\pi a}\,\sin\beta\cos\beta \ . \tag{8.18}$$

It follows that

$$\cot\beta = \frac{K_{II}}{K_I}, \tag{8.19}$$

and, thus, the inclination angle β might be used as a measure of the ratio of normal and shear load at the crack tip.

Firstly, we calculate the angle φ^* where the quantities $b_I^/ = |a| + s$ and $b_v^/$ take their maximum values

$$\frac{\partial b_I^/}{\partial\varphi} = 0 \quad \Rightarrow \quad \varphi_I^*, \qquad \frac{\partial b_v^/}{\partial\varphi} = 0 \quad \Rightarrow \quad \varphi_v^*. \tag{8.20}$$

We find

$$b_I \max = b_I^/(\varphi = \varphi_I^*), \qquad b_v \max = b_v^/(\varphi = \varphi_v^*) \tag{8.21}$$

and define these values as critical failure parameters b_{Ic} and b_{vc}, respectively. A pure mode I loading occurs at $\beta = \pi/2$, and in b_{Ic} and b_{vc}, the stress intensity factor K_I is equal to the fracture toughness K_{Ic}. With this "normalization" we can plot b_{Ic} and b_{vc} in a $(K_{II}/K_{Ic}, K_I/K_{Ic})$ -diagram together with various mixed-mode fracture criteria found in the literature (Kordisch 1982). As shown in Fig. 23 (a), the newly proposed criteria fit well into already existing ones.

Secondly, the crack kinking angle φ_0 can also be determined and compared to other approaches (see Fig. 23 (b)) with good agreement. Details of the calculation can be found in Kienzler (1993) and in Kienzler and Herrmann (2002).

In concluding this Chapter, it may be mentioned that local failure criteria based on the Eshelby tensor are shown to yield useful results for a crack under mixed-mode loading. For distributed defects (damage), similar criteria appear to be suitable, but, of course, the specific form of such a criterion based on the Eshelby tensor can be established only with the aid of systematic experiments.

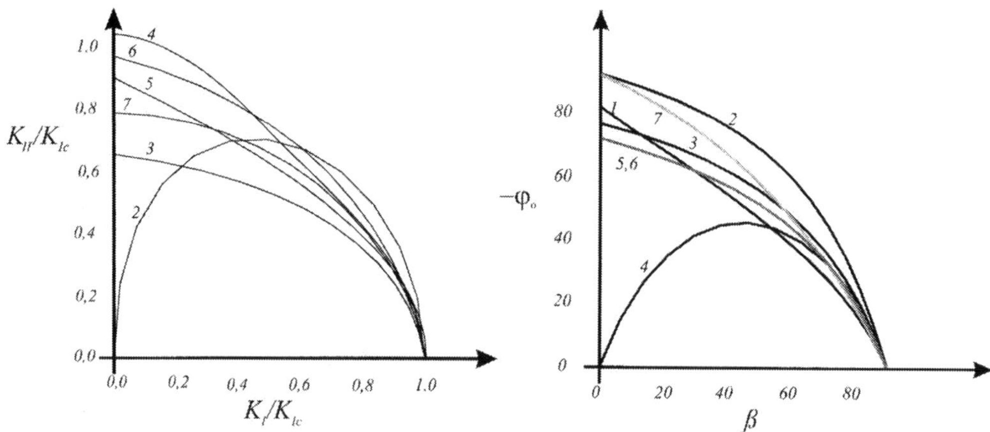

Fig. 23: Comparison of criteria for mixed-mode fracture (a) and crack-kinking angles (b). Legend (references are given in Kordisch (1982): 1 energy-density Sih, 2 tangential stress Erdogen and Sih, 3 energy-release rate Hussain, Pu and Underwood, 4 energy-release rate Hellen and Blackburn, 5 principal stress Finnie and Saith, 6 principal value b_{Ic}, 7 equivalent value b_{vc}

9 Fracture Mechanics in Strength - of - Materials

It is a remarkable circumstance that some elements of fracture mechanics, such as energy-release rates and stress-intensity factors, might be developed not on the basis of continuum theories, but on the basis of the much older and simpler theories of strength - of - materials. Here, bars in tension - compression are treated. A detailed outline of the theory and applications to cracked beams in bending, cracked shafts in torsion, arches, plates and shells with cracks may be fond in Kienzler and Herrmann (2000).

Let us consider a uniform elastic bar of unspecified length l and cross-section A subjected to end loads N_0 as sketched in Figure 24.

Equilibrium of forces in the x-direction requires

$$N' = 0 \Rightarrow N = N_0 = const., \tag{9.1}$$

where N is the resultant internal axial force and primes indicate differentiation with respect to the axial coordinate $x_1 = x$.

Let u be the displacement of any cross section, then u' is the strain ε

$$u' = \varepsilon . \tag{9.2}$$

Hooke's law is expressed as

$$N = EA\varepsilon . \tag{9.3}$$

a.)

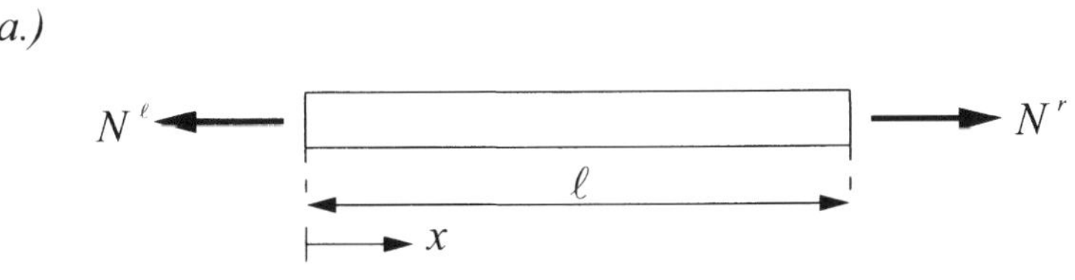

b.)

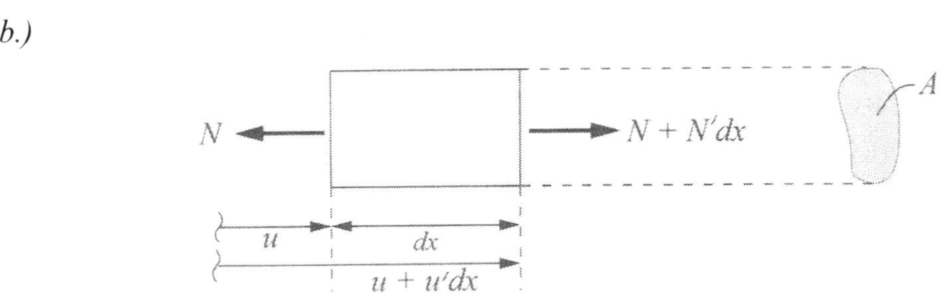

Fig. 24: Bar under tension-compression (a) and deformed in finitesimal element of a bar (b), cross-section A

The product EA is called the axial stiffness and is assumed to be constant for the present. The strain energy $\overline{W}$ per unit length of the bar is given by

$$\overline{W} = \frac{1}{2} EAu'^2 = \frac{1}{2} \frac{N^2}{EA} . \tag{9.4}$$

Conservation laws for bars in tension - compression may be obtained in various ways as sketched in Chapter 5. We simply deduce the relations by transition von $3D$ to $1D$ as follows

$$3D \quad \rightarrow \quad 1D$$

$$b_{ij} = W\delta_{ij} - \sigma_{ik} u_{k,i} \quad \rightarrow \quad B = \overline{W} - Nu' , \tag{9.5}$$

$$b_{ij,i} = 0 \quad \rightarrow \quad B' = 0 . \tag{9.6}$$

Equations (9.1) and (9.6) are the one-dimensional physical and material "conservation laws", respectively. Let us now explore the physical significance of the material force B by considering a bar containing a jump in axial stiffness EA at an arbitrarily fixed position $x = \xi$ given by

$$EA(x) = \begin{bmatrix} EA^{(1)} = const. \; for \; x < \xi \\ \\ EA^{(2)} = const. \; for \; x > \xi \end{bmatrix}, \tag{9.7}$$

(see Figure 25, state *I*. At the transition point $x = \xi$, we can distinguish between continuous and discontinuous variables.

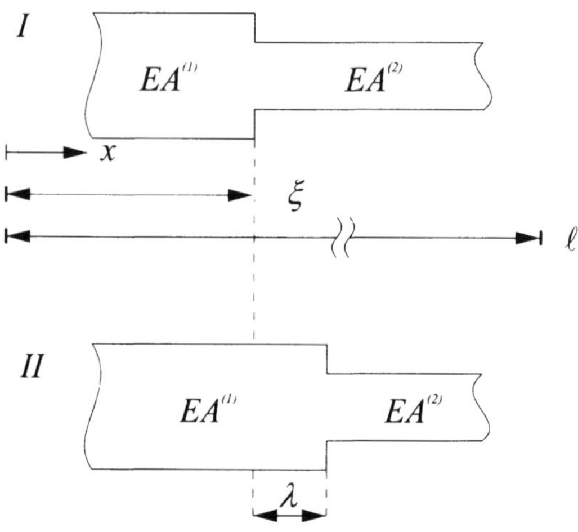

Fig. 25: Bar with jump in axial stiffness

Obviously, N and u are continuous, while EA and u' are discontinuous. The expression for the material force B given by (9.5) might be rearranged by (9.2) - (9.4) as

$$B = -\frac{1}{2}\frac{N^2}{EA}. \tag{9.8}$$

Therefore, it follows that the material force B is discontinuous. The jump term $[B]$ in B is easily calculated to be

$$[B] = B^+ - B^- = -\frac{1}{2}N^2(\xi)[C], \tag{9.9}$$

where $[C]$ is the jump in the compliance $C = \dfrac{1}{EA}$

$$[C] = \frac{1}{EA^{(2)}} - \frac{1}{EA^{(1)}}. \tag{9.10}$$

To provide a physical interpretation of $[B]$, a bar (length ℓ) is considered to be composed of two sections, $EA^{(1)}$ (length ξ) and $EA^{(2)}$ (length $\ell-\xi$) subjected to end forces N_0. The potential Π^i of the internal forces, i. e., the strain energy of the system, is given by

$$\Pi^i = \int_0^\ell W dx = \frac{1}{2}\int_0^\ell \frac{N_0^2}{EA} dx\,, \tag{9.11}$$

and, due to Clapeyron's theorem (2.17), the potential of the external forces follows to be

$$\Pi^a = -2\Pi^i\,. \tag{9.12}$$

Thus the complete potential energy Π is given by

$$\Pi = \Pi^i + \Pi^a = -\Pi^i\,. \tag{9.13}$$

We wish now to calculate the change of energy when the cross section $x=\xi$ is translated by a small amount λ (Figure 25 state *II*). Using (9.11) and (9.13), the result is

$$\begin{aligned} \lambda = 0: \qquad & \Pi_I = -\frac{1}{2}\left(\frac{N_0^2\xi}{EA^{(1)}} + \frac{N_0^2(\ell-\xi)}{EA^{(2)}}\right), \\ \lambda \neq 0: \qquad & \Pi_{II} = -\frac{1}{2}\left(\frac{N_0^2(\xi+\lambda)}{EA^{(1)}} + \frac{N_0^2(\ell-\xi-\lambda)}{EA^{(2)}}\right). \end{aligned} \tag{9.14}$$

The change of energy due to this material translation (see Chapter 3) turns to be

$$\Delta\Pi = \Pi_I - \Pi_{II} = \frac{1}{2}\lambda N_0^2[C] = -\lambda[B]\,. \tag{9.15}$$

The quantity $\lambda[B]$ may be, therefore, interpreted as the work of the material (concentrated) force $[B]$ in the material translation λ. The force $[B]$ acts at the cross section $x=\xi$ and points horizontally in the direction of the stiffer material (Kienzler, 1993). The force is likely to lower the amount of Π, by weakening the beam, i. e., by shifting the discontinuity such that the weaker part of the beam increases (removal of mass).

With

$$\overline{\mathcal{G}} = -\lim_{\lambda\to 0}\frac{\Pi(\xi+\lambda)-\Pi(\xi)}{\lambda} = \frac{-\partial\Pi}{\partial\lambda} = +[B]\,, \tag{9.16}$$

$[B]$ may be identified as the energy-release rate $\overline{\mathcal{G}}$ due to a translation of the cross section $x=\xi$ in x-direction. Equations (9.15) and (9.16) are also valid for arbitrary loading and boundary conditions. Here, N_0 merely has to be replaced by $N(\xi)$. For the calculation of $[B]$

it is necessary to know only $N(\xi)$ and $[C]$.

Next we consider a bar (length ℓ) containing at $x=\xi$ a segment of length 2c with reduced stiffness (Figure 26, state *I*).

$$EA(x)=\begin{cases} EA^{(1)}=const.\ for\ 0\le x<\xi-c\ and\ \xi+c<x\le\ell, \\ EA^{(2)}<EA^{(1)}\ for\ \xi-c<x<\xi+c. \end{cases} \tag{9.17}$$

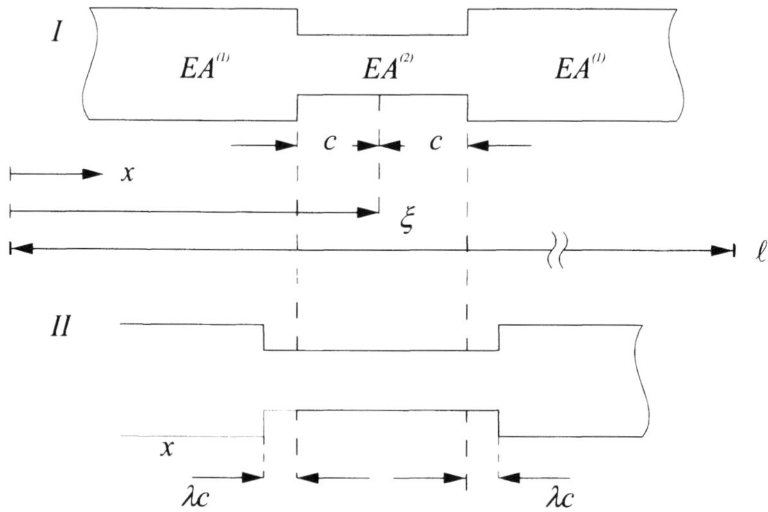

Fig. 26: Bar with partly reduced stiffness

Again, the bar is subjected to pure tension $N_0=const.$ The change of total energy, now, due to a selfsimilar expansion $c\to c(1+\lambda)$ (Figure 26, state *II*) is calculated in the same way as above yielding

$$\Delta\Pi=-\lambda cN^2[C]=2\lambda c[B]. \tag{9.18}$$

The energy-release rate due to the transformation $c\to c(1+\lambda)$ is

$$\overline{\mathcal{G}}=-\frac{\partial\Pi}{\partial(\lambda c)}=-2[B]. \tag{9.19}$$

In the following , we will assume that the length c in Figure 26 is small in comparison to the length ℓ. Such that the bar with partly reduced stiffness might be regarded as a structure with two symmetric edge cracks.

As mentioned in Chapter 7, energy-release rates play an essential role in fracture mechanics. The energy-release rate $\mathscr{G}$ due to crack extension $a \to a + \Delta a$ is closely related in linear elasticity to stress-intensity factors K (see eq. (7.2))

$$\overline{\mathscr{G}} = \frac{K^2}{E^*} . \tag{9.20}$$

In Kienzler and Herrmann (1986) it was assumed and in Kienzler and Herrmann (2000) thoroughly discussed that the energy-release rate $\mathscr{G}$ for crack extension is equal to that for crack widening, i. e., $c \to c(1+\lambda)$. Although $\overline{\mathscr{G}}$ corresponds to a virial acting perpendicular to the crack driving force, it was postulated that

$$\mathscr{G} = \overline{\mathscr{G}} / d . \tag{9.21}$$

For dimensional reason ($\mathscr{G}$ is defined in plane elasticity whereas $\overline{\mathscr{G}}$ is defined in bar theory) a measure of thickness d needs to be introduced.

Combining (9.9), (9.10). (9.19) - (9.21) leads to a remarkably simple formula to calculate stress-intensity factors for bars with cracks under tension.

$$K = N\sqrt{\frac{1}{d}\left(\frac{1}{A_2} - \frac{1}{A_1}\right)} . \tag{9.22}$$

As a first example, consider a bar of rectangular cross-section $d\ x\ h$ with symmetrical edge cracks of length a under pure tension (Figure 27).

The cross-section areas are

$$A^{(1)} = dh ,$$

$$A^{(2)} = d(h - 2a) .$$

With (9.22) the stress-intensity factors for the bar under tensile loading are given by

$$K = \frac{N}{d\sqrt{h}} g_N\left(\frac{a}{h}\right) \tag{9.22a}$$

with

$$g = \left(\frac{a}{h}\right) = \sqrt{\frac{1}{1 - 2\frac{a}{h}}} - 1 . \tag{9.22b}$$

The results are compared graphically with those of Benthem and Koiter (1973) in Figure 27. The agreement is quite satisfactory.

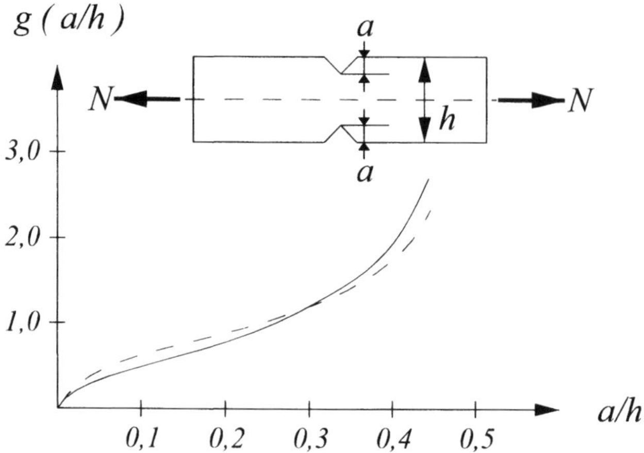

Fig. 27: Stress-intensity factor vs. dimensionless crack length for a bar under simple tension with symmetrical edge cracks (— (9.22), Benthem and Koiter, 1973)

As a second example, consider a bar of rectangular cross-section dxh with a centered crack of length $2a$ under pure tension/bending (Figure 29). As in the example above, the cross-sectional aras are

$$A^{(1)} = dh \qquad \text{and} \qquad A^{(2)} = d(h - 2a).$$

It is, therefore, not possible to distinguish between edge cracks and a center crack for the tension loading. Equation (9.22) is applicable likewise. The result is compared graphically with those of Benthem and Koiter (1973) in Figure 28.

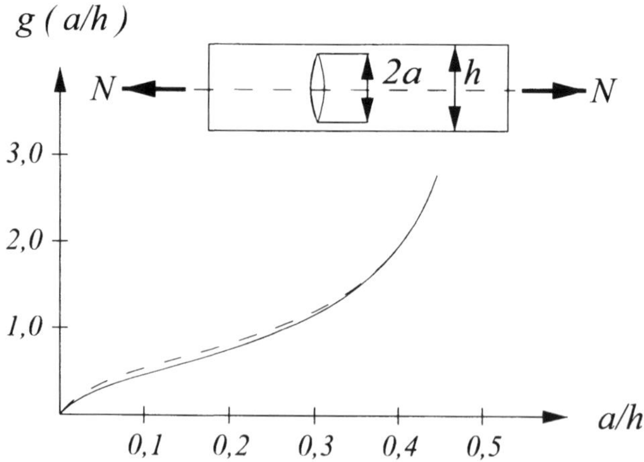

Fig. 28: Stress-intensity factor vs. dimensionless crack length for a bar under pure bending with center crack (— (9.22), ···Benthem and Koiter, 1973)

In Kienzler and Herrmann (2000) further applications are given and it may be concluded that the far-reaching and rather unexpected applicability of theories of strength-of-materials to cracked structural elements of a great variety confirms the power of these theories which rest mainly on their simplicity and accuracy as compared to theories of elastic continua.

References

Bakker, A. (1984). The three-dimensional J-integral. Ph.D. Thesis, Technical University Delft.

Benthem, J.P., and Koiter, W.T. (1973). Asymptotic approximations to crack problems. In: Sih, G.C. (ed.), *Mechanics of Fracure I.* Noordhoff, Leyden 131-178.

Budiansky, B. and Rice, J.R. (1973). Conservation laws and energy-release rates. *J. Appl. Mech.* **40**,201-203

Chien, N. (1992). Conservation laws in non-homogeneous and dissipative mechanical systems. Ph. D. Thesis, Stanford University.

Dundurs, J. and Mura, T. (1964). Interaction between an edge dislocation and a circular inclusion. *J. Mech. Phys. Sol.* **12**, 177-189.

Eshelby, J.D. (1975 a). The elastic energy-momentum tensor. *J. Elast.* **5**, 321-335.

Eshelby, J.D., (1975b). The calculation of energy release rates. In: Sih, G.C., van Elst, H.C. and Broek, D. (eds.), *Prospects of Fracture Mechanics.* Noordhoff, Leyden 69-84.

Freund, L.B. (1972). Energy flux into the tip of an extending crack in an elastic solid. *J. Elast.* **2**, 341-349

Freund, L. B. (1993). *Dynamic Fracture Mechanics.* Cambridge University Press, Cambridge, U. K.

Fung, Y.C. (1963). *Foundations of solid mechanics.* Prentice-Hall, Englewood Cliffs, N. J.

Gelfand, I.M. and Fomin, S.V. (1963). *Calculus of Variations.* Prentice-Hall, Englewood Cliffs, N. J.

Golebiewska-Herrmann, A. and Herrmann, G. (1981). On energy-release rates for a plane crack. *J. Appl. Mech.* **48**, 525-528.

Griffith, A.A. (1920). The phenomena of rupture and flow in solids. *Phil. Trans. Roy. Soc. London* **A 211**, 163-198.

Günther, W. (1962). Über einige Randintegrale der Elastomechanik. *Abh. Braunschw. Wiss. Ges.* **14**, 53-72.

Honein, T. Chien, N. and Herrmann, G. (1991). On conservation laws for dissipative systems. *Phys. Lett.* **A 155**, 223-224.

Irvin, G.R. (1958). Fracture. In: Flüge, S. (ed.). *Handbuch der Physik.* Springer, Berlin **VI**, 551-590.

Kanninen, M.F. and Popelar, C.H. (1985). *Advanced Fracture Mechanics.* Clarendon Press, Oxford.

Kienzler, R. (1993). *Konzepte der Bruchmechanik.* Vieweg, Braunschweig.

Kienzler, R. and Duan, Z.P. (1987). On the distribution of hoop stresses around circular holes in elastic sheets. *J. Appl. Mech.* **54**, 110-114.

Kienzler, R. and Herrmann, G. (1986). An elementary theory of defective beams. *Acta Mech.* **62**, 37-46.

Kienzler, R. and Herrmann, G. (1997). On the properties of the Eshelby tensor. *Acta Mech.* **125** 73-91.

Kienzler, R. and Herrmann, G. (2000). *Mechanics in material space*. Springer, Berlin.

Kienzler, R. and Herrmann, G. (2002). Fracture criteria based on local properties of the Eshelby tensor. *Mech. Res. Comm.* **29**, 521-527.

Kienzler, R. and Kordisch, H. (1990). Calculation of J_1 and J_2 using the L and M integrals. *Int. J. Fracture* **43**, 213-225.

Kordisch, H. (1982). Untersuchungen zum Verhalten von Rissen unter überlagerter Normal- und Scherbeanspruchung. Dissertation, Universität Karlsruhe.

Noether, E. (1918). Invariante Variationsprobleme. *Nachr. Ges. Wiss. Gottingen, Math.-Phys. Kl.* **2**, 235-257.

Olver, P.J. (1993). *Applications of Lie Groups to Differential Equations*, 2nd ed. Graduate Texts in Mathematics, No 107. Springer, New York.

Paris, P.C. and Sih, G.C. (1965). Stress analysis of cracks. In: *Fracture Toughness Testing and its Applications*, ASTM STP 381. American Society for Testing and Materials, Philadelphia 30-81.

Rice, J.R. (1968). A path independent integral and the approximate analysis of strain concentration by notches and cracks. *J. Appl. Mech.* **35**, 379-386.

Riedel, H. (1984). Die dritte Dimension in der Bruchmechanik. Tagungsband der 16. Sitzung des Arbeitskreises Bruchvorgänge im DVM. Karlsruhe 7-14.

Schweins, (1849). Fliehmomente oder die Summe $(xX + yY)$ bei Kräften in der Ebene, und $(xX + yY + zZ)$ bei Kräften im Raume. *Crelles J. reine angew. Math.* **38**. 77-88.

Sih, G.C. (1973). *Hanbook of Stress Intensity Factors*. Lehigh University.

Stark, J.P. (1976). *Solid State Diffusion*. Wiley, New York.

Tada, H., Paris, P.C. and Irwin, G.R. (1973). *The Stress Analysis of Cracks Handbook*. Hellertown, Pennsylvania.

Timoshenko, S.P., and Goodier, J.N. (1970). *Theory of Elasticity*, 3rd ed. McGraw-Hill, New York.

Volterra, V. (1907). Sur l'équilibre des corps élastiques multiplement connexes. *Annls. Scient. Éc. norm. sup. Paris* **24**, 401-517.

Williams, M.L. (1957). On the stress distribution at the base of a stationary crack. *J. Appl. Mech.* **24**, 109-144.

Phase Separation in Binary Alloys

- Modeling Approaches

Peter Fratzl and Richard Weinkamer

Max-Planck Institute of Colloids and Interfaces,
Department of Biomaterials,
14424 Potsdam Germany

Abstract. The physical principles and modern modeling approaches for phase separation in binary alloys are reviewed. A fundamental distinction between the different simulation models is their description of the moving interface during the separation process. The interface between the phases is either atomically rough, or a diffuse interface with a smooth transition between the phases, or finally a sharp geometric interface with a discontinuous jump of the bulk properties of one phase to the other. Advantages and disadvantages of the these microscopic, mesoscopic and macroscopic modeling approaches are presented. Concerning examples given some emphasis is put on the influence of elastic interactions caused by a misfit between the phases on the transformation process.

List of Notations

a	(cubic) lattice constant
B	elastic anisotropy, defined in (3.18)
$\widetilde{B}(\boldsymbol{k})$	elastic potential
$c,\ c_B$	concentration of solute B atoms (defined as lattice site fraction)
$\widetilde{c}(\boldsymbol{k})$	Fourier transform of the concentration $c(\boldsymbol{r})$ defined in (2.28) and (4.21)
c_A	concentration of solvent A atoms
c_α	equilibrium concentration of the matrix phase α
c_β	equilibrium concentration of the precipitate phase β
c_R	concentration at the curved interface between matrix and precipitate
$\overline{c}$	average concentration of the solid solution
c_∞	concentration of the supersaturated matrix
D	diffusion constant
E	energy
f	free energy density, i.e., free energy per unit volume
F	(Helmholtz) total free energy

F_{chem}	chemical contribution to the total free energy F
F_{int}	interfacial contribution to the total free energy F
$F_{elastic}$	contribution of the coherent strain energy to the total free energy F
g	thermodynamic potential defined in (2.16)
$\tilde{g}$	spherically averaged pair correlation function
G	shear modulus
$I(\boldsymbol{k},t)$	intensity, defined in (2.30)
j	atomic flux
J	ordering energy, defined in (2.5) and (3.6), respectively
k_B	Boltzmann´s constant
K	bulk modulus
$\hat{K}$	nucleation rate
M	mobility
$\boldsymbol{n}$	unit vector normal to the interface Γ
N	number of lattice sites or atoms, respectively
N_{XY}	number of NN bonds of atoms, $X,Y = A,B,V$
p_i^{α}	occupation parameter, defined in (3.2)
R	size (radius) of a precipitate
$\overline{R}$	average precipitate size
R_D	inter-precipitate spacing
R_A	size of the domain of precipitates with the same orientation
S	entropy
$S(\boldsymbol{k},t)$	structure function, defined in (3.19)
T	temperature
T_c	critical temperature of the phase transition
U	asymmetry parameter, defined in (2.5) and (3.6), respectively
$\boldsymbol{u}$	displacement of the atoms from their regular lattice position
u_{ij}	components of the strain tensor defined in (4.18)
$\boldsymbol{v}$	velocity of the interface
V	volume of the system
$W(\boldsymbol{x}_l \rightarrow \boldsymbol{x}_{l'})$	transition probability for passing from state $\boldsymbol{x}_l$ to $\boldsymbol{x}_{l'}$
Z	coordination number of the lattice
δ	lattice misfit
ε_{XY}	binding energy between a NN pair of atoms, $X,Y = A,B,V$
μ	chemical potential
σ	specific interfacial energy
σ_i, $\sigma(\boldsymbol{r})$	spin variable, defined in (3.3)
χ	interfacial energy parameter used in mesoscopic models
Ω	atomic volume

1. Introduction

Most metallic alloys are hardened by the introduction of defects such as precipitates formed in a phase separation process. The composition, size and distribution of these precipitates, which are usually not in thermodynamic equilibrium, are decisive for the actual mechanical properties of the alloy. For the design of new metallic materials, it is therefore essential to control the kinetics of nucleation, growth, and coarsening, as well as the stability and dissolution of precipitates under various conditions of temperature and applied load. All these processes involve movements of the interface between matrix and precipitate. A large research effort is therefore devoted to the development of suitable theoretical models to simulate the movement of interfaces due to thermodynamic or mechanical driving forces.

Interfaces can be described theoretically in different ways (Fratzl, et al., 1999), depending on the desired level of resolution (see Fig. 1). *(i)* The first approach consists in a macroscopic modeling of sharp moving interfaces. The thermodynamic phases are treated separately and the interface defines suitable boundary conditions for their development. The movement of the interface is described in such models by a velocity field $\boldsymbol{v}(\boldsymbol{r})$, which is calculated by solving appropriate differential equations. Macroscopic models are only briefly discussed in Section 5 since they are treated in detail elsewhere in this book (Fischer and Simha, 2004). *(ii)* In mesoscopic models, the alloy is described by continuous phase fields (e.g., the chemical composition $c(\boldsymbol{r})$) and the interface corresponds to the continuous variation of the phase field from its value in one phase to its value in the other. Such an interface is smooth and has a finite width. The original ideas are based on Cahn and Hilliards theory of spinodal decomposition (Cahn, 1961, Cahn and Hilliard, 1958, 1959) and have been developed considerably in recent years to allow large-scale simulations of complex alloys (Lochte, et al., 2000, Onuki and Nishimori, 1991, Orlikowski, et al., 1999, Vaithyanathan, et al., 2002, Wang, et al., 1993, Wang, et al., 2001) Recent reviews on spinodal decomposition can be found in (Binder and Fratzl, 2001, Chen, 2002, Raabe, 1998, Thornton, et al., 2003, Wang, et al., 1996), some basic principles will be discussed in Section 4. *(iii)* The most precise description of an interface is the atomic level. Lattice positions within the crystalline solid are occupied by different atomic species and the interface is atomically rough. An important class of microscopic models is based on the kinetic Ising model which has been used since the early 1970's to describe phase separation in alloys (Binder and Fratzl, 2001, Bortz, et al., 1974, Marro, et al., 1975, Rao, et al., 1976). The model consists of two types of atoms, A and B atoms, on a fixed lattice with periodic boundary conditions. The kinetics were described by stochastic exchanges of A and B atoms with a probability specified by a Hamiltonian describing the interactions between the atoms. These models are quite successful in describing nucleation and growth of precipitates (Sur, et al., 1977). They also led to the discovery of dynamical scaling (Fratzl, et al., 1991, Lebowitz, et al., 1982, Marro, et al., 1979) predicted on theoretical grounds (Binder and Stauffer, 1974). Recent reviews of atomistic modeling can be found in (Bellon, 2003, Binder and Fratzl, 2001, Fratzl, et al., 1999, Thornton, et al., 2003, Weinkamer, et al., 2004), Section 3 introduces the underlying principles and gives two examples. A short overview of the physical principles of phase separation (at the micro-, meso-, and macroscopic level) is given in Section 2.

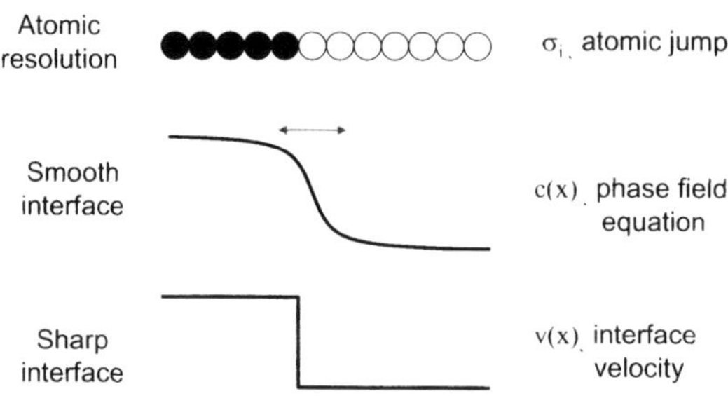

Figure 1. Description of an interface at different levels of resolution. At the microscopic level (atomic resolution), the alloy is described by lattice sites i with an occupation variable σ_i which takes different values depending on the atomic species on site i. The mobility is provided by atomic jumps. At the mesoscopic level, the interface is smooth and corresponds to the continuous change of a phase field $c(\boldsymbol{r})$ from one to the other phase. The phase field varies in time according to a (partial) differential equation (kinetic phase field equation). At the macroscopic level, the interface is a sharp boundary between the two phases. The change of the alloy morphology is described by the velocity of the interface, $\boldsymbol{v}(\boldsymbol{r})$, which is governed by thermodynamic or mechanical driving forces.

2. Physics of phase separation

2.1 The regular solid solution model

The basic model of a binary alloy consists of a lattice, whose N sites are occupied by two different types of atoms, A and B atoms, as well as vacancies V (see Fig. 2). For the purpose of these section, we can neglect the contribution of the vacancies to the thermodynamic functions. A full treatment of this model (including vacancies) will be given in Section 3.1. In the model of the regular solution, introduced by Becker (Becker, 1938), the concentrations of the atoms are fixed, $c_A + c_B = 1$ (since the vacancies are neglected). Therefore the system contains a number Nc_B of B atoms, and a number $Nc_A = N(1-c_B)$ of A atoms. The chemical short-range interaction between atoms are described phenomenologically as effective pair interactions between nearest neighbor (NN) atoms. ε_{AB} denotes the energy of a NN bond between an A and a B atom, ε_{AA} the binding energy for an A-A pair and ε_{BB} for an B-B pair.

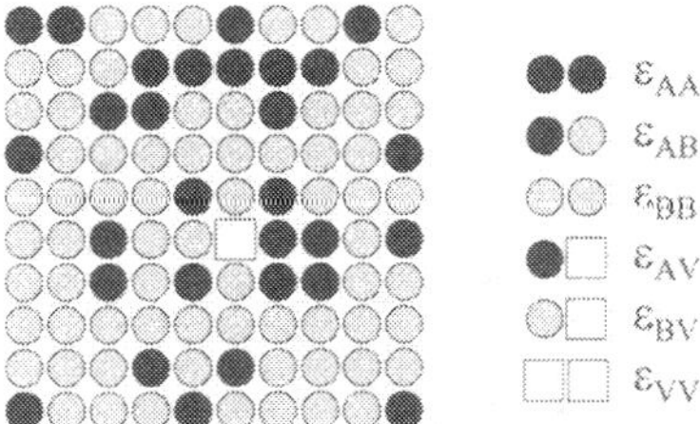

Figure 2. Definition of effective pair interaction energies in a binary alloy containing A and B atoms as well as vacancies. For the purpose of calculating the thermodynamic functions, the contribution of the vacancies can usually be neglected (since the vacancy concentration is very small). This is not true for the description of alloy kinetics (see Section 3).

We consider the case of an alloy at constant temperature T, constant volume V and constant concentrations. The equilibrium condition for the alloy is then given by the minimum of the Helmholtz free energy F,

$$F = E - TS \ , \tag{2.1}$$

where E denotes the total energy of the system and S the entropy. Simply counting and adding the energies between the atoms gives the total energy E for the regular solution (RS) model alloy,

$$E^{RS} = N_{AA}\varepsilon_{AA} + N_{BB}\varepsilon_{BB} + N_{AB}\varepsilon_{AB} \tag{2.2}$$

with N_{AB} the number of A-B bonds, and analog definitions for N_{AA} and N_{BB}. The total number of B atoms within nearest neighbor bonds can be calculated in two ways. Since each atom is contained within Z nearest neighbor bonds (Z being the coordination number of the lattice), there is a total of $Z\,N\,c_B$ of B atoms within bonds. The other way is to say that there are two B atoms within each B-B bond and one within each A-B bond, which leads to the relations

$$\begin{aligned} 2N_{AA} + N_{AB} &= ZNc_A = ZN(1-c_B) \\ 2N_{BB} + N_{AB} &= ZNc_B \end{aligned} \ . \tag{2.3}$$

Note that the total number of bonds is just $N_{AA} + N_{BB} + N_{AB} = ZN/2$. Using (2.3) to eliminate N_{AA} and N_{BB} leads to

$$\begin{aligned} E^{RS} &= \frac{1}{2}NZ((1-c_B)\varepsilon_{AA} + c_B\,\varepsilon_{BB}) - \frac{1}{2}(\varepsilon_{AA} + \varepsilon_{BB} - 2\varepsilon_{AB})N_{AB} = \\ &= \frac{1}{2}NZ\varepsilon_{AA} + 2NZUc_B - 2J\,N_{AB} \end{aligned} \ , \tag{2.4}$$

where U and the *ordering energy* J are defined as

$$U = \frac{1}{4}(\varepsilon_{BB} - \varepsilon_{AA}) \qquad \text{and} \qquad J = \frac{1}{4}(\varepsilon_{AA} + \varepsilon_{BB} - 2\varepsilon_{AB}) \; . \tag{2.5}$$

For $J > 0$, the total energy E^{RS} decreases when N_{AB} increases, therefore A atoms prefer B atoms in the nearest neighbor shell and vice versa leading to an *ordering* tendency in the system. In the case $J < 0$, E^{RS} is lowest when a large number of A-A- and B-B-bonds are formed. The system tends to separate into two pure phases consisting only either of A or B atoms (Fig. 3). Opposed to these tendencies to ordering or *phase separation* is the entropy term in the free energy (2.1), which becomes more important with increasing temperature. The configurational entropy is a maximum for a statistical arrangement of the atoms.

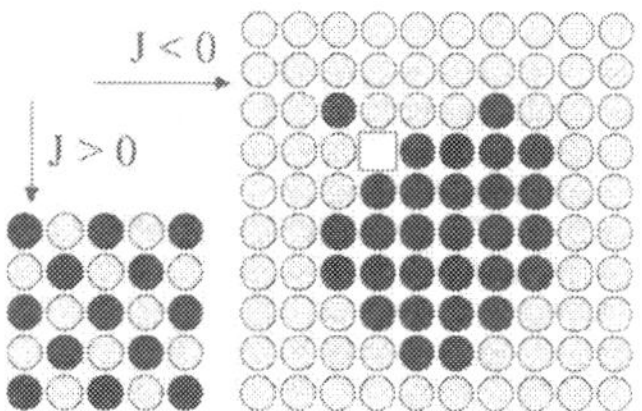

Figure 3. Configurations which reduce the interaction energy E^{RS} (eq. 2.4) with respect to a random distribution of A and B atoms on the lattice. A positive value of the ordering energy J (eq. 2.5) leads to intermetallic ordering with alternating A and B atoms on the lattice, a negative J to phase separation into A-rich and B-rich domains.

In the following, we consider only the case of phase separation for which $J < 0$. While the expression (2.4) for the internal energy of the alloy is valid for any arrangement of A and B atoms, the calculation of the entropy requires some additional assumptions to be tractable analytically. In particular, it is assumed that the atoms are statistically distributed on the lattice. This is a rather crude approximation, since it neglects all fluctuations of the composition within the alloy. For high temperatures it is a reasonable assumption.

For a lattice with randomly distributed A and B atoms, the configurational entropy, also called *mixing entropy*, can be calculated using Boltzmann's formula $S^{RS} = k_B \ln W$, where W counts the possibilities to arrange a number Nc_B of B atoms on N lattice sites:

$$S^{RS} = k_B \ln \frac{N!}{(Nc_B)!(N(1-c_B))!} \cong -k_B N\left(c_B \ln c_B + (1-c_B)\ln(1-c_B)\right) \; . \tag{2.6}$$

Stirling's formula, $\ln M! \cong M \ln M$ (for large M), was employed. Moreover, for a random solid solution $N_{AA} = \frac{1}{2}NZc_A^2$, $N_{BB} = \frac{1}{2}NZc_B^2$ and $N_{AB} = NZc_A c_B$. Hence, the free energy per lattice site can be written as

$$F^{RS}/N = \varepsilon_0 + 2ZUc - 2ZJc(1-c) + k_B T\big(c\ln c + (1-c)\ln(1-c)\big) \equiv \Omega f(c), \qquad (2.7)$$

where we have used the abbreviations $c \equiv c_B$ and $\varepsilon_0 \equiv Z\varepsilon_{AA}/2$; Ω denotes the atomic volume, therefore $f(c)$ is a free energy density, i.e., a free energy per unit volume. From this, it follows immediately that

$$\Omega\frac{\partial f}{\partial c} = 2Z(U-J) + 4ZJc + k_B T\ln\frac{c}{1-c} \quad \text{and} \quad \Omega\frac{\partial^2 f}{\partial c^2} = 4ZJ + \frac{k_B T}{c\,(1-c)} \qquad (2.8)$$

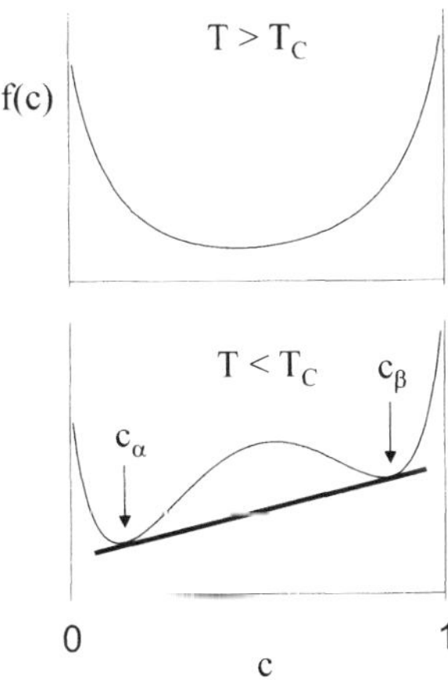

Figure 4. Free energy per lattice site *f(c)* (given by eq. 2.7) plotted as a function of *c* for two typical temperatures. Above T_c^{RS} (eq. 2.9), the function is convex, below it has two minima. The special values c_α and c_β are NOT the minima of the free energy function but are defined as the two contact points to the common tangent (thick straight line) which represents the convex envelope of the function *f(c)*.

The function $c\,(1-c)$ is a parabola with its maximum at $c = 1/2$, which implies that $c\,(1-c) \le 1/4$, for all values of c. By virtue of (2.8), it follows that $f(c)$ is a convex function (that is, $\partial^2 f/\partial c^2 \ge 0$ for all c) as long as the temperature is larger than a critical value T_c^{RS} given as (where it must be remembered that $J < 0$)

$$T_c^{RS} = \frac{Z|J|}{k_B}. \qquad (2.9)$$

For temperatures lower than the critical temperature T_c^{RS}, the function $f(c)$ is non-convex and has two separate minima. This is shown schematically in Figure 4.

Since the free energy must always be a convex function at equilibrium, it becomes clear that the homogeneous mixture of A and B atoms is thermodynamically unstable below the critical temperature T_c^{RS}. Hence, phase separation occurs below T_c^{RS}.

A closer analysis of the model is possible in some cases without the restrictive hypothesis that A and B atoms are always randomly distributed (that is, without neglecting concentration fluctuations, see Section 3 for the application of Monte Carlo methods). The binary alloy on the two-dimensional square lattice is one of the very few examples where an exact solution of the model is known (Onsager, 1944). The exact value of the critical temperature is

$$T_c^{exact} = \frac{2}{\ln(1+\sqrt{2})} \frac{|J|}{k_B} \approx 2.27 \frac{|J|}{k_B} . \tag{2.10}$$

Considering that $Z = 4$ for the square lattice, the prediction of the regular solution approximation (eq. 2.9) overestimates the critical temperature by a factor of almost 2 ! Figure 5 shows a computer simulation of the fluctuations occurring at a temperature of $2.5\,|J|/k_B$, that is just above the exact critical temperature. The solid solution is, indeed, stable at this temperature (despite the fact that it lies below the critical temperature predicted by the regular solution approximation). The large fluctuations in the local composition of the alloy are decreasing considerably the total free energy, thus stabilizing the solid solution. This explains why the exact critical temperature is much lower than the prediction from the regular solution model. Despite this defect, the regular solution model shows a qualitatively correct behavior and it is worth to analyze the predictions of the model a bit further, keeping in mind, however, that the predictions will only be qualitative but never quantitative.

Figure 5. Monte Carlo computer simulation of the composition fluctuations above the critical temperature as compared to a random distribution of the atoms (which is a basic assumption in the regular solution model). The model corresponds to a square lattice with equal number of A (black) and B atoms (white).

2.2 Stability of the solid solution in the mean-field approximation

The main simplification in the regular solution model is to assume a completely random distribution of the A and B atoms on the lattice. This is obviously not a appropriate description for a microstructure consisting of A-rich and B-rich regions. A more sophisticated approach – referred to as "coarse graining" - is to subdivide the lattice into representative volume elements W and assume the mean-field approximation of a random arrangement of the atoms only in each of these volumes (Fig. 6). It should be noted, that this new approximation still neglects fluctuations with a wavelength smaller than the coarse-graining size. Hence, one cannot expect that the values of the critical temperature, for instance, will be predicted correctly within this mean-field approximation.

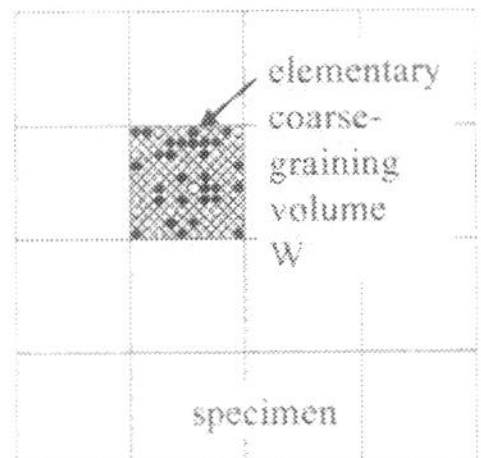

Figure 6. In the coarse-graining approximation, the specimen is subdivided into elementary volumes W, which are large compared to the inter-atomic distance, but small enough to describe variations of the composition at the microscopic level. Typically, one can imagine the edge length of W to be in the order of (several) ten nanometers. Within each elementary volume, the alloy is assumed to have a composition c and to behave like a regular solution with this composition (mean-field approximation). Hence, the coarse-grained model "inherits" the main defect of the regular solution approximation, which is an overestimation of the critical temperature. Indeed, fluctuations, with a wavelength smaller than the edge-length of W are suppressed even in the coarse-graining approximation.

The result is a "coarse-grained" free energy which depends on the concentration c, but c being now a continuous function of the position, $c = c(\boldsymbol{r})$, where $\boldsymbol{r}$ is the center of gravity of the volume element W. The volume W has to be large enough to allow a transition to a continuum description, but small enough to resolve the changes of the concentration at interfaces, for example. Since the width of such interface can be a few nanometers only, it is not always possible to fulfill both coarse-graining conditions.

Let us consider a coarse-grained binary alloy of A and B atoms. In a given volume element W_i at the position $\boldsymbol{r}_i$ within the specimen, the number of lattice sites is denoted by n_i and the lattice site fraction occupied by B atoms is called $c(\boldsymbol{r}_i)$. Using the result from the regular solution (2.7), the free energy per lattice site f is (within W_i)

$$\Omega f(c(\boldsymbol{r}_i)) = 2ZUc(\boldsymbol{r}_i) - 2ZJc(\boldsymbol{r}_i)(1-c(\boldsymbol{r}_i)) + k_B T(c(\boldsymbol{r}_i)\ln c(\boldsymbol{r}_i) + (1-c(\boldsymbol{r}_i))\ln(1-c(\boldsymbol{r}_i))) + const \tag{2.11}$$

If $V = \sum W_i$ is the volume of the system, the total free energy can be written as

$$F = \sum_i n_i \Omega f(c(\mathbf{r}_i)), \tag{2.12}$$

or in a continuous formulation,

$$F = \int_V f(c(\mathbf{r}))\, d\mathbf{r} \quad . \tag{2.13}$$

Due to the conservation of matter within the system, there is the constraint that

$$\int_V c(\mathbf{r})\, d\mathbf{r} = V\,\bar{c} = const \quad , \tag{2.14}$$

where $\bar{c}$ is the average composition of the alloy. Minimizing (2.13) under the constraint (2.14) is equivalent to minimizing

$$G = \int_V \left[f(c(\mathbf{r})) - \mu\, c(\mathbf{r}) \right] d^3\mathbf{r} \quad , \tag{2.15}$$

where μ is a Lagrange multiplier. Obviously, the homogeneous solid solution (for which $c(\mathbf{r}) = \bar{c}$, for all $\mathbf{r}$) minimizes (2.15), if the integrand

$$g(c) = f(c) - \mu\, c \tag{2.16}$$

has a minimum at $c = \bar{c}$. This can only be true if

$$\left.\frac{\partial g}{\partial c}\right|_{c=\bar{c}} = 0, \quad \text{or equivalently} \quad \mu = \left.\frac{\partial f}{\partial c}\right|_{c=\bar{c}} . \tag{2.17}$$

With this choice of the Lagrange multiplier μ, usually called *chemical potential*, the function $g(c)$ has always an extremum at the average composition $c = \bar{c}$. For temperatures larger than T_c, $f(c)$ is a convex function and, therefore, $g(c)$ is convex as well. Hence, $g(c)$ is always minimal at $c = \bar{c}$, which confirms the stability of the solid solution above T_c. For lower temperatures, a more detailed stability analysis can be made with the help of Figure 7.

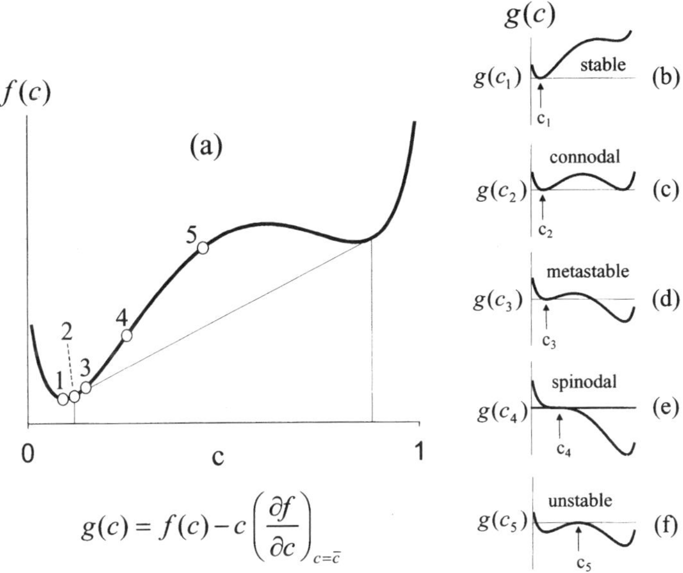

Figure 7. (a) Schematic drawing of the free energy density *f* as a function of composition *c*. The function *g(c)* is plotted on the right side (graphs (b) to (f)) for five different values of the average composition, c_1 to c_5 (also indicated on the left by the numbers 1 to 5). The point 2 indicates one of the contact points with the double tangent corresponding to the convex envelope of *f(c)*. The point 4 indicates the spinodal (where the second derivative of *f(c)* is zero).

It shows schematically the function $f(c)$ as well as $g(c)$ (eq. 2.16) for five different values of the average alloy composition $\bar{c} = c_i$, (i = 1,..., 5). At the point 1 ($\bar{c} = c_1$), the function $g(c)$ has an absolute minimum at c_1. Hence, the solid solution with the average composition $\bar{c} = c_1$ is stable. The point 2 indicates one of the contact points to the double tangent corresponding to the convex envelope of $f(c)$. At this composition, there is the limit of stability of the solid solution (also called "*connodal*"). Indeed, for average concentrations somewhat larger than c_2 (say at c_3), the function $g(c)$ does not have an absolute minimum at $\bar{c}$, but only a relative minimum (see Fig. 7d). Hence, the solid solution is not stable anymore, it is metastable. This is true for all concentrations up to $\bar{c} = c_4$, where the curvature of $f(c)$ changes its sign from positive to negative. The point $\bar{c} = c_4$ is called "*spinodal*". Beyond this concentration, e.g., at the point 5, the function $g(c)$ has a maximum rather than a minimum at $\bar{c}$: the solid solution is unstable.

From this discussion, it follows that three regions (separated by c_2 and by c_4) can be recognized. At small concentrations ($\bar{c} < c_2$), the solid solution is stable, at larger concentrations ($c_2 < \bar{c} < c_4$), it is metastable, and beyond the spinodal it is unstable. As shown in Figure 8 (top), 4 special points can be identified. First, there are the two contact points of the double-tangent (convex envelope of $f(c)$). Second, there are the two inflection points of $f(c)$ corresponding to the spinodal (where the second derivative vanishes). Since $f(c)$ depends on temperature, the values of these four special points will also be temperature dependent. This can be used to construct the phase diagram of the regular solution model (Fig. 8, bottom). The dependence of the connodal compositions on temperature, as defined by the double-tangent construction, give the connodal line (full line in Fig. 8, bottom) and the inflection points the spinodal line (broken line in Fig. 8, bottom). Outside the connodal, the solid solution is stable. Inside, there is a miscibility gap, which is subdivided into a metastable and an unstable region.

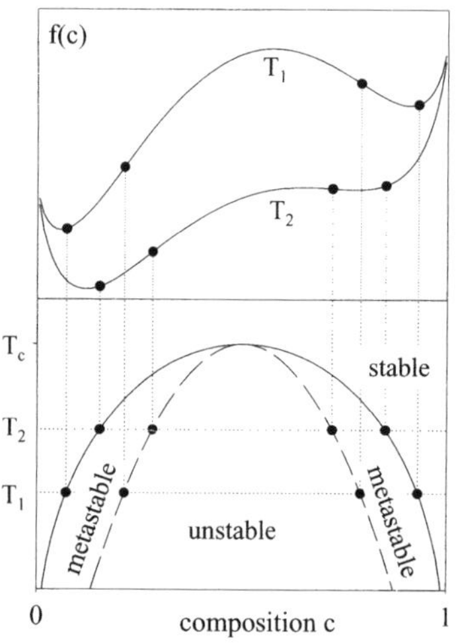

Figure 8. Construction of the miscibility gap from the shape of the free energy curve of the regular solution model. In the lower part of the graph, the full line corresponds to the contact points of the free energy curve to the common tangent forming the convex envelope. This line delimits the miscibility gap of the alloy. The broken line corresponds to the inflection points of the free energy function defining the spinodal. Inside the spinodal, the regular solution is unstable, while it is metastable between the connodal and the spinodal.

It should be mentioned at this point, that the existence of a sharp spinodal line is a result of the mean-field approximations of the regular solution model. In alloys where short-range interactions dominate, the transition from the metastable to the unstable region is gradual and a sharp spinodal line does not exist (for details, one may refer to (Binder and Fratzl, 2001)).

Moreover, the free energy function $f(c)$ as defined by (2.7) completely neglects the influence of coherency stresses between the precipitates and the matrix. The origin of the additional energy contribution from elastic interactions is shown in Figure 9. Let us call c_α and c_β the two equilibrium compositions of matrix and precipitates, respectively. Without the influence of elastic interaction energies, these compositions are given by the connodal line of Figure 8. However, if the equilibrium lattice spacing is different in the α- and in the β-phase (see Fig. 9), the coherent embedding of a β-precipitate in an α-matrix results in an elastic relaxation of the lattice in both matrix and precipitate. The elastic energy associated with this relaxation is added to the free energy $f(c)$, which in turn may affect the values of c_α and c_β. Indeed, the addition of the elastic energy changes the shape of $f(c)$ and, thus, the contact points of the double-tangent.

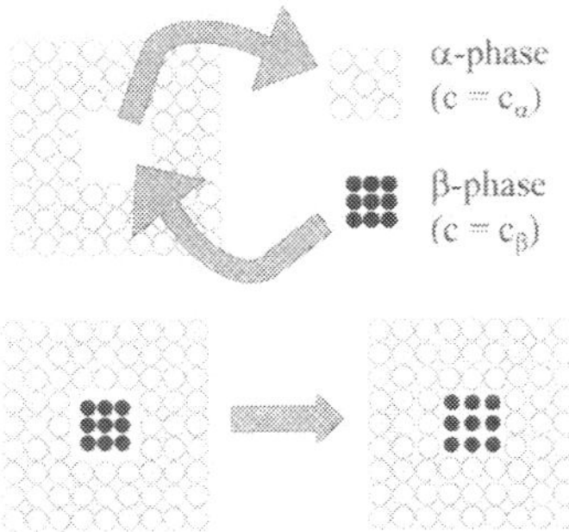

Figure 9. Method to calculate the elastic energy of a coherent inclusion (after (Eshelby, 1957, 1959, Khachaturyan, 1983). A region inside the (elastically undeformed) matrix is removed virtually and replaced by the same number of atoms but corresponding to the β-phase rather than the α-phase. Since the lattice spacing is different in the β-phase, elastic strains appear to accommodate the difference. The elastic energy associated with this deformation has to be added to the free energy of the alloy.

The effects of coherent misfit strains on the phase separation have been reviewed in detail in (Binder and Fratzl, 2001). Anisotropic precipitate shapes and arrangements and modified growth laws typically result from such coherency stresses.

In many cases coherency stresses may reach the critical value for movement or generation of dislocations. Under such conditions, misfit dislocations may accumulate at the interface between the two phases and, thus, release (part of) the misfit stresses. In extreme cases, the interface may become totally incoherent so that no misfit stresses remain (Fig. 10).

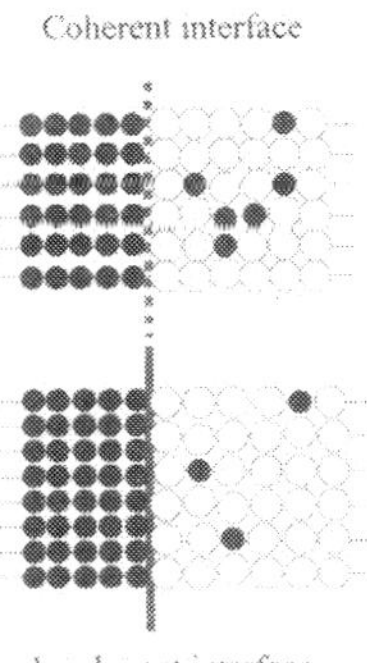

Figure 10. Difference between a coherent and an incoherent interface. The lattice planes can be followed across the coherent interface and the lattice spacing is changed locally by elastic distortion. Across the incoherent interface, lattice planes are discontinuous but lattice spacings are relaxed to their equilibrium values.

The last remark in this section concerns the influence of elastic coherency on the miscibility gap (Cahn, 1961, 1968). If the single phase domains are large enough and separated by incoherent interfaces, the miscibility gap is given by the usual double tangent construction (incoherent miscibility gap in Figure 11). This miscibility gap is usually the one found in phase diagrams. As mentioned before, coherent elastic misfit stresses contribute to the free energy of the system and shift the connodal line ("coherent connodal" in Fig. 11), which is relevant for (homogeneously nucleated) coherent precipitates. This means that the homogeneous nucleation of coherent precipitates might be substantially suppressed due to elastic misfit stresses. In the region marked "b" in Figure 11, nucleation is only possible heterogeneously by the formation of incoherent interfaces with the matrix (at grain boundaries for instance) (see Section 2.3.2).

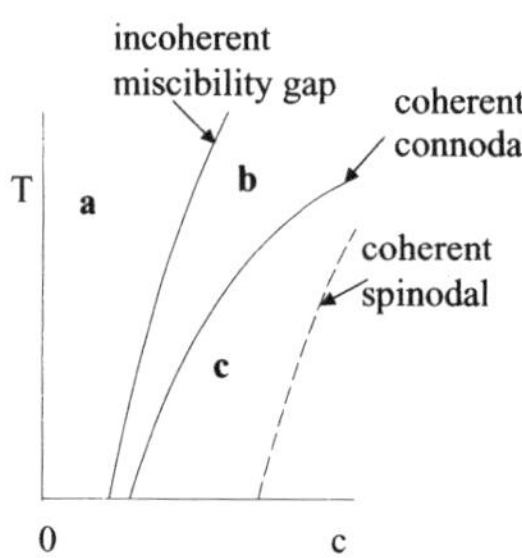

Figure 11. The incoherent miscibility gap describes the phase equilibrium in the absence of elastic misfit stresses. The coherent connodal is shifted due to the contribution of elastic strain energy to the total free energy of the alloy. The coherent spinodal is obtained by a construction such as in Figure 8, taking the elastic strain energy into account.

2.3 Early stages of phase separation

2.3.1 Homophase and heterophase fluctuations

We consider a solid mixture *A-B* with a miscibility gap. The classical experiment is to quench the system from an equilibrium state in the single phase region to a state below the connodal line (see Fig. 8). Since inside the miscibility gap the solid solution is not stable, the alloy will separate into two phases. However, the stability analysis in the preceding section does not give any information on how the process of phase separation starts and proceeds. Starting from an initially homogeneous mixture of atoms, there are two possibilities for small fluctuations leading to the formation of new precipitates (Fig. 12). Either there is a small deviation in the concentration (Fig. 12a) or a small size fluctuation with a substantial deviation in concentration (Fig. 12b).

In the first case shown in Figure 12a, small-amplitude statistical fluctuations (which may be spread over large volumes) are supposed to develop. This means that the concentration slightly deviates from its average value $\bar{c}$ in large regions of the specimen. In cases where $\bar{c}$ is located in the metastable region of the phase diagram (e.g., $\bar{c} = c_3$ in Fig. 7), the total free energy of the system is increased as long as the amplitude of the fluctuation is not too large (this can be concluded from Figure 7d or the left part of Figure 12b, where $g(c)$ has clearly a minimum for $c = \bar{c}$). Hence, such fluctuations will always be damped and phase separation can not proceed by this mechanism. However, Figure 7f (or the left part of Fig. 12a) clearly shows that the situation is completely different in the unstable region of the phase diagram, where the free energy is clearly lowered by any concentration fluctuation (since $g(c)$ has a maximum at $c = \bar{c}$ in this case). Hence, the non-equilibrium solid solution is unstable with respect to the formation of such spatially extended fluctuations. This unmixing reaction initiated via the spontaneous formation and subsequent growth of concentration fluctuations is termed *spinodal decomposition* and can be described in the framework of the Cahn-Hilliard theory (see Section 2.3.3).

Figure 12b shows the second possibility for a fluctuation. The deviation from the average concentration $\bar{c}$ is large from the beginning, but the initial spatial extension of the fluctuation is very small. Coming back to Figure 7d (or the left part of Fig. 12b), it becomes clear that for such a fluctuation (provided the deviation in concentration is large enough to overcome the energy barrier given by the maximum in $g(c)$) the total free energy is actually reduced, since $g(c)$ presents a second minimum which is lower than the one at $c = \bar{c}$. Hence, such fluctuations can grow even in the metastable region of the phase diagram by classical nucleation and growth (see Section 2.3.2).

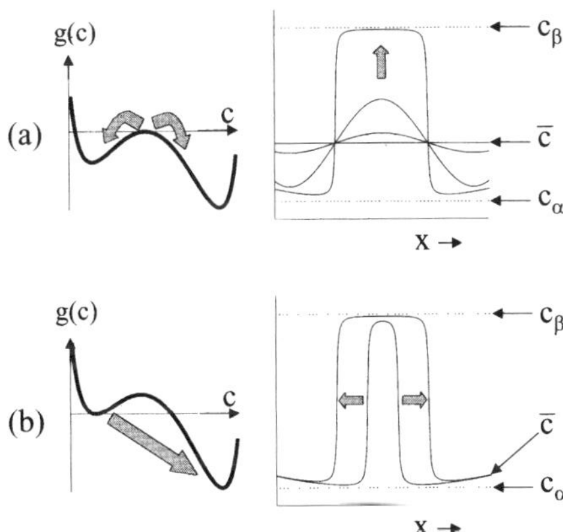

Figure 12. (a) The left side reproduces the free energy curve from Figure 7f which has a maximum at the average concentration (corresponding to the unstable region of the phase diagram Fig. 8). Hence each concentration fluctuation leads to a reduction in free energy and will therefore grow. **(b)** The left side reproduces the free energy curve from Figure 7d which has a (relative) minimum at the average concentration (corresponding to the metastable region of the phase diagram Fig. 8). The consequence is that a concentration fluctuation with a small amplitude such as in (a) leads to an increase in free energy and will therefore be unable to grow. However, fluctuations with a large excursion in concentration but a small size (right graph in (b)) lead to a reduction of the free energy, as soon as the concentration difference is large enough to overcome the energy barrier in *g(c)*.

2.3.2 Homogeneous and heterogeneous nucleation

The classical nucleation theories calculate the energy needed for the formation of a droplet of the new phase β in the matrix α. The main assumptions are, first, that the droplet composition is close to its equilibrium value from the beginning (to allow a reduction of free energy as shown in Figure 12b). Second, the interface between matrix and precipitate is assumed to be sharp (in the sense of Fig. 1). Considering a coherent droplet with radius R (which may also give rise to some elastic coherency strains), the change in the total free energy of the system is,

$$\Delta F(R) = \frac{4\pi}{3\Omega} R^3 (\Delta g_{chem} + \Delta g_{elast}) + 4\pi R^2 \sigma \ , \tag{2.18}$$

where Δg_{chem} and Δg_{elast} are the changes in the chemical and the elastic thermodynamical potential, respectively. The use of g ensures the conservation of the concentration in the system (see Section 2.2). $\Delta g_{chem} + \Delta g_{elast}$ essentially corresponds to the difference between the two minima in the free energy curve shown in Fig. 12b, for instance. Obviously $\Delta g_{chem} + \Delta g_{elast}$ is negative inside the miscibility gap. σ denotes the specific interfacial energy between matrix and precipitate. Figure 13 shows the typical R-dependence of $\Delta F(R)$. At small R, the interface energy term dominates and ΔF increases proportional to R^2. At larger R, the volume energy term dominates and ΔF decreases proportional to R^3. Such a curve displays a maximum at $R \equiv R^*$,

$$R^* = \frac{2\sigma\Omega}{-(\Delta g_{chem} + \Delta g_{elast})} \ . \tag{2.19}$$

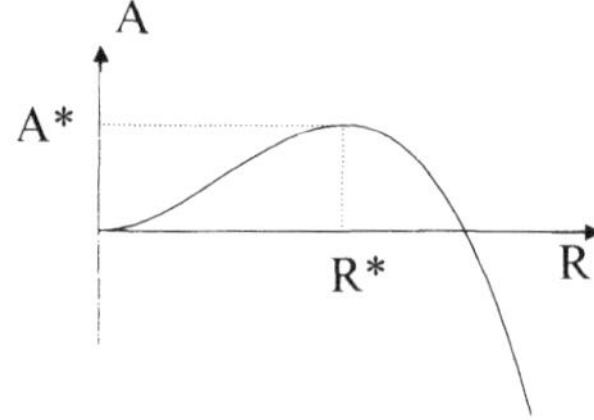

Figure 13. Free energy for the nucleation of a droplet in a solid solution according to eq. (2.18).

Only droplets with an radius R larger than the critical radius R^*, $R > R^*$, are predicted to grow continuously. To escape the metastable state requires a fluctuation to become a critical droplet, i.e., to overcome the activation barrier for nucleation,

$$A^* = \Delta F(R^*) = \frac{16\pi}{3} \frac{\sigma^3 \Omega^2}{(\Delta g_{chem} + \Delta g_{elast})^2} \ . \tag{2.20}$$

Assuming that the droplets grow or shrink only via the condensation or evaporation of single solute atoms, the number of critical droplets per unit of time, i.e., the nucleation rate $\hat{K}$, is given by

$$\hat{K} = \hat{K}_0 \exp(A^* / k_B T) \ . \tag{2.21}$$

Because a sound derivation of the prefactor $\hat{K}_0$ is still missing, the comparison with experimental results is very limited (Gunton, 1999). The most blatant approximation of classical nucleation theories is that macroscopic properties are attributed to a droplet which can contain only a few atoms. In Equation (2.18) the bulk values for Δg_{chem}, Δg_{elast} and σ are used. Also the change in the composition of the matrix as the result of the growth of supercritical nuclei is usually neglected. Homogeneous nucleation as a whole is an idealization because it does not take into accounts defects or other preferred nucleation sites in the system. There are a number of proposals for the prefactor in the literature which can be found in textbooks, e.g., (Cahn and Haasen, 1983, Wagner, et al., 2001).

2.3.3 Cahn-Hilliard theory

The theory by Cahn and Hilliard (Cahn, 1961, Cahn and Hilliard, 1958, 1959) starts from Equation (2.13) by adding a surface energy term in analogy to the one in (2.18),

$$F = F_{chem} + F_{int} = \int_V \left[f(c(\boldsymbol{r},t)) + \frac{1}{2}\chi^2 (\nabla c(\boldsymbol{r},t))^2 \right] d\boldsymbol{r} \ . \tag{2.22}$$

The second term in this expression integrates the regions where the concentration gradient is large. In a two-phase mixture, $(\nabla c)^2$ is typically zero (or very small) inside the single-phase domains and large only at interfaces. Hence, the integral will be essentially proportional to the total amount of interface in the system, which identifies F_{int} clearly as an interface energy term. It is interesting to note, that this interface is NOT sharp since it can be described by a smooth function corresponding to the gradient squared of the concentration. Hence, the Cahn-Hilliard theory represents the prototype of the mesoscopic models (called phase field models, Section 4) as introduced in Figure 1.

The driving force for phase separation is the reduction of (2.22) under the constraint of mass conservation (2.14). Considering the composition $c(\boldsymbol{r},t)$ to be dependent on space as well as on time (to allow a temporal evolution of the alloy microstructure), a kinetic equation can be formulated using the variational derivative of (2.22) (Binder and Fratzl, 2001)

$$\frac{\partial c(\boldsymbol{r},t)}{\partial t} = M \nabla^2 \left(\frac{\partial f}{\partial c}(\boldsymbol{r},t) - \chi^2 \nabla^2 c(\boldsymbol{r},t) \right) . \tag{2.23}$$

where M is a mobility coefficient (see Section 4.1 for further details). This is a non-linear partial differential equation, often called Cahn-Hilliard-Equation (CHE), which can in principle be solved numerically. This is the basis of so-called phase-field models (see Section 4). Here we shall only examine the most crude approximation which is the linearized CHE.

A linearization is possible by assuming that in the most early stages of phase separation, the deviation of the local concentration from its initial value in the homogeneous alloy, $c(\boldsymbol{r},t)-\bar{c}$, is very small. Then one can approximate $\partial f/\partial c$ by a linear expansion:

$$\frac{\partial f}{\partial c} \approx \left.\frac{\partial f}{\partial c}\right|_{c=\bar{c}} + (c-\bar{c})\left.\frac{\partial^2 f}{\partial c^2}\right|_{c=\bar{c}} . \tag{2.24}$$

Insertion into Equation (2.23) gives the linearized CHE,

$$\frac{\partial c(\boldsymbol{r},t)}{\partial t} = M\left(\left.\frac{\partial^2 f}{\partial c^2}\right|_{c=\bar{c}} \nabla^2 c(\boldsymbol{r},t) - \chi^2\nabla^4 c(\boldsymbol{r},t)\right) . \tag{2.25}$$

which can be solved by Fourier transformation,

$$\frac{\partial \tilde{c}(\boldsymbol{k},t)}{\partial t} = -Mk^2\left(\left.\frac{\partial^2 f}{\partial c^2}\right|_{c=\bar{c}} + \chi^2 k^2\right)\tilde{c}(\boldsymbol{k},t) \equiv -R(k)\tilde{c}(\boldsymbol{k},t) , \tag{2.26}$$

yielding

$$\tilde{c}(\boldsymbol{k},t) = \tilde{c}(\boldsymbol{k},0)e^{-R(k)t} , \tag{2.27}$$

where

$$\tilde{c}(\boldsymbol{k},t) = \int_V d\boldsymbol{r}\, c(\boldsymbol{r},t)e^{i\boldsymbol{kr}} . \tag{2.28}$$

The Fourier transform $\tilde{c}(\boldsymbol{k},t)$ represents a concentration wave with a wavelength $\lambda = \frac{2\pi}{k}$. The solution (2.27) predicts a damping of fluctuations with wave vectors k, such that $R(k) > 0$, but an exponential growth for the case $R(k) < 0$. It is important to realize that $R(k) < 0$ is only possible if $\left.\partial^2 f/\partial c^2\right|_{c=\bar{c}} < 0$, i.e., inside the unstable region of the miscibility gap. In analogy to the critical radius R^* in the classical nucleation theory (2.19), there is a critical wavelength λ^* for the linearized CHE,

$$\lambda^* = \frac{2\pi}{k^*} = 2\pi \frac{\chi}{\sqrt{-\left.\frac{\partial^2 f}{\partial c^2}\right|_{c=\bar{c}}}} \quad . \tag{2.29}$$

Only fluctuations with a wavelength larger than λ^* are able to grow. Subcritical fluctuations (with a wavelength smaller than λ^*) will dissolve again. This is visible in Figure 14, where the intensity

$$I(\boldsymbol{k},t) = \left|\tilde{c}(\boldsymbol{k},t)\right|^2 \tag{2.30}$$

is plotted as a function of $k = |\boldsymbol{k}|$ for different times t. For $k > k^*$ the intensity decreases (corresponding to the damping of fluctuations), while for $k < k^*$ the intensity increases exponentially. The intensity stays fixed at the point $k = k^*$.

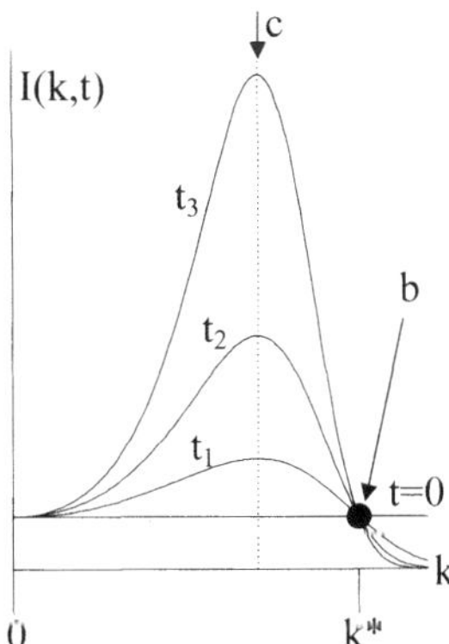

Figure 14. $I(k,t)$ defined by (2.30) plotted as a function of k for different times t using the solution (2.27) of the linearized CHE. k^* is defined by (2.29).

The intensity $I(\boldsymbol{k},t)$ is an interesting quantity since it can be measured by small-angle scattering of X-rays or neutrons (see, e.g., (Fratzl, 2003)). A qualitatively rather similar behavior of the intensity was found for phase separation in many alloys (Binder and Fratzl, 2001) with, however, some important differences: usually, there is neither a fixed point nor an exponential growth (see Fig. 17). This is attributed to the linear approximation in the CHE which does not apply even in the earliest measurable states for metal alloys (Binder and Fratzl, 2001). A more detailed discussion of the (non-linear) CHE and of similar equations can be found in Section 4.

2.4 Late stages of phase separation

2.4.1 Growth and coarsening

After nucleation (by any of the mechanisms described in the previous section) a growing nucleus with concentration c_β is embedded in the still supersaturated matrix with concentration $c_\infty > c_\alpha$. The mobility of the interface is assumed high enough to allow the solute concentration c_R at the curved interface to attain local equilibrium. When the particles are small the growth rate is not controlled by the supply of atoms to the interface by diffusion, but by the rate the atoms cross the interface. Mass balance at the interface leads to

$$(c_\beta - c_R)\frac{dR}{dt} = -j = D\left.\frac{\partial c}{\partial r}\right|_{r=R} \approx -\frac{D}{R}(c_R - c_\infty) \ , \tag{2.31}$$

where j denotes the flux of atoms across the interface. Solving the diffusion equation around the precipitate, gives

$$c(r) = c_\infty + (c_R - c_\infty)\frac{R}{r} \qquad \text{for } r > R \tag{2.32}$$

which is plotted in Figure 15.

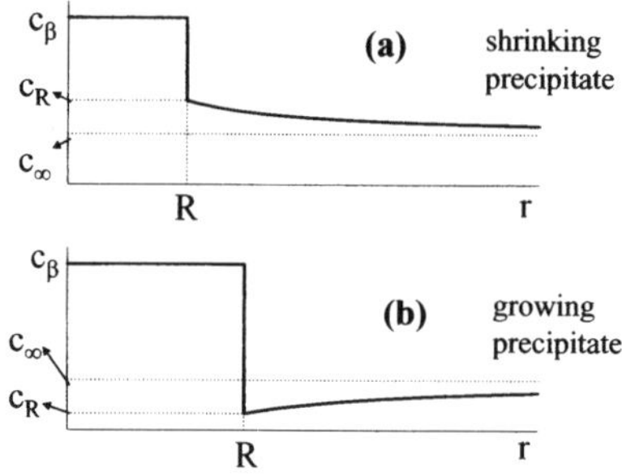

Figure 15. Concentration profiles around growing and shrinking precipitates in the later stages of phase separation.

Assuming local equilibrium at the interface according to (2.19) and neglecting the elastic misfit interactions, it follows that

$$\frac{2\sigma\Omega}{R} = -\Delta g = g_R(c_R) - g_R(c'_R), \text{ where } g_R(c) = f(c) - c\left.\frac{\partial f}{\partial c}\right|_{c=c_R} \tag{2.32}$$

and c'_R the position of the second minimum of g (see Fig. 12b, for instance). In a situation where c_R is close to the connodal line, we have $c_R \approx c_\alpha$ and $c'_R \approx c_\beta$, c_α and c_β being the concentrations defined by the double tangent construction. A development of (2.32) to first order in $c_R - c_\alpha$ gives:

$$g_R(c_R) - g_R(c'_R) \approx (c'_R - c_R)\left(\left.\frac{\partial f}{\partial c}\right|_{c=c_R} - \left.\frac{\partial f}{\partial c}\right|_{c=c_\alpha}\right) \approx (c_R - c_\alpha)(c_\beta - c_\alpha)\left.\frac{\partial^2 f}{\partial c^2}\right|_{c=c_\alpha}, \tag{2.33}$$

where we have used that

$$f(c_\alpha) - c_\alpha \left.\frac{\partial f}{\partial c}\right|_{c=c_\alpha} = f(c_\beta) - c_\beta \left.\frac{\partial f}{\partial c}\right|_{c=c_\beta} \quad \text{and} \quad \left.\frac{\partial f}{\partial c}\right|_{c=c_\alpha} = \left.\frac{\partial f}{\partial c}\right|_{c=c_\beta}. \tag{2.34}$$

A further simplification occurs at low temperatures when $c_\alpha << 1$. Then it follows from Equation (2.8) that

$$\left.\frac{\partial^2 f}{\partial c^2}\right|_{c=c_\alpha} \approx \frac{k_B T}{c_\alpha}, \tag{2.35}$$

which gives the classical (linearized) Gibbs-Thomson equation:

$$c_R - c_\alpha \approx \frac{2\sigma\Omega\, c_\alpha}{k_B T(c_\beta - c_\alpha)\, R}. \tag{2.36}$$

Insertion of the Gibbs-Thomson equation into (2.31) results in the classical differential equation studied by Lifshitz, Slyozov and Wagner (LSW) (Lifshitz and Slyozov, 1961, Wagner, 1961).

$$\frac{dR}{dt} = \frac{D}{R}\frac{c_\infty - c_\alpha}{c_\beta - c_\alpha}\left(1 - \frac{R^*(c_\infty)}{R}\right) \quad \text{with} \quad R^*(c_\infty) = \frac{2\sigma\Omega}{k_B T(c_\beta - c_\alpha)(c_\infty - c_\alpha)} \tag{2.37}$$

Both R and c_∞ are time-dependent in this expression. They are related, however, via the mass conservation since the total volume of all precipitates at any time just corresponds to the reduction of the supersaturation from $\bar{c} - c_\alpha$ to $c_\infty - c_\alpha$. The analysis of this equation shows that, under the assumption of a constant supersaturation $c_\infty - c_\alpha$, an isolated particle with a size $R >> R^*$ grows according to a power law,

$$R(t) \propto (Dt)^{1/2}. \tag{2.38}$$

At later stages, when the growth of precipitates is replaced by coarsening, the mass conservation leads to a change in the growth law. The classical LSW-theories, which are derived under the assumption of very dilute alloys, predict

$$\overline{R}(t) \propto (Dt)^{1/3} \tag{2.39}$$

where $\overline{R}(t)$ is the average precipitate size at time t (Voorhees, 1985). Moreover, the supersaturation decreases as

$$c_x(t) - c_\alpha \propto \frac{1}{\overline{R}(t)} \tag{2.40}$$

Finally, the particle size distribution follows a scaling law according to

$$h(R,t) = \frac{1}{\overline{R}(t)} \tilde{h}\left(R / \overline{R}(t)\right) \tag{2.41}$$

where $\tilde{h}$ is a function independent of time. The actual shape of this function was predicted (in the limit of infinite dilution) in (Lifshitz and Slyozov, 1961, Wagner, 1961). The resulting function is usually too narrow to account for realistic precipitate size distributions (Voorhees, 1985, 1992). The LSW theory has been extended to somewhat larger volume fractions of precipitates and the scaling property still holds in many cases (at least when elastic misfit interactions are not too large (Fratzl, et al., 1999)) but the shape of the rescaled size distributing is considerably changed (Voorhees, 1985).

Up to now nucleation, growth and coarsening were presented as well separated and consecutive processes on the time scale. Since experiments showed the limited validity of this assumption, different approaches have been proposed which cover nucleation, growth and coarsening as concomitant processes. Binder and coworkers proposed a cluster-dynamics approach which considers the coagulation and splitting of clusters (Binder, et al., 1978, Mirold and Binder, 1977). Predictions about the time evolution of the mean precipitate size, the number density of precipitates, the supersaturation and the nucleation rate over the whole course of phase separation is given by the theory of Langer and Schwartz (Langer and Schwartz, 1980). A modification of this theory using the non-linearized Gibbs-Thomson equation (see (2.36)) gives more accurate results (Kampmann and Wagner, 1984). Kampmann and Wagner further proposed an algorithm termed the Numerical model, which allows to compute the precipitate size distribution without any approximations (Kampmann and Wagner, 1984, Wagner, et al., 2001).

2.4.2 Dynamical scaling

Independently of the assumptions in the previous section, dynamic scaling has been shown to hold for many two-phase mixtures (when misfit interactions are negligible) (Binder and Fratzl, 2001, Binder and Stauffer, 1974, Fratzl, et al., 1991). The idea of scaling is shown in

Figure 16. The time evolution of the microstructure is self-similar, therefore simply blowing up a part of the microstructure gives a reasonable temporal description of its evolution. A reformulation of this idea is that without a scale it is impossible to decide whether coarsening just started or already proceeded for a long time because the microstructure always "looks the same". A single length scale is enough for the description of the system. A striking verification of the self-similar coarsening of the precipitate structure is obtained by small-angle scattering experiments.

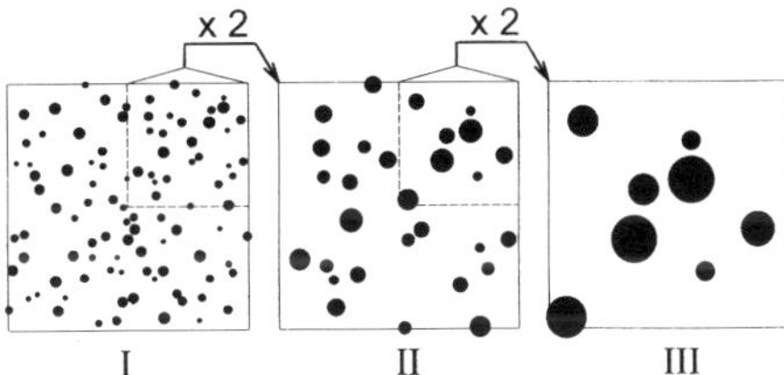

Figure 16. Self-similarity of the microstructure is the basis of dynamical scaling. Simply taking one quarter of a picture and blowing it up to the original size adds up to a time series of the coarsening process.

Figure 17 (left) shows the scattering intensity $I(k,t)$ at small angles for different annealing times. A peak in $I(k,t)$, which originates from the typical spacing between the precipitates, develops with increasing time t. Since this spacing increases, the position of the peak shifts to smaller values of k in reciprocal space (compare Figure 17 to Figure 14 to catch the shortcomings of the linearized Cahn-Hilliard equation). After renormalization by the position and the height of the maximum, all the $I(k,t)$ data for different times superimposes (Fig. 17, right). In mathematical terms that means that $I(k,t)$ can be written as,

$$I(k,t) = I_{\max}(t)\,\widetilde{F}(k / k_{\max}(t)) \; , \tag{2.42}$$

where $\widetilde{F}$ is a time-independent scaling function, plotted as a continuous line on the right-hand side of Figure 17 (Binder and Fratzl, 2001, Fratzl, et al., 1991). Equation (2.42) is just the translation of the self-similarity of the microstructure into measurable quantities. Both the size and the spacing of the precipitates increase at the same rate with time.

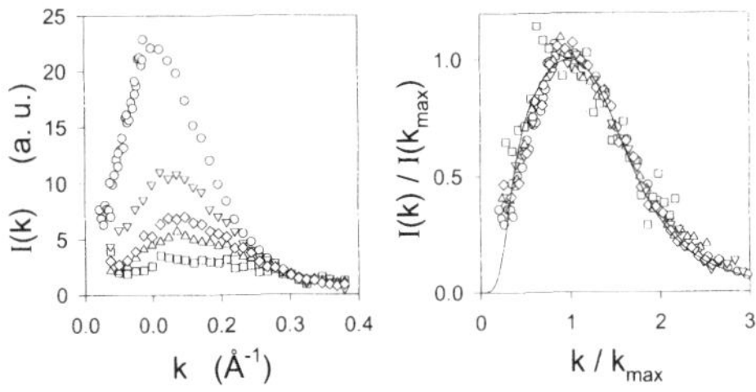

Figure 17. Intensity measured in a small-angle neutron scattering (SANS) experiment in an Al-Zn-Mn alloy as a function of annealing time (left). The same data but renormalized by the position, $k_{\max}$, and the height, $I_{\max} = I(k_{\max})$, of the peak (from (Blaschko and Fratzl, 1983))

3. Monte Carlo simulation of phase separation

3.1 The kinetic Ising Model including vacancies

A major assumption in the model of the regular solid solution (Section 2.1) and in the derivation of the free energy (2.7) was that the atoms are statistically arranged on the lattice. Now we study the general model and how this model can be used in Monte Carlo (MC) simulations to study diffusional phase transformation. Due to the dynamical interpretation of the simulation results, we obtain information not only about the equilibrium, but about the dynamics of the phase transformation.

Like in the model of the regular solid solution (Section 2.1) the computer model for a binary alloy used in kinetic Monte Carlo simulations consists of a lattice, whose sites are occupied by two different types of atoms, A and B atoms (Fig. 2). Since most of alloys crystallize in a body-centered cubic (bcc) or face-centered cubic structure (fcc), these are the lattices preferentially chosen. Simulations are performed with a fixed number of N_A A atoms and N_B B atoms. The system can be allowed to contain lattice defects, namely N_V vacancies. To avoid finite size effects, the number of lattice sites, $N = N_A + N_B + N_V$, and therefore the linear dimensions of the lattice have to be clearly larger than a typical length characterizing the emerging microstructure, e.g., the size of the formed precipitates. A rule of thumb is that the lattice size is at least six times larger than the precipitate size (Heermann, et al., 1996). Although $N > 10^6$ is feasible on modern computers for a study of the nucleation, growth and coarsening of precipitates, the size of precipitates that can be studied with atomistic MC simulations is therefore still restricted to about 10 *nm*. To keep the influence of the finite size of the lattice small, periodic boundary conditions are usually chosen.

3.1.1 Hamiltonian of the ABV model

The chemical short-range interactions between atoms are again described phenomenologically as pair interactions between nearest neighbor (NN) atoms, e.g., ε_{AB} denoting the energy of an NN-pair of an A and B atom (see Fig. 2). The description of the energetics is here restricted to interactions between nearest neighbor atoms, but an inclusion of interactions between atoms in more distant neighbor shells is straightforward (Weinkamer, et al., 1998). Also the model can easily be extended to alloys with more than two components. We consider here a model for a binary alloy including vacancies. The Hamiltonian of this so-called ABV model (Yaldram and Binder, 1991) is then given by:

$$H = N_{AA}\varepsilon_{AA} + N_{BB}\varepsilon_{BB} + N_{AB}\varepsilon_{AB} + N_{AV}\varepsilon_{AV} + N_{BV}\varepsilon_{BV} + N_{VV}\varepsilon_{VV} = \\ = \sum_{<ij>}^{NN} \sum_{\alpha,\beta}^{A,B,V} \varepsilon_{\alpha\beta} p_i^\alpha p_j^\beta \quad , \qquad (3.1)$$

where $<ij>$ denotes the sum over all NN-atom pairs. The occupation parameter p_i^α is defined as

$$p_i^\alpha = \begin{cases} 1 & \textit{if an atom of typ } \alpha \textit{ occupies the site } i \\ 0 & \textit{else} \end{cases} , \tag{3.2}$$

$\alpha = A, B$ or V. Equation (3.1) and the introduction of the occupation parameter corresponds to a transition from a global energetical description to a local description which considers only sums over NN pairs. This transition is completed by introducing a spin variable σ_i,

$$\sigma_i = \begin{cases} -1 & \textit{if an A atom occupies site } i \\ 1 & \textit{if a B atom occupies site } i \\ 0 & \textit{if a vacancy V occupies site } i \end{cases} . \tag{3.3}$$

Inserting the elementary relations

$$\begin{aligned} p_i^A &= 1/2(-\sigma_i + \sigma_i^2) \\ p_i^B &= 1/2(\sigma_i + \sigma_i^2) \\ p_i^V &= 1 - \sigma_i^2 \end{aligned} \tag{3.4}$$

in (3.1), results in the following form of the Hamiltonian, i.e. of the total energy (Vives and Planes, 1993),

$$E = \tilde{E}_0 + K \sum_{<ij>}^{NN} \sigma_i^2 \sigma_j^2 + J \sum_{<ij>}^{NN} \sigma_i \sigma_j + U \sum_{<ij>}^{NN} (\sigma_i^2 \sigma_j + \sigma_i \sigma_j^2) + M \sum_{<ij>}^{NN} \sigma_i^2 + R \sum_{<ij>}^{NN} \sigma_i \tag{3.5}$$

where,

$$\begin{aligned} K &= \frac{1}{4}(\varepsilon_{AA} + \varepsilon_{BB} + 2\varepsilon_{AB}) + \varepsilon_{VV} - \varepsilon_{AV} - \varepsilon_{BV} , \\ J &= \frac{1}{4}(\varepsilon_{AA} + \varepsilon_{BB} - 2\varepsilon_{AB}) , \\ U &= \frac{1}{4}(\varepsilon_{BB} - \varepsilon_{AA}) - \frac{1}{2}(\varepsilon_{BV} - \varepsilon_{AV}) , \\ M &= \frac{Z}{2}(\varepsilon_{AV} + \varepsilon_{BV} - 2\varepsilon_{VV}) , \\ R &= \frac{Z}{2}(\varepsilon_{BV} - \varepsilon_{AV}) , \\ \tilde{E}_0 &= \frac{NZ}{2}\varepsilon_{VV} . \end{aligned} \tag{3.6}$$

$\tilde{E}_0$ is constant and since the number of A and B atoms is fixed, resulting in

$$\sum_{<ij>}^{NN} \sigma_i^2 = N_A + N_B \quad \text{and} \quad \sum_{<ij>}^{NN} \sigma_i = N_A - N_B , \tag{3.7}$$

also the last two terms of (3.5) are independent of the atomic configuration. Another simplification of the total energy E is obtained when considering the case of a single vacancy, $N_V = 1$, which typically corresponds to a vacancy concentration $c_V \approx 10^{-6}$, which is realistic for a lot of alloys. Then,

$$\sum_{<ij>}^{NN} \sigma_i^2 \sigma_j^2 = Z(\frac{N}{2} - 1) , \tag{3.8}$$

and summarizing all the terms independent of the atomic configuration into the new constant $\tilde{E}'_0$, the Hamiltonian can be rewritten as,

$$E = \tilde{E}'_0 + J \sum_{<ij>}^{NN} \sigma_i \sigma_j + U \sum_{<ij>}^{NN} (\sigma_i^2 \sigma_j + \sigma_i \sigma_j^2) . \tag{3.9}$$

From the five independent pair interactions, ε_{AA}, ε_{AB}, ε_{BB}, ε_{AV}, and ε_{BV}, only two independent parameters, J and U, remain. Their role within the model can be better understood when we go back to a global description of the Hamiltonian (Weinkamer and Fratzl, 2003),

$$E = E'_0 - 2 (J N_{AB} + U N_{BV}) \tag{3.10}$$

with E'_0 denoting another constant. The ordering energy J describes the interaction between the atoms and, therefore, determines the phase diagram of the system. For $J < 0$ the systems tends to separate into A-rich and B-rich phases, $J > 0$ leads to ordering on the lattice. The presence of a single vacancy, i.e., a vacancy concentration $c_V \approx 10^{-6}$, does not significantly change the phase diagram (Fratzl and Penrose, 1994). The interaction of the vacancy with the atoms is specified by the asymmetry parameter U (Kikuchi and Sato, 1972). $U > 0$ means that the vacancy prefers B atoms in its NN-shell. The ordering energies $J^{(k)}$ – the superscript *(k)* denotes the k-th neighbor shell for the case of interactions beyond the nearest neighbor shell – can be determined by means of diffuse scattering experiments (Schweika, 1998) and using the so-called inverse Monte Carlo method (Gerold and Kern, 1987). It was shown that the obtained $J^{(k)}$ are not unique and that many-body interactions have to be included for a good reproduction of real phase diagrams (Wolverton, et al., 1997). Values for the asymmetry parameter U can be estimated from the difference of the pure metal cohesive energies, the vacancy formation energies, or the vacancy migration energies (Roussel and Bellon, 2001). These estimates often lead to quite large ratios, $|U/J| > 1$.

3.1.2 Dynamics

For the dynamical interpretation of a MC simulation, it is important to use a realistic mechanism to model the elementary changes in the atomic configuration. In early MC simulations using models, which did not include vacancies, the time evolution of the system was modeled by exchanges between atoms of different type (Kawasaki dynamics, abbreviated K-dynamics). In alloys it is well-known that changes in the atomic configuration occur via a jump of an atom into a neighboring vacancy. The first simulations using vacancy dynamics (V-dynamics) were performed in the early nineties (Yaldram and Binder, 1991, 1991). An unrealistically high vacancy concentration of 16% chosen in these works concealed the new physics which is associated with the introduction of vacancies, and which is presented below.

3.1.3 Exchange probabilities

The general principles of MC simulations can be found in (Binder, 1997, Binder and Heermann, 2002) and need not be repeated here. For our problems the state of the system is defined by a complete set of the spin variables, $\boldsymbol{x}_l = \{\sigma_i, i = 1,..., N\}$. In a kinetic MC simulation the aim is to construct a Markov chain of states $... \rightarrow \boldsymbol{x}_l \rightarrow \boldsymbol{x}_{l+1} \rightarrow \boldsymbol{x}_{l+2} \rightarrow ...$, corresponding to a temporal trajectory in phase space, where on the long run the different states appear with the statistical weights given by the Boltzmann distribution. A sufficient condition for achieving this distribution is that the transition probabilities $W(\boldsymbol{x}_l \rightarrow \boldsymbol{x}_{l'})$ for passing from state $\boldsymbol{x}_l$ to $\boldsymbol{x}_{l'}$ satisfy the detailed balance condition,

$$\frac{W(\boldsymbol{x}_l \rightarrow \boldsymbol{x}_{l'})}{W(\boldsymbol{x}_{l'} \rightarrow \boldsymbol{x}_l)} = \exp[-\Delta E/k_B T] \,, \tag{3.11}$$

with k_B Boltmann's constant and ΔE the "energy difference" (see below). Since Equation (3.11) does not define the probability $W(\boldsymbol{x}_l \rightarrow \boldsymbol{x}_{l'})$ uniquely, different "rules" have been proposed, most prominent the Metropolis rule,

$$W(\boldsymbol{x}_l \rightarrow \boldsymbol{x}_{l'}) = \begin{cases} \dfrac{1}{\tau_0} \exp[-\Delta E/k_B T] & if \;\; \Delta E > 0 \\ \dfrac{1}{\tau_0} & if \;\; \Delta E \le 0 \end{cases} \,, \tag{3.12}$$

The factor $1/\tau_0$ corresponds to an attempt frequency and defines the physical time in the simulation. Our physical understanding of how diffusion proceeds in crystalline solids allows only a very limited number of "moves" of the system, i.e., only very few $W(\boldsymbol{x}_l \rightarrow \boldsymbol{x}_{l'})$ are non-zero. More precisely, only exchanges between a vacancy and a neighboring atom are allowed (vacancy dynamics). The choice of the correct elementary allowed moves of the system is crucial for the results obtained and can therefore be seen as the Achilles heel of a kinetic MC simulation.

In Equations (3.11) and (3.12) ΔE denotes the jump-energy of the exchange vacancy/atom. Usually ΔE is calculated as the difference between the energy of the final state and the initial state, $\Delta E = E_{final} - E_{initial}$. An alternative is to introduce an energy barrier, which separates these two states and which the atom has to overcome in changing sites. The jump-energy ΔE is then taken to be the barrier energy minus the energy of the initial state (Abinandanan, et al., 1998, Bellon, 2003). Typically a constant energy barrier is considered, which is greater than all possible states of the system, and ΔE is calculated in a broken bond approximation (Abinandanan, et al., 1998). The influence of the jump-energy evaluation is often non-trivial (Weinkamer, et al., 2004), as was demonstrated for the case of diffusion in an ordered alloy (Weinkamer, et al., 1998), and deserves attention. Since a direct comparison between MC simulations and experimental results have high priority (Pareige, et al., 1999, Weinkamer, et al., 1999), all attempts to make the modeling more realistic are of great importance.

3.1.4 Time incrementation

Following the standard Metropolis algorithm an exchange between the vacancy and a randomly chosen NN-atom is carried out with the probability given by Equation (3.12). The time unit is then usually defined as one Monte Carlo step (MCS), i.e., one attempted exchange per lattice site. The Metropolis algorithm has the disadvantage that the rejection rate drastically increases at low temperatures, e.g., when the vacancy is trapped in a low energy configuration . This waste of computing time can be avoided by employing the residence time algorithm of (Bortz, et al., 1975), where it is ensured that the vacancy performs a jump at each attempt. The incrementation of the total time corresponding to an exchange depends then on all the probabilities of possible vacancy jumps. A "trapped" vacancy would be freed again at the expense of a large time increment. The mean residence time and the simulation results are unaffected by the way the time is incremented (Novotny, 1995, Weinkamer, et al., 1998). The idea of residence algorithms was extended to higher order algorithms, which consider in the time incrementation also the possibility that the vacancy jumps straight back to its previous position (Athenes, et al., 1997, Rautiainen and Sutton, 1999). The conversion of computer time measured in MCS to physical time measured in seconds is accomplished by giving τ_0 the value necessary to make the diffusivity of the model equal to the experimentally measured value of the diffusivity (Abinandanan, et al., 1998, Weinkamer, et al., 1999).

3.2 Influence of the vacancy dynamics on phase separation

The influence of the vacancy mechanism on the simulation results of phase separation in the kinetic Ising model is demonstrated giving two examples: First we compare results obtained with Kawasaki and vacancy dynamics, where the single vacancy in the system does not energetically distinguish between the two different types of atoms, A and B. This corresponds to the case $U = 0$ for the asymmetry parameter in Equation (3.6), which further simplifies the Hamiltonian of Equation (3.5) and (3.9) to the standard Ising form,

$$E = E_0 + J \sum_{<ij>}^{NN} \sigma_i \sigma_j = E_0' - 2J N_{AB} \; . \qquad (3.13)$$

In a second step the influence of U is studied choosing values $U \neq 0$. The two different results can be best demonstrated with different microstructure morphologies: an interconnected and intertwined structure at a solute concentration $c = 0.5$ and isolated precipitates at low concentration, respectively (compare Fig. 25).

3.2.1 The vacancy dynamics influences the coarsening rate of the precipitates

Starting with the 1970s (Bortz, et al., 1974), phase separation in a binary alloy was studied using the kinetic Ising model with Kawasaki dynamics – abbreviated in the following as K-dynamics and using a subscript K. Modeling the elementary diffusion step more realistically as an exchange between vacancy and atom (V-dynamics) brought as a premature conclusion that the actual diffusion mechanism is only of minor importance for the process of phase separation (Yaldram and Binder, 1991, 1991). Confirmed by numerous simulations, this is true as far as the developing domain structure is concerned. Snapshot pictures, or the spherically averaged pair correlation function and its Fourier transform, the structure function $S(\boldsymbol{k},t)$, which is experimentally accessible with scattering methods, do not allow one to discriminate whether K- or V-dynamics were used in the simulations ($S(\boldsymbol{k},t)$ defined in (3.19) and the measured intensity $I(\boldsymbol{k},t)$ are proportional to each other). For later stages the kinetic Ising model showed dynamical scaling (Section 2.4.2) (Fratzl, et al., 1991, Marro, et al., 1979) with identical scaling functions for both dynamics. The spherically averaged pair correlation function $\tilde{g}(r)$ can be written as (see (2.42) for comparison),

$$\tilde{g}(r,t) = g_0(T)\, h\big(r / R(t)\big) \; , \qquad (3.14)$$

where R, the first zero of $\tilde{g}(r)$, is a measure of the domain size of the two-phase structure; T denotes the temperature. The scaling function h was not only independent of the dynamics used, but also the same for different lattices (sc = simple cubic, bcc, fcc) and almost unchanged for different temperatures (Fratzl, et al., 2000).

Although the morphologies were all but identical, the coarsening rates were quite different for simulations using either K- or V-dynamics. In all the plots in Figure 18 the domain size R is shown as a function of $t^{1/3}$. Especially at low temperatures the rate of coarsening was significantly enhanced in the case of V-dynamics. For temperatures close to T_c, however, this difference vanished. The straight lines correspond to fits using the formula proposed by Huse (Huse, 1986),

$$R(t) = R_0 + (\lambda t)^{1/3} \; . \qquad (3.15)$$

The quality of the fit was excellent for all different temperatures and lattices.

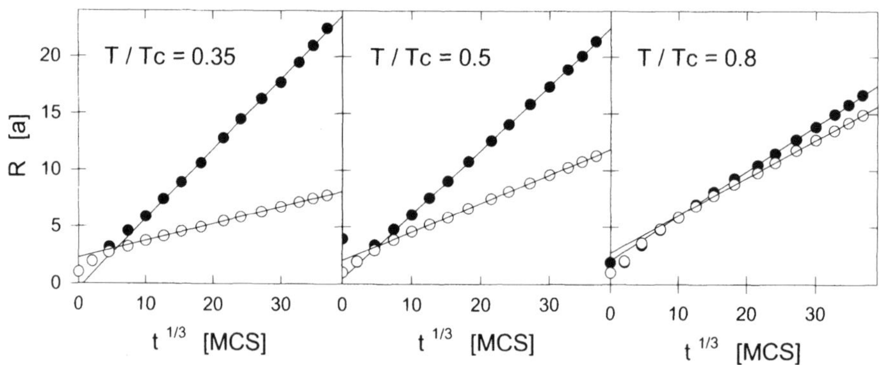

Figure 18. First zero R of the spherically averaged pair correlation function $\tilde{g}(r)$ for a simple cubic lattice and three different temperatures, plotted as a function of $t^{1/3}$, where t is time in MCS. R is given in units of the cubic lattice constant a, $c = 0.5$. Black disks represent vacancy and circles Kawasaki dynamics. The straight lines are fits with equation (3.15) for all data points $t \geq 2000$ MCS.

The difference between the two dynamics is most noticeable when comparing the temperature dependence of the prefactor λ. For K-dynamics λ_K decreased with decreasing temperature. Since coarsening proceeds via a diffusion of monomers for K-dynamics, this behavior can be well understood as due to the drastic reduction of monomers at low temperatures. For V-dynamics, however, λ_V increased with decreasing temperature. Qualitatively this can be understand due to the fact that the vacancy is located preferentially at domain interfaces, because there it breaks an energetically unfavorable bond between an A and a B atom. At the interface the vacancy supports interface diffusion and leads to the enhanced rate of phase separation. At higher temperatures the binding to the interface gets smaller, and the vacancy "loses time" roaming in one of the two phases. A more profound understanding of the simulation results was attempted in (Fratzl, et al., 2000). The importance of interface diffusion is reasonable in the case when the two-phase microstructure is interconnected and most of the interface can be reached via paths on the interface itself. This is surely the case for an equal concentration of A and B atoms, $c = 0.5$. Surprisingly the effect of an enhanced coarsening rate in the case of V-dynamics turned up also in the case of isolated precipitates, although not so pronounced. An explanation for this is given in the following section.

3.2.2 Coarsening by coagulation of precipitates or via evaporation and condensation of atoms?

The importance of the asymmetry parameter U (Eqs. (3.6) and (3.9)) on the kinetic path of the model alloy evolution was realized only recently: for the case of diffusion (Weinkamer, et al., 1999), nucleation (Soisson and Martin, 2000), phase separation (Athenes, et al., 2000, Roussel and Bellon, 2001, Weinkamer and Fratzl, 2003) and ordering (Athenes, et al., 1996, Porta, et al., 1999). The value of U affects the preferred location of the vacancy. For $U = 0$ the vacancy prefers a nearest-neighbor shell occupied roughly by the same number of A and B atoms, i.e., the interface between matrix and precipitates. In simulations with $U = J$ the vacancy is preferentially inside the precipitates, while for $U = -J$ the vacancy stays most of the time inside the matrix. For higher temperatures the vacancy path becomes less restricted to either the precipitates, the matrix or the interface.

In the framework of the Lifshitz-Slyozov-Wagner (LSW) theory (Lifshitz and Slyozov, 1961, Wagner, 1961), it is assumed that coarsening occurs via a migration of single precipitate atoms, which evaporate from smaller precipitates and finally condensate on larger ones – therefore it is often referred to as evaporation-condensation mechanism. The result is a growth of larger precipitates at the expense of smaller ones. An alternative mechanism is that precipitates move as a whole. When they meet, they coalesce, again with the result of a coarsening microstructure. That this latter coagulation mechanism can act during coarsening, was demonstrated in a simulation with the following artificial setting (and $U = 0$): a single, small precipitate of B atoms was put in a box with only A atoms, the boundary sites of the box filled with fixed B atoms. The small precipitate moved around and eventually disappeared by coagulation with the boundary before it was evaporated (Fratzl and Penrose, 1997).

To get a quantitative measure of the efficiency of the two different mechanisms during coarsening, MC simulations have been started from a microstructure, prepared by a preliminary simulation, comprising more than 200 precipitates, the largest containing close to 4000 atoms. Monitoring the size evolution of all the precipitates $n_i(t)$ in the system (Roussel and Bellon, 2001), plots like those shown in Figure 19 were obtained for all four different dynamics under investigation, K-dynamics and V-dynamics with $U = 0, \pm J$ (Weinkamer and Fratzl, 2003). During the time step between two measurements, the size of each precipitate changed either only by a small amount of less than about 12 atoms, or rather dramatically by more than 50 atoms. Small changes in the size of a precipitate are due to the evaporation and condensation of monomers and small clusters (LSW mechanism). Whereas the events involving large changes and which are displayed as jumps in the trajectories in Figure 19, has to be interpreted as coalescence events.

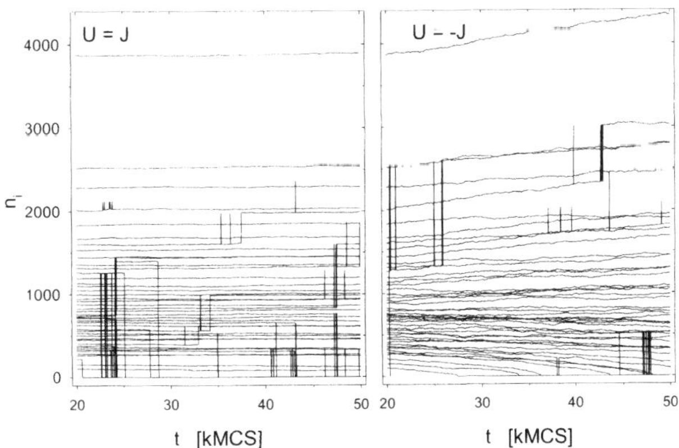

Figure 19. Size of individual precipitates as a function of time using vacancy dynamics with $U = J$ (left) and $U = -J$ (right) at $T/T_c = 0.35$ starting from the same initial configuration. For clarity trajectories of only every fifth precipitate when ordered by size are drawn.

The total increase of the mean precipitate size $\langle \Delta n \rangle$ could therefore be separated into contributions from the LSW mechanism and from the coagulation mechanism, $\langle \Delta n \rangle = \langle \Delta n \rangle_{LSW} + \langle \Delta n \rangle_{COAG}$.

Figure 20 shows the results for all four cases with a rather smooth contribution for the LSW mechanism, but a jagged contribution for the coagulation mechanism. To quantify the predominance of a coarsening mechanism, we computed the ratio $\varepsilon \equiv \langle \Delta n \rangle_{COAG} / \langle \Delta n \rangle$ or a mean value $\bar{\varepsilon}$ respectively, since ε slowly decreases with larger precipitate sizes.

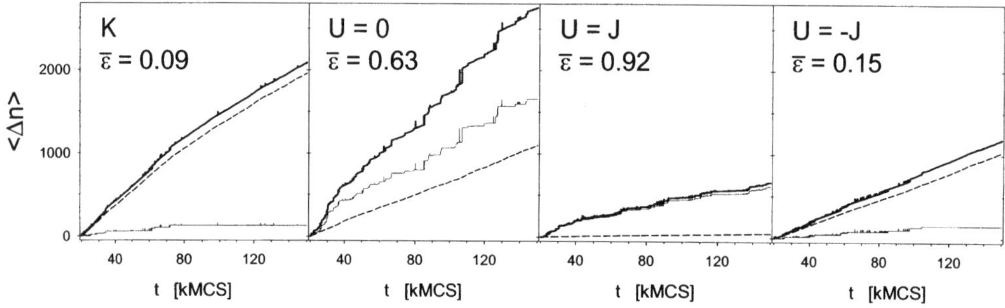

Figure 20. Time evolution of $\langle \Delta n \rangle$, the increase of the mean precipitate size $\langle n \rangle$ with respect to its initial value, for Kawasaki dynamics and vacancy dynamics with $U = 0$ and $U = \pm J$ (thick line). The dashed (smoother) line shows the contribution of the LSW mechanism to the overall increase, the thin (jagged) line the contribution of the coagulation mechanism, $\langle \Delta n \rangle = \langle \Delta n \rangle_{LSW} + \langle \Delta n \rangle_{COAG}$ (reprinted with permission from (Weinkamer and Fratzl, 2003))

This change of the predominant coarsening mechanism with different values for U can be understood as the result of a completely different vacancy path. In the case of $U = J$ the vacancy stays most of the time inside the precipitate, just leaving and reentering the precipitate at varying locations, which leads to a overall shift of the whole precipitate. The higher mobility of the precipitate enhances the probability of a coalescence event, resulting in a predominant coagulation mechanism. In contrast, for $U = -J$ the vacancy path is mainly confined to sites in the matrix, where it enhances the mobility of the monomers and small clusters of precipitate atoms, which are dissolved in the matrix. As a result coarsening occurs by the LSW mechanism. The special case of a vacancy attracted to the interface, $U = 0$, showed significant distributions for both coarsening mechanisms, $\bar{\varepsilon}_{U=0} = 0.63$, leading to the highest coarsening rate of all cases under investigation.

The finding of two different coarsening mechanisms active during phase separation in a model alloy naturally gives rise to the question, whether the predominant coarsening mechanism can be determined experimentally. Since conventional scattering techniques do not allow a discrimination between the mechanisms, the use of coherent X-rays from high-brilliance synchrotron sources was proposed (Weinkamer and Fratzl, 2003) performing a X-ray photon correlation spectroscopy (XPCS) experiment (Lengeler, 2001). Measuring the time-correlations of intensity fluctuations – so-called speckles – the XPCS technique allows investigations of dynamic processes on a nanometer scale. Performing a computer XPCS experiment and analyzing the intensity fluctuations using fluctuation analysis (FA) (Kantelhardt, et al., 2001) showed that a measurable quantity – the fluctuation exponent - as a function of length of the scattering vector, $|\boldsymbol{k}|$ was different depending on the predominant coarsening mechanism (Weinkamer and Fratzl, 2003). In a recent XPCS experiment the same differences were found in experiments on the alloys Al-6 at.%Ag and Al-9 at. % Zn under conditions comparable to the simulation. These measurements indicate that coarsening in Al-6 at.%Ag proceeds predominately via the classical LSW mechanism, while in Al-9 at. % Zn the coagulation mechanism is prevailing (Stadler, et al., 2003).

A last interesting effect due to the asymmetry parameter U should be mentioned. For the modified solubility of solute B atoms in the A-rich matrix (or of A atoms in the B precipitates) due to the curved interface, Roussel and Bellon found good agreement with the Gibbs-Thomson equation (2.36) in the case $U = 0$ after short simulation times. For $U \neq 0$, however, the solubility deviated significantly from the theoretical value (Roussel and Bellon, 2001).

3.3 An extended Ising model including elastic interactions

Up to now only chemical interactions between the atoms were considered. The system is then completely determined by the set $\{\sigma_i, i = 1,..., N\}$, which defines, how the rigid lattice is occupied by the different types of atoms. For the investigation of the influence of a lattice misfit between the A-rich and B-rich phases on phase separation, we used an extended atomistic model, where also long-range elastic interactions are included (Fratzl and Penrose, 1995). Displacements $\boldsymbol{u}(\boldsymbol{r})$ of the atoms from sites of the undistorted lattice $\boldsymbol{r}$ are allowed. They are assumed to be small enough that a one-to-one mapping between atoms and these regular lattice sites still holds, i.e. the lattice remains coherent. Furthermore, the small displacements justify the use of the harmonic approximation for the interaction between the atoms. In the three-dimensional model each pair of NN-atoms can be thought to be connected by springs with a longitudinal, L, and two different transverse spring constants, T_1 and T_2, allowing for noncentral forces (Gupta, et al., 2001). The spring constants are independent of the types of atoms they connect (homogeneous elasticity). The lattice misfit and the resulting elastic interactions are due to difference in radii of the two types of atoms, $R_A > R_B$. The Hamiltonian of the model system can be written as a quadratic form of the spin variables $\sigma(\boldsymbol{r})$ and the displacements $\boldsymbol{u}(\boldsymbol{r})$, corresponding to a second-order-expansion treatment of compositional and displacive disorder (Asta and Foiles, 1996, Cook and Defontaine, 1969). The relaxation time of the lattice distortions is assumed much shorter than the diffusion time of the atoms, i.e. the system is always in mechanical equilibrium. Using the Fourier transform of the spin variable, $\tilde{\sigma}(\boldsymbol{k}) = \sum_{\boldsymbol{r}} \sigma(\boldsymbol{r}) e^{i\boldsymbol{k}\boldsymbol{r}}$, the Hamiltonian can then be written as (Khachaturyan, 1983),

$$E = \widetilde{E}_0 + \frac{1}{2N} \sum_{k \neq 0} \widetilde{B}(\boldsymbol{k}) \left| \widetilde{\sigma}(\boldsymbol{k}) \right|^2 . \tag{3.16}$$

where the sum extends over the first Brillouin zone of the lattice, $\widetilde{H}_0$ a constant. The "elastic potential" $\widetilde{B}(\boldsymbol{k})$ can be calculated for any type of lattice (Cook and Defontaine, 1969, Gupta, et al., 2001). It is important to note that although the interactions are now long-range and anisotropic, they are still describable as pairwise interactions between atoms (Fratzl and Penrose, 1995), i.e.,

$$E = \widetilde{E}_0' + \frac{1}{2} \sum_{r} \sum_{r'} (J(\boldsymbol{r} - \boldsymbol{r}'))\, \sigma(\boldsymbol{r})\, \sigma(\boldsymbol{r}') . \tag{3.17}$$

Model parameters like the spring constants can be experimentally obtained (e.g., by neutron scattering) or calculated using more accurate descriptions of the alloy energetics (e.g., the embedded-atom-method (Asta and Foiles, 1996)). The effects of an additional external stress can also be modeled within this framework to study directional coarsening and rafting (Section 3.4.2). To observe non-trivial effects the model has to be modified by including a weak dependence of the spring constants on the atoms connected by the springs (Laberge, et al., 1997).

An important parameter defining the elastic properties of the system is the elastic anisotropy B, which is defined as

$$B = \frac{c_{11} - c_{12} - 2c_{44}}{2} . \tag{3.18}$$

In our simulations we chose a negative elastic anisotropy corresponding to elastically soft $<100>$ directions ($<10>$ direction in the 2D case). In particular, for the three-dimensional model Gupta et al. took the experimentally obtained Born-van Karman parameters of copper to specify the values for the spring constants, L, T_1 and T_2. Since introducing a vacancy in a model which allows for lattice distortions is not straightforward, the simulations investigating the influence of elastic interactions have been performed using Kawasaki dynamics (Gupta, et al., 2001).

We want to mention an alternative model approach, the so-called Discrete Atom method (Lee, 1998), which assumes central forces, but considers non-homogeneous elasticity. The assumption of harmonic forces can also be suspended (Ikeda and Matsuda, 1993). A comprehensive overview on modeling alloys with misfits including also continuum models can be found in (Fratzl, et al., 1999).

3.4 Influence of elastic interactions on phase separation

3.4.1 Varying the lattice misfit

The lattice misfit δ is defined as $\delta \equiv (a_A - a_B)/a_A$, where a_A and a_B are the lattice constants of the pure A and B phases. As a first simulation result is was shown that changes in the value of δ , involve also changes of the phase diagram, especially of the critical temperature T_c . If the elastic interactions, as measured by the size of δ are increased at fixed value for the parameter J which specify the "chemical" part of the interaction, the effect is to increase the critical temperature for phase separation. This was demonstrated by simulations and the same effect can be seen in the mean-field approximation (Fratzl and Penrose, 1996, Gupta, et al., 2001). For order-disorder transitions, on the other hand, the critical temperature is decreased by the elastic interaction, as is the tricritical temperature. This was shown by simulations of (Fischer and Nielaba, 2000).

Figure 21 summarizes some results of 3D simulations concerning the influence of different misfits and different concentration of precipitate atoms on the morphology of the microstructure (Weinkamer, et al., 2004). For no misfit (Fig. 5, left column) the isolated precipitates are round and irregularly arranged at low concentration and form an interconnected structure at higher concentrations. For $\delta = 1\%$ the precipitates are clearly orientated perpendicular to the elastically soft <100>-directions. In the case $c = 0.2$ the precipitates have the form of platelets. They are grouped together in domains of platelets with the same orientation. At higher concentration ($c = 0.5$) only a single domain is observed, built up of a series of plates. For intermediate misfit of $\delta = 0.5\%$ in the microstructure some alignment of the precipitates is combined with more or less irregular features. In agreement with earlier simulations in 2D (Fratzl and Penrose, 1996), the elastic interactions seem more effective for higher concentrations, leading to a more regular microstructure.

To study the length scales in the system the structure function $S(\boldsymbol{k},t)$ can be used,

$$S(\boldsymbol{k},t) = \frac{1}{N}\left|\sum_{\boldsymbol{r}} \sigma(\boldsymbol{r},t) e^{i\boldsymbol{kr}}\right|^2 , \tag{3.19}$$

i.e., the squared Fourier transform of the atom configuration defined by the spin variable $\sigma(\boldsymbol{r},t)$, N being the number of lattice sites. Since $S(\boldsymbol{k},t)$ is measured in a single crystal small angle scattering (SAS) experiment, Figure 22 corresponds to the scattering pattern obtained with an area detector in a SAS experiment. In the case of no misfit the SAS pattern shows spherical symmetry. For the case $\delta = 1\%$ and $c = 0.2$ (middle row on the right in Fig. 21), the alignment of the precipitates is reflected in an anisotropic SAS pattern (Fig. 22, left), with streaks along the horizontal and vertical directions. Also higher order maxima can be observed.

The task to define a typical length scale is not straightforward because the shapes of the precipitates and their alignment and relative positions are not spherically symmetric. The spherically averaged structure function, $S(|\boldsymbol{k}|,t)$ (or one of its higher moments) usually displays a peak, which shifts during coarsening towards smaller values of $|\boldsymbol{k}|$ (Fig. 17, left). From the position of the peak a mean distance R between the precipitates can be derived as,

$R(t) \equiv 2\pi / k_{\max}(t)$. Interestingly, spherically averaging $S(\boldsymbol{k},t)$ in Figure 22 results in a function which displays two peaks (Fig. 22, right). Both peaks come from contributions along the <100> directions and one peak is not simply a higher order peak of the other. There is a peak at $k = 2\pi / R_D$, where R_D is the distance between the precipitates, but there is a second peak at a position $k = 2\pi / R_A$; the value of R_A is about the same as the size of the domains of parallel precipitates that are visible in Figure 21.

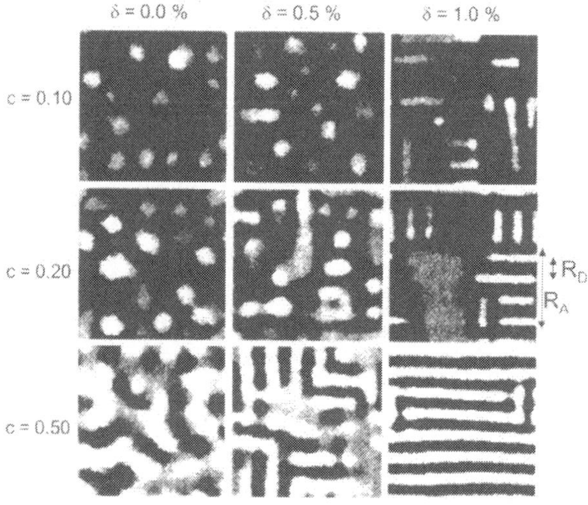

Figure 21. Snapshots of the precipitate morphology for three different misfits δ and three different concentrations of solute atoms c at $t = 6000$ MCS and $T = 5\ J/k_B$. The two-dimensional plots were obtained by taking a cross-section of a finite thickness equal to 6 lattice constants and averaging along the thickness. Light regions denote regions of a high concentration of solute A atoms. On the right the two different length scales R_D, the inter-precipitate spacing, and R_A, the precipitate domain size, are plotted (from (Weinkamer, et al., 2004))

Studying the time evolution of $R_D(t)$ gives a slowing down of the kinetics with increasing misfit. Plotted on the natural time scale, $t^{1/3}$, the data cannot be fitted by a straight line (as shown in Figure 18, for instance). For $\delta > 1$ coarsening comes virtually to rest at late times. Also the precipitate domains, $R_A(t)$, grow in time. Since their size is close to the length of the whole system, finite size effects will have a considerable influence on the result. But making just a qualitative statement, we observe in our simulations for large lattice misfit a coarsening process on two length scales.

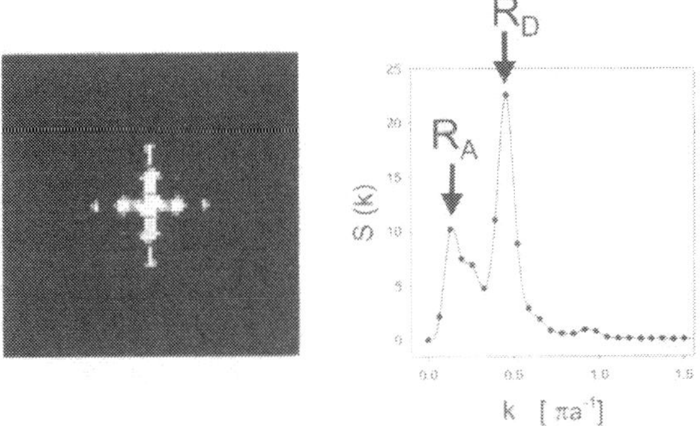

Figure 22. (100) plane of the structure function $S(\boldsymbol{k})$ (left) for misfit $\delta = 1\%$ and concentration $c = 0.2$, corresponding to the microstructure shown in Figure 21. The spherically averaged structure function $S(|\boldsymbol{k}|)$ displays two peaks corresponding to two different length scales in the system; a denotes the cubic lattice constant (from (Weinkamer, et al., 2004)).

3.4.2 Applying an additional external load – precipitate rafting

From real alloys it is known that the interference of an externally applied load with an internal stress state due to a lattice misfit between the phases causes new effects during phase separation. Most strikingly the microstructure "follows" the orientation impressed by an uniaxial load. Early MC simulations in 2 dimensions demonstrated that the emerging precipitate morphology exhibited a preferred orientation. Wavy stripes developed aligned parallel to the axis of the uniaxial tensile load (Laberge, et al., 1995, 1997). The drawback of these simulations is that there is no direct conclusion from a stripe-like pattern in two dimensions to the microstructure in 3D.

Predicting the rafting morphology in an alloy with negative anisotropy needs the consideration of *(i)* the sign of the misfit, i.e., is the lattice spacing of the precipitate phase larger or smaller, *(ii)* the sign of the elastic inhomogeneity, i.e. is the precipitate phase harder or softer, and *(iii)* the sign of the applied load, i.e. tensile or compressive. In 3D simulations the case of a larger lattice misfit ($\delta = 1\%$) and larger elastic constants in the precipitates was investigated (Gupta, et al., 2001). In agreement with theoretical considerations (Socrate and Parks, 1993) cylindrical precipitates along the loading direction are formed when applying a tensile load and plate-like precipitates perpendicular to the loading direction in the case of a compressive load (Gupta, et al., 2001) (Fig. 23). In the framework of our model, the change of one of the three signs mentioned above, e.g., changing the sign of the lattice misfit, would result in a switch from the cylindrical to the plate-like morphology or vice versa. It is notable that the precipitates are far from being perfect plates or cylinders, but they are partly interconnected, showing holes and other irregularities (Fig. 23).

The application of an external load leads from the very beginning of phase separation to a distinction of the loading direction. Reversing the external stress from tensile to compressive or vice versa, caused only very slow changes in the precipitate microstructure. The parallel-cylinder morphology remained fairly stable when a compressive load was applied (Gupta, et al., 2001). The time evolution of the precipitate size was not only very similar regardless of the sign of the applied load, but also of whether an additional load was applied at all. Like in the case without external stress the kinetics slowed down significantly (Gupta, et al., 2001).

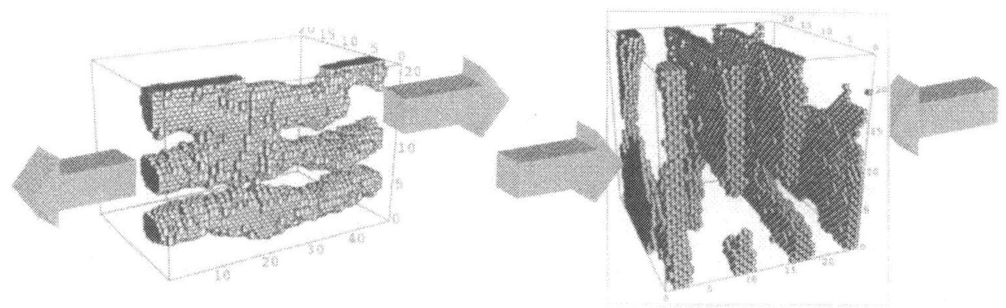

Figure 23. Snapshots of the precipitate morphology for an alloy evolved from a random configuration at $t = 4000$ MCS for external tensile stress and compressive stress as indicated by the arrows ($c = 0.2$ and $T = 5\ J/k_B$). Lattice sites occupied by solute *B* atoms are represented by cubes. *A* atoms and *B* atoms dissolved in the matrix are not drawn. In each case only a small part of the whole system is shown.

An explanation of this reduced growth rate of the precipitates is based on the regularity of the emerging microstructure. In the case of a tensile load, the cylindrical precipitates are arranged in a regular two-dimensional array, forming a quadratic mesoscopic precipitate lattice. For a compressive load the plate-like precipitates form an one-dimensional array. Once such a regular arrangement of precipitates was formed it is hard to imagine how coarsening should proceed at all. For example, how is the number of precipitates reduced in the tensile case, say from 49 cylindrical precipitates arranged in 7×7 array to 36 forming an 6×6 array. Simulations showed that the mesoscopic precipitate lattices are not perfect, but contain mesoscopic lattice defects (Weinkamer, et al., 2000). In analogy to defects in ordinary crystals, the defects can be classified as "meso-vacancies", "meso-interstitials" and "meso-dislocations". These defects of the precipitate lattice are involved in complex dynamical processes, which decrease the number of precipitates and consequently increase the lattice constant of the mesoscopic lattice: Meso-dislocations can move against each other and annihilate, meso-interstitials can merge with a precipitate occupying a regular site of the meso-lattice. The series of plate-like precipitates in the compressive case exhibited a mesoscopic analog to a stacking fault in crystal lattices (Weinkamer, et al., 2000).

3.4.3 Ordered Precipitates

Choosing antiferromagnetic nearest neighbor interactions, $J^{(1)} > 0$, and ferromagnetic next nearest neighbor interactions, $J^{(2)} < 0$, results in a ordering tendency. The inclusion of elastic interactions into a two-dimensional model (Nielaba, et al., 1999) leads to the development of a stripe-like morphology for the segregated ordered and disordered regions. A new morphological feature is the presence of narrow channels subdividing the ordered plates. These channels can be explained by the possibility of anti-phase boundaries (APB´s) separating the two variants of the ordered phase. Since it is commonly found that neighboring precipitates are ordered on different sublattices, the narrow channels correspond to wetted APB´s. The great difference in the morphology with 30% ordered phase in the disordered matrix compared to the "reverse case" with 30% disordered phase is as well attributed to the presence of APB´s.

The effect of an additional external stress is to cause rafting of the ordered precipitates. Because of the APB´s the stripes are then segmented. The simulation results are best compared to the technologically important class of nickel base superalloys where the precipitates are also ordered. Figure 24 shows transmission electron micrographs of Ni-Al-Mo alloys annealed with and without external stress and simulation results. Despite the great difference in the length scale, the great similarity between the real and the computed morphologies is striking.

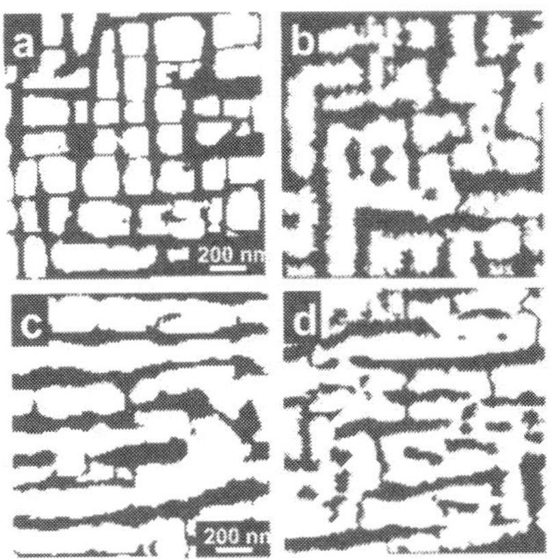

Figure 24. Comparison between transmission electron micrographs and simulation snapshots. The disordered phase is shown black, while all ordered variants are shown in white. Plots (a) and (c) on the left are TEM images of Ni-Al-Mo alloy (from (Paris, et al., 1997)) without external load (a) and with an external load along the vertical [010]-direction (c). The simulations were performed at a concentration $c = 0.35$ and temperature $T = 0.567\,J/k$. Snapshots are taken after 10^6 MCS, without (b) and with (d) including an external stress (from (Fratzl, et al., 1999)).

3.5 Concluding remarks

The kinetic Ising model and its extensions to the ABV model (Section 3.1) or to include an elastically deformable lattice (Section 3.3) is able to predict a number of important details of phase separation which depend on processes on the atomic level. While these microscopic models are probably too detailed to perform large-scale simulations of real alloys, they provide basic knowledge complementary to the results which may be obtained by mesoscopic or macroscopic models of phase separation, which are presented below in Sections 4 and 5, respectively.

4 Phase field modeling of phase seperation

4.1 Phase field kinetic equations

A phase-field model describes the whole microstructure of a system, i.e. the compositional/structural domains and the interfaces, by using a set of order-parameter fields. These "phase fields" are spatially inhomogeneous, but continuous across the interfacial regions. Hence phase field models belong to the class of diffuse interface models. Typical examples for phase fields are the composition field, $c(\boldsymbol{r},t)$, which defines in a coarsed grained approach the concentration of solute atoms at position $\boldsymbol{r}$ at time t (see Section 2.2), or the order field $\eta(\boldsymbol{r},t)$ (see Section 4.2). Field variables can be conserved, i.e., they have to satisfy the local conservation condition, or they are non-conserved. The composition field $c(\boldsymbol{r},t)$ is conserved,

$$\int_V c(\boldsymbol{r},t)d\boldsymbol{r} = const \ , \tag{4.1}$$

while the order field $\eta(\boldsymbol{r},t)$ is an example of a non-conserved field variable.

The problem, which has to be solved, is to find field kinetics equations, which describe the temporal evolution of the phase fields. The free energy that tends to its minimum value is the driving force of the microstructural evolution. Two crucial steps are, first, to determine the phase fields ψ_i, $i = 1,\ldots,p$, which are necessary for a complete description of the microstructure. Secondly, the free energy functional F as a function of the phase fields ψ_i has to be specified,

$$F = F[\psi_1(\boldsymbol{r},t), \psi_2(\boldsymbol{r},t), \ldots] \ . \tag{4.2}$$

The time-dependent Ginzburg-Landau kinetic equations describe the evolution of the phase fields using a linear response ansatz,

$$\frac{\partial \psi_i(\boldsymbol{r},t)}{\partial t} = \hat{M}_{ij} \frac{\delta F}{\delta \psi_j(\boldsymbol{r},t)} \ , \tag{4.3}$$

where δ denotes the functional derivative and $\hat{M}_{ij}$ is the kinetic coefficient matrix. Summation over repeated indices is implied. Although the equations are linear with respect to the driving force $\frac{\delta F}{\delta \psi_j(\boldsymbol{r},t)}$, they are, of course, nonlinear with respect to the field variables $\psi_i(\boldsymbol{r},t)$. For conserved quantities like $c(\boldsymbol{r},t)$ the kinetic parameter $\hat{M}$ is proportional to the Laplacian, $\nabla^2 \equiv \frac{\partial^2}{\partial x^2} + \frac{\partial^2}{\partial y^2} + \frac{\partial^2}{\partial z^2}$, $\hat{M} = M\nabla^2$ (Wang, et al., 1996), so that (4.3), becomes,

$$\frac{\partial c(\boldsymbol{r},t)}{\partial t} = M\nabla^2 \frac{\delta F}{\delta c(\boldsymbol{r},t)} , \tag{4.4}$$

with M a mobility.

For the case of coherent phase transformations, three contributions to F has to be considered,

$$F = F_{chem} + F_{int} + F_{elast} ,$$

with F_{chem} the "chemical" bulk free energy, F_{int} the total interfacial free energy and F_{elast} the coherent strain energy.

4.2 Phase Separation

Starting with a simple example in this section the complexity of the problem will be enhanced stepwise in the following sections. We consider phase separation into a phase with a high concentration of solute atoms and a matrix rich of solvent atoms in the absence of coherent strains in the system. Only F_{chem} and F_{int} contribute then to free energy function F. Since the bulk free energy density f is a function of composition, which itself varies with position in the specimen, the chemical contribution reads,

$$F_{chem} = \int_V f(c(\boldsymbol{r},t))d\boldsymbol{r} . \tag{4.5}$$

Below the critical temperature, $f(c)$ has two minima corresponding to the composition of the two phases. In the phase field approach an interface is characterized by a strong gradient of $c(\boldsymbol{r})$. The introduction of an interface tension corresponds to penalize rapid spatial variations of $c(\boldsymbol{r})$. A common ansatz is therefore $F_{int} \propto \int_V (\nabla c(\boldsymbol{r},t)^2 d\boldsymbol{r}$, where the power 2 just makes the interfacial free energy independent of the sign of the gradient of $c(\boldsymbol{r})$. The energy functional reads

$$F = F_{chem} + F_{int} = \int_V \left[f(c(\boldsymbol{r},t)) + \frac{1}{2}\chi^2 (\nabla c(\boldsymbol{r},t)^2 \right] d\boldsymbol{r} . \tag{4.6}$$

Inserting F in (4.4), which means essentially performing the variational derivative $\frac{\delta F}{\delta c(\boldsymbol{r},t)}$, results in the celebrated Cahn-Hilliard equation (Cahn, 1961, Cahn and Hilliard, 1958, 1959) (see also (2.23) in Section 2.3.3),

$$\frac{\partial c(\boldsymbol{r},t)}{\partial t} = M\nabla^2\left(\frac{\partial f}{\partial c}(\boldsymbol{r},t) - \chi^2\nabla^2 c(\boldsymbol{r},t)\right) . \tag{4.7}$$

When the interest of the investigation is primarily in a qualitative analysis of the sequence of different microstructures and not in its details like the exact transformation rates, a simple phenomenological ansatz for the free energy is sufficient. The approximation of the free energy only has to provide a correct topology as a function of the phase fields. This means essentially that the minima of the free energy are taken for values of the phase field variables which correspond to the equilibrium states of the different phases. In the simplest case of a isostructural phase separation, where the composition field $c(\boldsymbol{r})$ is the only phase field, a polynomial for the local free energy density can be used in accord with Landau´s theory of phase transitions,

$$f(c(\boldsymbol{r},t)) = -\frac{1}{2}A_1(c(\boldsymbol{r},t)-c_0)^2 + \frac{1}{4}A_2(c(\boldsymbol{r},t)-c_0)^4 , \tag{4.8}$$

where A_1, A_2 and c_0 are phenomenological constants, $A_1, A_2 > 0$ below the critical temperature. The constants have to be fitted from theory or experiment. The specific free energy density f given above has a "double-well" form, with two minima at

$$c_{1,2} = c_0 \pm \sqrt{\frac{A_1}{A_2}} , \tag{4.9}$$

which are close to the composition of the precipitate phases α and the matrix phase β in equilibrium (see Section 2.2). Inserting (4.8) into Equation (4.7) leads to a partial differential equation which describes the time evolution of the composition field $c(\boldsymbol{r})$ during phase separation,

$$\frac{\partial c(\boldsymbol{r},t)}{\partial t} = M\nabla^2\left(-A_1(c(\boldsymbol{r},t)-c_0) + A_2(c(\boldsymbol{r},t)-c_0)^3 - \chi^2\nabla^2 c(\boldsymbol{r},t)\right) . \tag{4.10}$$

For a proper mathematical formulation of the problem, additionally the boundary and initial conditions must be specified. To reduce surface effects usually periodic boundary conditions are used. The initial condition is naturally the state of the system before the quench into the miscibility gap, i.e., a homogeneous solid solution with the concentration $\bar{c}$. Using this initial condition, $c(\boldsymbol{r},t=0) = \bar{c} = const$, leads to $\frac{\partial c}{\partial t} = 0$, and therefore results only in the solution $c(\boldsymbol{r},t) = \bar{c}$ for all times. The reason for yielding only a trivial solution is the entirely deterministic character of the field kinetic equations. With these equations a adequate description of growth and coarsening is possible, but not of processes where thermal fluctuations play an important role like the nucleation of new precipitates (see Section 2.3.2). To overcome this shortcoming the field kinetic equations can be complemented by adding a stochastic noise term $\xi(\boldsymbol{r},t)$ of Langevin type into the right-hand parts of the kinetic equations, e.g.,

$$\frac{\partial c(\boldsymbol{r},t)}{\partial t} = M\nabla^2\left(\frac{\partial f}{\partial c}(\boldsymbol{r},t) - \chi^2\nabla^2 c(\boldsymbol{r},t)\right) + \xi(\boldsymbol{r},t) \ . \qquad (4.11)$$

The noise term $\xi(\boldsymbol{r},t)$ is Gaussian distributed and its correlation properties satisfy the requirements of the fluctuation-dissipation theorem (Chandler, 1987). With the noise-term extension equation (4.11) is often referred to as Cahn-Hilliard-Cook equation (Cook, 1970).

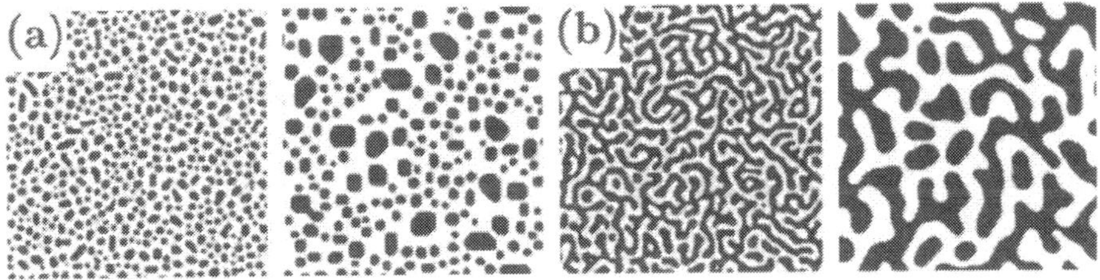

Figure 25. Phase separation in the absence of elastic interactions ($G_1 = 0, B_1 = 0, B_0 = 0$ in the nomenclature of equation (4.24)): two snapshots at early and late times for each of the two concentrations, $c = 0.35$ on the left, $c = 0.5$ on the right (reprinted with permission from (Orlikowski, et al., 1999); copyright by the American Physical Society).

The partial differential equation (4.10) can be simply integrated forward in time and the goal of a numerical solution is to track the time evolution of $c(\boldsymbol{r},t)$ with some desired accuracy. The straightforward approach is to discretize the system in space and time by choosing points with equal spacing Δt along the time axis and spacings Δx, Δy, Δz along the spatial axes. The concentration field $c(\boldsymbol{r},t)$ is then represented by the values $c(i\,\Delta x, j\,\Delta y, k\,\Delta z, n\,\Delta t) \equiv c^n_{i,j,k}$ at the grid points. The simplest choice to present the time derivative in terms of a differential quotient is the forward Euler differencing,

$$\frac{\partial c(\boldsymbol{r},t)}{\partial t} \approx \frac{c^{n+1}_{i,j,k} - c^n_{i,j,k}}{\Delta t} \ , \qquad (4.12)$$

leading to a system of linear equations in the variables $c^n_{i,j,k}$. Depending on the problem a more accurate algorithm should be used. A number of numerical tricks are available to solve partial differential equations efficiently (Press, 1997), e.g., in the case of considering elastic interactions (Section 4.4) and homogenous elasticity it is advantageous to solve the Fourier representation of the kinetic field equations (see eq. (4.20)) (Wang, et al., 1996). The final result is the temporal evolution of the discretized composition field $c^n_{i,j,k}$ as shown in Figure 25 for two different concentrations of solute atoms for the two-dimensional case.

4.3 Ordering

Increasing the complexity of the problem we consider now the precipitation of an ordered phase from a disordered phase. The degree of order is usually described by order parameters, $0 \le |\eta| \le 1$, with $|\eta| = 1$ for a full ordered structure and $\eta = 0$ for a random arrangement of the atoms. In the case of a *bcc* lattice and B2 order a single order parameter is sufficient. Order in the important case of an *fcc* lattice and $L1_2$ order has to be described by three order parameters (Inden, 2001). Taking as an example the precipitation of a B2 ordered phase and translating it into the language of phase field modeling, the system is described now by two phase fields. Beside the composition field $c(\boldsymbol{r},t)$, which is a conserved quantity, the order field $\eta(\boldsymbol{r},t)$ describes the degree of order with $\eta = \pm 1$ for the two fully ordered variants. The order field is non-conserved since order can emerge from an initially totally unordered state. The field kinetic equations for both phase fields are therefore

$$\frac{\partial c(\boldsymbol{r},t)}{\partial t} = M\nabla^2 \frac{\delta F}{\delta c(\boldsymbol{r},t)}$$
$$\frac{\partial \eta(\boldsymbol{r},t)}{\partial t} = L \frac{\delta F}{\delta \eta(\boldsymbol{r},t)} \quad . \tag{4.13}$$

In the course of phase separation, two different diffuse interfaces emerge. The first are interfaces between the ordered precipitates and the disordered matrix, the second ones are antiphase boundaries, which separate the two variants of the ordered phases. Since there is a relation between values of η and c, e.g., for $\eta = 1$ c has to be equal to 0.5, the two phase fields have to be coupled in the functional of the local free energy density. The simplest Landau expansion with the right topology of the free energy surface has the following form (Wang, et al., 1996)

$$f[c(\boldsymbol{r},t),\eta(\boldsymbol{r},t)] = \frac{1}{2}A_1(c(\boldsymbol{r},t)-c_0)^2 - \frac{1}{2}A_2(c(\boldsymbol{r},t)-c_1)(\eta(\boldsymbol{r},t))^2 - \frac{1}{4}A_3(\eta(\boldsymbol{r},t))^4 + \frac{1}{6}A_4(\eta(\boldsymbol{r},t))^6 \tag{4.14}$$

where $A_1 - A_4$ are positive constants below the critical temperature and c_0 and c_1 are close to the equilibrium composition of the disordered matrix and the ordered precipitate phase, respectively (Fig. 26).

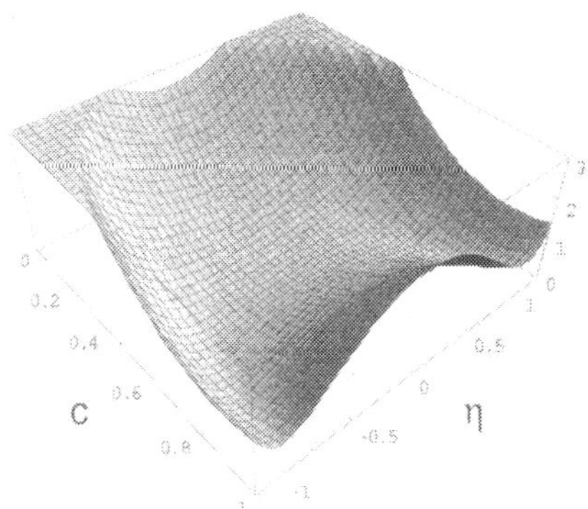

Figure 26. Free energy density as a function of the concentration c and the order parameter η according to eq. (4.14).

For the B2 type ordering the free energy should be invariant with respect to a transition $\eta \to -\eta$, which is the reason that $f[c,\eta]$ contains only even powers of η. This is not the case for the $L1_2$ type ordering. In a study of the ordered precipitation with $L1_2$ order the following functional was used combining the composition field $c(\boldsymbol{r},t)$ and the three order fields $\eta_i(\boldsymbol{r},t)$, $i = 1,2,3$ (Vaithyanathan and Chen, 2000):

$$f[c,\eta] = \frac{1}{2}A_1(c-c_0)^2 - \frac{1}{6}A_2(c-c_1)\sum_{i=1}^{3}\eta_i^2 - \frac{1}{3}A_3\prod_{i=1}^{3}\eta_i + \frac{1}{24}A_4\sum_{i=1}^{3}\eta_i^4 + \frac{1}{24}A_5\sum_{\substack{i,j=1\\ i\neq j}}^{3}\eta_i^2\eta_j^2 \tag{4.15}$$

where the constants have the same characteristics like in equation (4.14).

It should be noted that in this phase field description of ordered systems unphysical states are not excluded. For instance in the last example, all η_i can be equal to 1 at the same position $\boldsymbol{r}$. Such a state is only penalized by a high local value of the bulk free energy density. This "forgiving" description of the physical system makes the phase field approach very powerful. The examples of ordered precipitation given above, however, expose a weak point of the method. Specifying the permitted order parameter fields to write down the field kinetic equations, the results of the phase transformation are a priori restricted to changes in these variables. The possibility remains that an important phase field was omitted and therefore transient states in the microstructure evolution cannot be described. This limitation, however, can be remedied by using a microscopic phase field kinetic model (Chen and Khachaturyan, 1991, Wang, et al., 1996).

4.4 Elastic Interactions

Elastic interactions caused by a lattice misfit between the matrix and the precipitating phase have a crucial influence on the shape of the precipitates, on their orientation and mutual arrangement, as well as on the kinetics of the phase transformation. Similar to the description used in the Monte Carlo simulations (Section 3.3), the system is described now by the composition field $c(\boldsymbol{r},t)$ and the vector field of the displacements, $\boldsymbol{u}(\boldsymbol{r},t)$. The free energy is the sum of three contributions, the chemical and interfacial part as before, and additionally an elastic contribution (Orlikowski, et al., 1999, Sagui, et al., 1998):

$$F[c,\boldsymbol{u}] = F_{chem} + F_{int} + F_{elast} = \iint_V \left(f(c(\boldsymbol{r},t)) + \frac{1}{2}\chi^2 (\nabla c(\boldsymbol{r},t))^2 + \alpha c(\boldsymbol{r},t)\nabla \boldsymbol{u}(\boldsymbol{r},t) + f_{elast}(\boldsymbol{u}(\boldsymbol{r},t)) \right) d\boldsymbol{r} \tag{4.16}$$

On the right-hand side under the integral the third term couples the composition field and the displacement field taking into account that the stress free state depends on the composition; α denotes a coupling constant. Considering a three-dimensional system with cubic symmetry the elastic free energy density is

$$f_{elast}(\boldsymbol{u}) = \frac{1}{2} K (\nabla \boldsymbol{u})^2 + G \sum_{i,j=1}^{3} (u_{ij} - \frac{\delta_{ij}}{3} \nabla \boldsymbol{u})^2 + B \sum_{i=1}^{3} (u_{ii} - \frac{1}{3} \nabla \boldsymbol{u})^2 , \tag{4.17}$$

where the strain tensor is defined as usual

$$u_{ij} = \frac{1}{2} \left(\frac{\partial u_i}{\partial x_j} + \frac{\partial u_j}{\partial x_i} \right) . \tag{4.18}$$

The bulk modulus K, the shear modulus G and the elastic anisotropy B are given as a function of the elastic constants c_{11}, c_{12}, and c_{44} as

$$\begin{aligned} K &= c_{12} + \frac{2}{3} c_{44} \\ G &= c_{44} \\ B &= \frac{1}{2}(c_{11} - c_{12}) - c_{44} \end{aligned} \tag{4.19}$$

Under the assumption of mechanical equilibrium, $\frac{\partial F}{\partial u_i} = 0$, the free energy functional (4.16) can be reduced to a functional of the composition field alone, i.e., $F[c(\boldsymbol{r},t),\boldsymbol{u}(\boldsymbol{r},t)] = F[c(\boldsymbol{r},t)]$. Before dealing with the general case two special cases are addressed.

In the first simplest case of an elastically isotropic and homogeneous system, i.e., $B = 0$ and $K, G = const$, the form of (4.6) for the free energy functional is regained. The only change is that the incoherent bulk free energy density $f(c(\boldsymbol{r},t))$ has to be replaced by the coherent bulk free energy density (see Section 2.2). This replacement results in a change of the phase diagram, but the influences on the simulation results are only quantitative.

The second special case is to assume homogeneous elasticity, but to consider the system to be elastically anisotropic ($K,G = const$ and $B \neq 0$). This is exactly the case the investigations using the Monte Carlo method concentrated on (Section 3.3 and 3.4). As shown by Khachaturyan (Khachaturyan, 1965, 1983), an analytic solution for the elastic free energy can be calculated where the result is best given in Fourier space (see also its discrete version eq. (3.16) for comparison),

$$F_{elast} = \frac{1}{2} \int_{1^{st}BZ} \frac{d\boldsymbol{k}}{(2\pi)^3} \widetilde{B}(\boldsymbol{k}) \,|\, \tilde{c}(\boldsymbol{k}) \,|^2 \tag{4.20}$$

where the integral is over the first Brillouin zone avoiding the singularity at the origin and with

$$\hat{c}(\boldsymbol{k}) = \int_V d\boldsymbol{r}\, c(\boldsymbol{r}) e^{i\boldsymbol{k}\boldsymbol{r}} \quad . \tag{4.21}$$

The elastic potential $\widetilde{B}(\boldsymbol{k})$ is fixed once the symmetry of the lattice and the elastic constants are given. Figure 27 shows simulation results with different signs for the elastic anisotropy B resulting in different elastically soft directions in the system and consequently a different orientation of the microstructure.

In the general case the material exhibits an elastic anisotropy ($B \neq 0$) and the elastic constants are different in the matrix and in the precipitate phase. In the framework of a phase field approach that means that K, G and B depend on concentration. For the following, a linear dependence on concentration is assumed, i.e.,

$$\begin{aligned} K &= K_0 + K_1 c \\ G &= G_0 + G_1 c \quad . \\ B &= B_0 + B_1 c \end{aligned} \tag{4.22}$$

The free energy function can then be rewritten in the form

$$F = \int_V \left[\tilde{\tilde{f}}(c(\boldsymbol{r},t)) + \frac{1}{2}\chi^2 (\nabla c(\boldsymbol{r},t))^2 + f_{lr}(c(r,t)) \right] d\boldsymbol{r} \quad ,$$

where $\tilde{\tilde{f}}$ denotes a modified bulk free energy density. The third term under the integral describes the long-range elastic interactions which are responsible for the qualitative changes in the evolution of the microstructure. The details of the derivation and the final result for f_{lr} can be found in (Orlikowski, et al., 1999). Here only an instructive form is given,

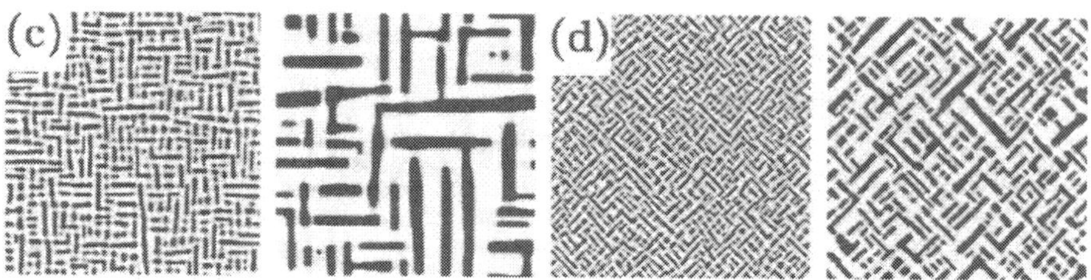

Figure 27. Phase separation in a homogeneous but anisotropic two-dimensional system ($G_1 = 0, B_1 = 0, B_0 = -1$ (left) and $B_0 = 1$ (right) in the nomenclature of equation (4.24)). Negative elastic anisotropy corresponds to elastically soft {100} -directions, $B > 0$ to elastically soft {111} -directions. Two snapshots for each value of B_0 corresponding to early and late times; $c = 0.35$ (from (Orlikowski, et al., 1999)).

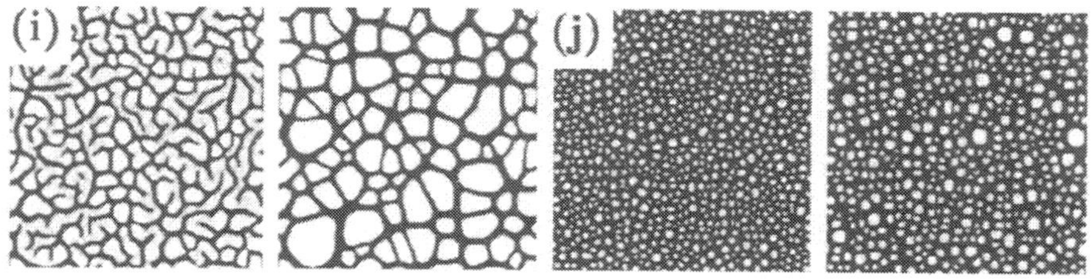

Figure 28. Phase separation in an elastically isotropic but inhomogeneous two-dimensional system ($G_1 = 0.15, B_1 = 0, B_0 = 0$ in the nomenclature of equation (4.24)). Two snapshots at early and late times for each of the two concentrations, $c = 0.35$ on the left, $c = 0.65$ on the right (from (Orlikowski, et al., 1999)).

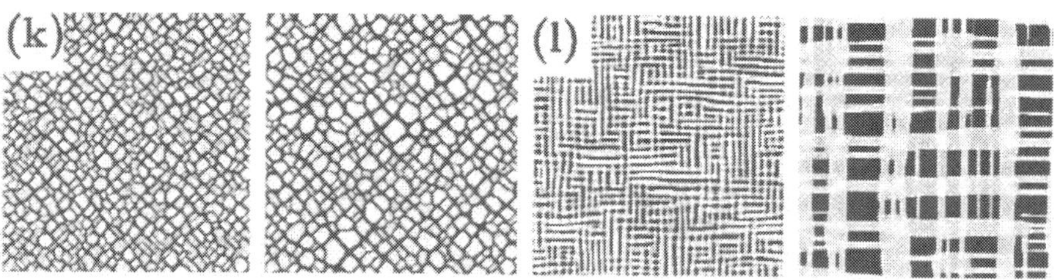

Figure 29. Phase separation in a both elastically inhomogeneous and anisotropic two-dimensional system ($G_1 = 0.5, B_1 = 0, B_0 = 0.4$ (left) and $G_1 = 0.25, B_1 = -1.4, B_0 = -1.4$ (right) in the nomenclature of equation (4.24)). Two snapshots for each set of parameters corresponding to early and late times; $c = 0.35$ (from (Orlikowski, et al., 1999)).

$$f_{lr}(c(\boldsymbol{r},t)) = \tilde{\delta}^2 \left(G_1 h_1(c(\boldsymbol{r},t)) + B_1 h_2(c(\boldsymbol{r},t)) + B_0 h_3(c(\boldsymbol{r},t))\right) , \quad (4.24)$$

where $\tilde{\delta}$ is a quantity corresponding to the lattice misfit defined at the beginning of Section 3.4.1; h_1, h_2 and h_3 are functions which do not depend on any elastic constants but only on the composition field. The first term on the right-hand side – the so-called Eshelby interaction - is due to the elastic inhomogeneity. The last term reflects the cubic anisotropy. The remaining second term was omitted in the original investigation of Onuki and Nishimori (Nishimori and Onuki, 1990, Onuki and Nishimori, 1991) and corresponds to a concentration dependence on the anisotropy.

Figures 27 to 29 give some examples of the variety of different microstructures that can occur during phase separation when elastic interactions play an important role. Note that in Figure 28 (left) the precipitate phase is the phase with the higher volume fraction. The following guidelines are useful in understanding the microstructures. Concerning the elastic anisotropy, the precipitates are orientated perpendicular to the elastically soft directions (Fig. 27). The elastic inhomogeneity is reflected by a wrapping of the elastically hard phase by the softer one (Fig. 28). In the case of a non-vanishing anisotropy and inhomogeneity both aspects are combined (Fig. 29).

4.5 Further extensions of the model

Four examples of the effect of different influences on the phase separation process and their modeling using phase field models are briefly presented below.

4.5.1 Externally applied stress

Figure 30 gives an example of a phase field simulation where the system is subjected to uniaxial and shear stress, respectively. As already mentioned in Section 3.4.2 qualitative changes are only expected when the system is elastically inhomogeneous. In the case of a uniaxial loading it is most obvious that the loading direction defines a new preferred direction.

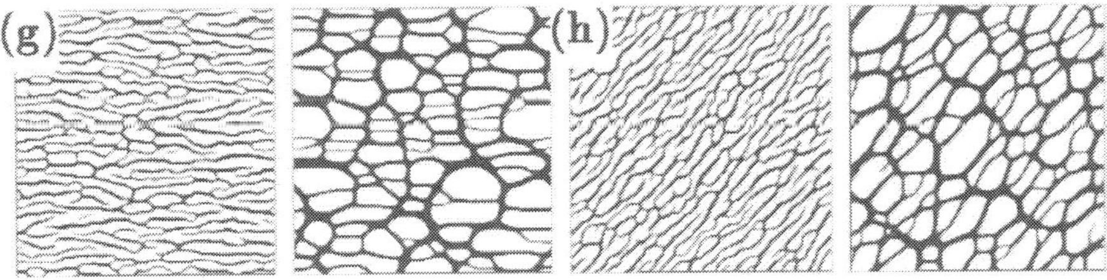

Figure 30. Phase separation in an elastically isotropic but inhomogeneous system ($G_1 = 0.25, B_1 = 0, B_0 = 0$ in the nomenclature of equation (4.24), see Figure 28 for the microstructure without external stress) which is subjected to an external load; uniaxial stress, $S_{xx} = 0.25$, on the left, and shear stress, $S_{xy} = 0.25$ on the right. Two snapshots for each loading condition corresponding to early and late times; $c = 0.35$ (reprinted with permission from (Orlikowski, et al., 2000); copyright by the American Physical Society).

4.5.2 Ordered precipitates

Combining the bulk free energy density of equation (4.15) and the elastic contribution of equation (4.20), phase separation in the technologically important class of nickel-based superalloys was studied (Chen, et al., 2001, Rubin and Khachaturyan, 1999). The new feature is that the merging between two precipitates can be prevented since it would create an energetically expensive antiphase boundary (see Section 3.4.3). The precipitates take a plate-like shape as observed in experimental investigation using electron microscopy or scattering experiments (Fig. 31).

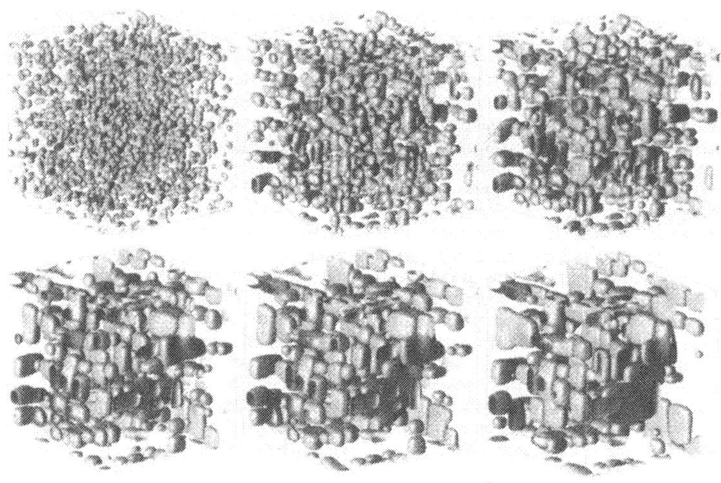

Figure 31. Temporal evolution of the microstructure during precipitation of L12 ordered precipitates in an unordered matrix. Homogenous elasticity is assumed in the three-dimenisonal phase-field simulation. (reprinted with permission from (Chen, et al., 2001)).

4.5.3 Defects

A recent major progress in the phase field method was the incorporation of defects into its models. With these models nucleation and growth near a dislocation (Hu and Chen, 2001) and the temporal evolution of many dislocations in an elastically anisotropic crystal during deformation have been investigated (Wang, et al., 2001). A study of dislocation dynamics near the free surface in thin films leads over to the final point (Wang, et al., 2003).

4.5.4 Surface-directed spinodal decomposition

Phase separation processes can be crucially influenced by the geometry of the sample due to interactions with the free surface. Theoretically this "defect" in form of a free surface is described in the theory of surface-directed spinodal decomposition (Binder, 1998, Binder and Fratzl, 2001). Figure 32 on the left shows a recent example, where this effect was observed in a metallic alloy on a macroscopic scale (Aichmayer, et al., 2003). The influence of the surface leads to the development of a concentric domain structure. Long-range elastic interactions are responsible for this macroscopic structure. A comparison with phase field calculations based on a standard Cahn-Hilliard-Cook equation (4.11), but including a surface potential which decreases in essence proportional to $1/R$, where R denotes the distance from the surface, shows good agreement with the experiment.

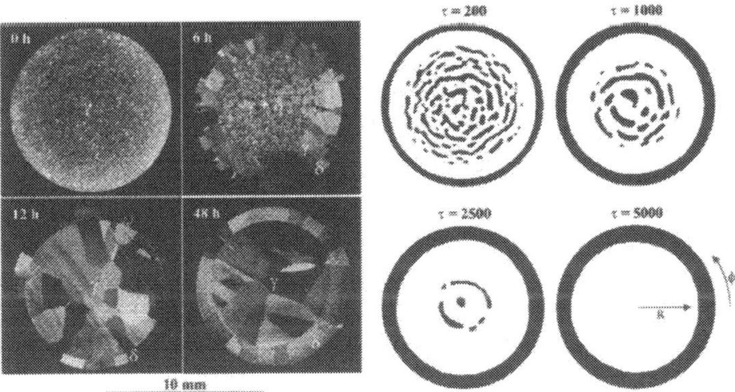

Figure 32. On the left, sample-shape dependent domain structure of cylindrical stell samples for annealing times as given in the pictures, annealing temperature $T = 1325°C$. δ denotes the ferrite, γ the austenite phase. On the right are simulation results using an extended Cahn-Hilliard-Cook equation at four dimensionless times τ . Regions with different sign of the phase field are marked either in black or in white; reprinted with permission from (Aichmayer, et al., 2003); copyright by the American Physical Society.

4.6 Concluding remarks

A disadvantage of the phase field model is the phenomenological description of a non-equilibrium and ill defined free energy density (see Section 2.2 and for example eqs. (4.8) and (4.14)). Recently attempts have been performed to make use of thermodynamical databases using the CALPHAD method (Zhu, et al., 2002) or to combine the phase field method with atomistic first-principle calculations and a mixed-space cluster expansion approach (Vaithyanathan, et al., 2002) to determine the driving forces for phase separation in real systems. Atom-probe experiments also show that the transition between phases can occur within a few nanometers. A realistic physical description resolving the interface would require then an enormous number of mesh points. The great strength of the phase field method is its general concept, which makes it applicable to a variety of different microstructure evolutions in materials science (Chen, 2002, Raabe, 1998, Wang, et al., 1996). Examples are displacive phase transformations (Wang and Khachaturyan, 1997), solidification and dendrite growth (Apel, et al., 2002, Boettinger, et al., 2002), and grain growth (Fan and Chen, 1997).

5. Macroscopic models

5.1 Phase separation with elastic interactions

In this section we briefly describe how phase separation is modeled in the framework of a sharp interface model. More about macroscopic modeling can be found in another article in this volume (Fischer and Simha, 2004). The interface Γ now represents the dividing surface where the properties change discontinuously from the bulk properties of one phase to those of the others. As an example we consider the growth of a number of isolated precipitates - the β phase – into the α matrix of a supersaturated binary alloy. The concentration within each phase varies by diffusion, which is modeled by the standard diffusion equation

$$\frac{\partial c(\boldsymbol{r},t)}{\partial t} = \nabla\big(D\nabla c(\boldsymbol{r},t)\big) . \tag{5.1}$$

As boundary conditions the concentrations on the two sides of the interface are required,

$$c(\boldsymbol{r}_i,t) = c_{\hat{\alpha}} \qquad \text{and} \qquad c(\boldsymbol{r}_i,t) = c_{\hat{\beta}} , \tag{5.2}$$

where $\boldsymbol{r}_i$ denote locations at the interface. Typically it is assumed that local equilibrium prevails, and the concentrations at the interface are calculated using a generalized Gibbs-Thomson equation (see eq. (2.36) for the classical linearized version). The evolution of the interface is given by the interfacial mass balance condition (see (2.31)),

$$(c_{\hat{\beta}} - c_{\hat{\alpha}})v_n = D_\alpha \nabla c\, n\big|_{r=r_i} - D_\beta \nabla c\, n\big|_{r=r_i} , \tag{5.3}$$

where $\boldsymbol{n}$ is a unit vector normal to Γ pointing into the α phase and v_n denotes the velocity of Γ in the direction of $\boldsymbol{n}$. The great difficulty in solving the Equations (5.1) and (5.3) stems from the fact that the shape of the boundary has to be determined as part of the solution. It was shown that constraining the shape of the precipitates to a special geometry, e.g. spheres or ellipsoids, can cause wrong results (Thornton, et al., 2003).

We present only an outline under which assumptions and using which tricks the problem can be tackled nowadays. Details can be found in (Akaiwa, et al., 2001). First, diffusion is restricted to take place in the matrix. In the quasistatic diffusion approximation it is assumed that the concentration field relaxes much faster than the motion of the interface and is therefore in equilibrium: Assuming a diffusion constant D independent of the concentration, equation (5.1) becomes the Laplace equation,

$$\nabla^2 c(\boldsymbol{r},t) = 0 . \tag{5.4}$$

An important simplification is that the free-boundary problem given above is reformulated to involve only quantities at the interface. This formulation of the problem in terms of the interface itself reduces the dimensionality of the problem by one and only the interface and not the entire space has to be meshed for the numerical solution. A modified Gibbs-Thompson condition (see (2.36)) which considers also elastic interactions, specifies the concentration at the interface,

$$c(\boldsymbol{r}_i) = c_\alpha + l_c\big(\kappa + \omega_{elast}\big) . \tag{5.5}$$

c_α denotes the concentration at a flat interface when the matrix is in equilibrium, l_c the capillary length defined as $l_c \equiv \dfrac{\sigma \Omega c_\alpha}{k_B T}$, κ the mean curvature and ω_{elast} the elastic contribution. The system is assumed to be elastically homogeneous and anisotropic.

The procedure of calculating the microstructure evolution during coarsening can be summarized as follows: *(i)* since mechanical equilibrium is assumed, solve the mechanical problem with the condition of coherency at the interface, *(ii)* specify the boundary conditions on the interface using equation (5.5), *(iii)* solve the quasistatic diffusion equation (5.4) with the above boundary conditions, and *(iv)* calculate the normal velocity of the interface using (5.3) respecting the mass conservation,

$$\sum_{j=1}^{M} \int_{\Gamma_j} v_n(\boldsymbol{r}_i) ds_j = 0 \ , \qquad (5.6)$$

where M is the total number of the precipitates and ds_j is the arc length element of the interface Γ_j of the j-th precipitate. Again we refer to (Akaiwa, et al., 2001) for all the technical details which have to be solved on this way.

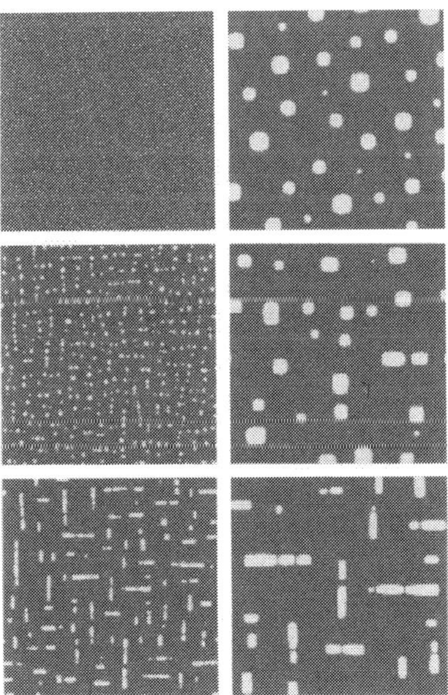

Figure 33. Microstructural evolution in a system with elastic interactions using boundary integral methods, $c = 0.1$. The left column shows the computational domain, while the right column is scaled with the average particle size (reprinted with permission from (Thornton, et al., 2001); copyright by the American Physical Society).

The macroscopic simulations presented here were started with 4000 particles and different initial conditions. In contrast to the results of MC simulations reported in Section 3.4.1 the elastic stress present in the system did not alter the growth exponent of 1/3 (Thornton, et al., 2001).

5.2 Concluding remarks

The sharp interface description of the microstructure becomes inadequate when typical length scales in the system become of the order of the interface thickness. Spindodal decomposition (see Section 2.3.3) is an important example where mesoscopic models (Section 4) offer a more appropriate approach. Another limitation of the sharp interface model as presented here is connected with topological changes such as precipitate coalescence during phase separation, which are extremely tricky to handle (Thornton, et al., 2003).

6. References

Abinandanan, T. A., Haider, F. and Martin, G. (1998). Computer simulations of diffusional phase transformations: Monte Carlo algorithm and application to precipitation of ordered phases. *Acta Materialia* 46: 4243-4255.

Aichmayer, B., Fratzl, P., Puri, S. and Saller, G. (2003). Surface-directed spinodal decomposition on a macroscopic scale in a nitrogen and carbon alloyed steel. *Physical Review Letters* 91: 015701/1-4.

Akaiwa, N., Thornton, K. and Voorhees, P. W. (2001). Large-scale simulations of microstructural evolution in elastically stressed solids. *Journal of Computational Physics* 173: 61-86.

Apel, M., Boettger, B., Diepers, H. J. and Steinbach, I. (2002). 2D and 3D phase-field simulations of lamella and fibrous eutectic growth. *Journal of Crystal Growth* 237: 154-158.

Asta, M. and Foiles, S. M. (1996). Embedded-atom-method effective-pair-interaction study of the structural and thermodynamic properties of Cu-Ni, Cu-Ag, and Au-Ni solid solutions. *Physical Review B* 53: 2389-2404.

Athenes, M., Bellon, P. and Martin, G. (1997). Identification of novel diffusion cycles in B2 ordered phases by Monte Carlo simulation. *Philosophical Magazine a-Physics of Condensed Matter Structure Defects and Mechanical Properties* 76: 565-585.

Athenes, M., Bellon, P. and Martin, G. (2000). Effects of atomic mobilities on phase separation kinetics: A Monte-Carlo study. *Acta Materialia* 48: 2675-2688.

Athenes, M., Bellon, P., Martin, G. and Haider, F. (1996). A Monte-Carlo study of B2 ordering and precipitation via vacancy mechanism in bcc lattices. *Acta Materialia* 44: 4739-4748.

Becker, R. (1938). Die Keimbildung bei der Ausscheidung in metallischen Mischkristallen. *Annalen der Physik* 32: 128-140.

Bellon, P. (2003). Kinetic Monte Carlo simulations in crystalline alloys: principles and selected applications. In Finel, A. e. a., Eds. *NATO Sci. Ser. II Math., Thermodynamics, Microstructures and Plasticity,* Kluwer Academic. 395-409.

Binder, K. (1997). Applications of Monte Carlo methods to statistical physics. *Reports on Progress in Physics* 60: 487-559.

Binder, K. (1998). Spinodal decomposition in confined geometry. *Journal of Non-Equilibrium Thermodynamics* 23: 1-44.

Binder, K., Billotet, C. and Mirold, P. (1978). Theory of Spinodal Decomposition in Solid and Liquid Binary-Mixtures. *Zeitschrift Fur Physik B-Condensed Matter* 30: 183-195.

Binder, K. and Fratzl, P. (2001). Spinodal decomposition. In Kostorz, G., Eds. *Phase Transformations in Materials,* Wiley-VCH. Chapter 6, 409-480.

Binder, K. and Heermann, D. W. (2002). Monte Carlo simulation in statistical physics : an introduction. Berlin ; New York: Springer.

Binder, K. and Stauffer, D. (1974). Theory for Slowing Down of Relaxation and Spinodal Decomposition of Binary-Mixtures. *Physical Review Letters* 33: 1006-1009.

Blaschko, O. and Fratzl, P. (1983). Experimental-Observation of a Time-Scaling Characteristic in Alloy Decomposition in the Alznmg System. *Physical Review Letters* 51: 288-291.

Boettinger, W. J., Warren, J. A., Beckermann, C. and Karma, A. (2002). Phase-field simulation of solidification. *Annual Review of Materials Research* 32: 163-194.

Bortz, A. B., Kalos, M. H. and Lebowitz, J. L. (1975). New Algorithm for Monte-Carlo Simulation of Ising Spin Systems. *Journal of Computational Physics* 17: 10-18.

Bortz, A. B., Kalos, M. H., Lebowitz, J. L. and Zendejas, M. A. (1974). Time Evolution of a Quenched Binary Alloy - Computer-Simulation of a 2-Dimensional Model System. *Physical Review B* 10: 535-541.

Cahn, J. W. (1961). On Spinodal Decomposition. *Acta Metallurgica* 9: 795-801.

Cahn, J. W. (1968). 1967 Institute of Metals Lecture Spinodal Decomposition. *Transactions of the Metallurgical Society of Aime* 242: 166-&.

Cahn, J. W. and Hilliard, J. E. (1958). Free Energy of a Nonuniform System .1. Interfacial Free Energy. *Journal of Chemical Physics* 28: 258-267.

Cahn, J. W. and Hilliard, J. E. (1959). Free Energy of a Nonuniform System .3. Nucleation in a 2-Component Incompressible Fluid. *Journal of Chemical Physics* 31: 688-699.

Cahn, R. W. and Haasen, P. (1983). Physical metallurgy. Amsterdam: North-Holland Physics Publishing.

Chandler, D. (1987). Introduction to modern statistical mechanics. New York: Oxford University Press.

Chen, L. Q. (2002). Phase-field models for microstructure evolution. *Annual Review of Materials Research* 32: 113-140.

Chen, L. Q. and Khachaturyan, A. G. (1991). Computer-Simulation of Structural Transformations During Precipitation of an Ordered Intermetallic Phase. *Acta Metallurgica Et Materialia* 39: 2533-2551.

Chen, L. Q., Wolverton, C., Vaithyanathan, V. and Liu, Z. K. (2001). Modeling solid-state phase transformations and microstructure evolution. *Mrs Bulletin* 26: 197-202.

Cook, H. E. (1970). Brownian Motion in Spinodal Decomposition. *Acta Metallurgica* 18: 297-&.

Cook, H. E. and De Fontaine, D. (1969). On Elastic Free Energy of Solid Solutions .I. Microscopic Theory. *Acta Metallurgica* 17: 915-924.

Eshelby, J. D. (1957). The Determination of the Elastic Field of an Ellipsoidal Inclusion, and Related Problems. *Proceedings of the Royal Society of London Series a-Mathematical and Physical Sciences* 241: 376-396.

Eshelby, J. D. (1959). The Elastic Field Outside an Ellipsoidal Inclusion. *Proceedings of the Royal Society of London Series a-Mathematical and Physical Sciences* 252: 561-569.

Fan, D. and Chen, L. Q. (1997). Computer simulation of grain growth using a continuum field model. *Acta Materialia* 45: 611-622.

Fischer, D. and Nielaba, P. (2000). Phase diagram of a model alloy with lattice misfit. *Physica a-Statistical Mechanics and Its Applications* 279: 287-295.

Fischer, F. D. and Simha, N. K. (2004). Thermodynamics and kinetics of phase and twin boundaries. In Eds. *This book ##*, Udine: CISM.

Fratzl, P. (2003). Small-angle scattering in materials science - a short review of applications in alloys, ceramics and composite materials. *Journal of Applied Crystallography* 36: 397-404.

Fratzl, P., Lebowitz, J. L., Penrose, O. and Amar, J. (1991). Scaling Functions, Self-Similarity, and the Morphology of Phase-Separating Systems. *Physical Review B* 44: 4794-4811.

Fratzl, P. and Penrose, O. (1994). Kinetics of Spinodal Decomposition in the Ising-Model with Vacancy Diffusion. *Physical Review B* 50: 3477-3480.

Fratzl, P. and Penrose, O. (1995). Ising-Model for Phase-Separation in Alloys with Anisotropic Elastic Interaction .1. Theory. *Acta Metallurgica Et Materialia* 43: 2921-2930.

Fratzl, P. and Penrose, O. (1996). Ising model for phase separation in alloys with anisotropic elastic interaction .2. A computer experiment. *Acta Materialia* 44: 3227-3239.

Fratzl, P. and Penrose, O. (1997). Competing mechanisms for precipitate coarsening in phase separation with vacancy dynamics. *Physical Review B* 55: R6101-R6104.

Fratzl, P., Penrose, O. and Lebowitz, J. L. (1999). Modeling of phase separation in alloys with coherent elastic misfit. *Journal of Statistical Physics* 95: 1429-1503.

Fratzl, P., Penrose, O., Weinkamer, R. and Zizak, I. (2000). Coarsening in the Ising model with vacancy dynamics. *Physica A* 279: 100-109.

Gerold, V. and Kern, J. (1987). The Determination of Atomic Interaction Energies in Solid-Solutions from Short-Range Order Coefficients - an Inverse Monte-Carlo Method. *Acta Metallurgica* 35: 393-399.

Gunton, J. D. (1999). Homogeneous nucleation. *Journal of Statistical Physics* 95: 903-923.

Gupta, H., Weinkamer, R., Fratzl, P. and Lebowitz, J. L. (2001). Microscopic computer simulations of directional coarsening in face-centered cubic alloys. *Acta Materialia* 49: 53-63.

Heermann, D. W., Li, Y. X. and Binder, K. (1996). Scaling solutions and finite-size effects in the Lifshitz-Slyozov theory. *Physica A* 230: 132-148.

Hu, S. Y. and Chen, L. Q. (2001). Solute segregation and coherent nucleation and growth near a dislocation - A phase-field model integrating defect and phase microstructures. *Acta Materialia* 49: 463-472.

Huse, D. A. (1986). Corrections to Late-Stage Behavior in Spinodal Decomposition - Lifshitz-Slyozov Scaling and Monte-Carlo Simulations. *Physical Review B* 34: 7845-7850.

Ikeda, H. and Matsuda, H. (1993). Effects of Difference in Elastic-Moduli between Constituents on Spinodal Decomposition Processes. *Materials Transactions Jim* 34: 651-657.

Inden, G. (2001). Atomic odering. In Kostorz, G., Eds. *Phase Transformations in Materials,* Wiley-VCH. Chapter 8, 519-581.

Kampmann, R. and Wagner, R. (1984). In Haasen, P., Gerold, V., Wagner, R. and Ashby, M. F., Eds. *Decomposition of Alloys: the Early Stages,* Oxford: Pergamon Press. 91-103.

Kantelhardt, J. W., Koscielny-Bunde, E., Rego, H. H. A., Havlin, S. and Bunde, A. (2001). Detecting long-range correlations with detrended fluctuation analysis. *Physica A* 295: 441-454.

Khachaturyan, A. G. (1965). On Theory of Modulated Structures in Binary Solid Solutions. *Soviet Physics Crystallography, Ussr* 10: 248-&.

Khachaturyan, A. G. (1983). Theory of structural transformations in solids. New York: Wiley.

Kikuchi, R. and Sato, H. (1972). Diffusion-Coefficient in an Ordered Binary Alloy. *Journal of Chemical Physics* 57: 4962-4979.

Laberge, C. A., Fratzl, P. and Lebowitz, J. L. (1995). Elastic Effects on Phase Segregation in Alloys with External Stresses. *Physical Review Letters* 75: 4448-4451.

Laberge, C. A., Fratzl, P. and Lebowitz, J. L. (1997). Microscopic model for directional coarsening of precipitates in alloys under external load. *Acta Materialia* 45: 3949-3962.

Langer, J. S. and Schwartz, A. J. (1980). Kinetics of Nucleation in near-Critical Fluids. *Physical Review A* 21: 948-958.

Lebowitz, J. L., Marro, J. and Kalos, M. H. (1982). Dynamical Scaling of Structure-Function in Quenched Binary-Alloys. *Acta Metallurgica* 30: 297-310.

Lee, J. K. (1998). Elastic stress and microstructural evolution. *Materials Transactions Jim* 39: 114-132.

Lengeler, B. (2001). Coherence in X-ray physics. *Naturwissenschaften* 88: 249-260.

Lifshitz, I. M. and Slyozov, V. V. (1961). The Kinetics of Precipitation from Supersaturated Solid Solutions. *Journal of Physics and Chemistry of Solids* 19: 35-50.

Lochte, L., Gitt, A., Gottstein, G. and Hurtado, I. (2000). Simulation of the evolution of GP zones in Al-Cu alloys: An extended Cahn-Hilliard approach. *Acta Materialia* 48: 2969-2984.

Marro, J., Bortz, A. B., Kalos, M. H. and Lebowitz, J. L. (1975). Time Evolution of a Quenched Binary Alloy .2. Computer-Simulation of a 3-Dimensional Model System. *Physical Review B* 12: 2000-2011.

Marro, J., Lebowitz, J. L. and Kalos, M. H. (1979). Computer-Simulation of the Time Evolution of a Quenched Model Alloy in the Nucleation Region. *Physical Review Letters* 43: 282-285.

Mirold, P. and Binder, K. (1977). Theory for Initial-Stages of Grain-Growth and Unmixing Kinetics of Binary-Alloys. *Acta Metallurgica* 25: 1435-1444.

Nielaba, P., Fratzl, P. and Lebowitz, J. L. (1999). Growth of ordered domains in a computer model alloy with lattice misfit. *Journal of Statistical Physics* 95: 23-43.

Nishimori, H. and Onuki, A. (1990). Pattern-Formation in Phase-Separating Alloys with Cubic Symmetry. *Physical Review B* 42: 980-983.

Novotny, M. A. (1995). A new approach to an old algorithm for the simulation of Ising-like systems. *Computers in Physics* 9: 46-52.

Onsager, L. (1944). Crystal statistics. I. A two-dimensional model with an order-disorder transition. *Physical Review* 65: 117-149.

Onuki, A. and Nishimori, H. (1991). Anomalously Slow Domain Growth Due to a Modulus Inhomogeneity in Phase-Separating Alloys. *Physical Review B* 43: 13649-13652.

Orlikowski, D., Sagui, C., Somoza, A. and Roland, C. (1999). Large-scale simulations of phase separation of elastically coherent binary alloy systems. *Physical Review B* 59: 8646-8659.

Orlikowski, D., Sagui, C., Somoza, A. M. and Roland, C. (2000). Two- and three-dimensional simulations of the phase separation of elastically coherent binary alloys subject to external stresses. *Physical Review B* 62: 3160-3168.

Pareige, C., Soisson, F., Martin, G. and Blavette, D. (1999). Ordering and phase separation in Ni-Cr-Al: Monte Carlo simulations vs three-dimensional atom probe. *Acta Materialia* 47: 1889-1899.

Paris, O., Fahrmann, M., Fahrmann, E., Pollock, T. M. and Fratzl, P. (1997). Early stages of precipitate rafting in a single crystal Ni-Al-Mo model alloy investigated by small-angle X-ray scattering and TEM. *Acta Materialia* 45: 1085-1097.

Porta, M., Vives, E. and Castan, T. (1999). Vacancy-assisted domain growth in asymmetric binary alloys: A Monte Carlo study. *Physical Review B* 60: 3920-3927.

Press, W. H. (1997). Numerical recipes in C : the art of scientific computing. Cambridge Cambridgeshire ; New York: Cambridge University Press.

Raabe, D. (1998). Computational materials science - the simulation of materials microstructures and properties. Weinheim: Wiley-VCH.

Rao, M., Kalos, M. H., Lebowitz, J. L. and Marro, J. (1976). Time Evolution of a Quenched Binary Alloy .3. Computer-Simulation of a 2-Dimensional Model System. *Physical Review B* 13: 4328-4335.

Rautiainen, T. T. and Sutton, A. P. (1999). Influence of the atomic diffusion mechanism on morphologies, kinetics, and the mechanisms of coarsening during phase separation. *Phys Rev B* 59: 13681-13692.

Roussel, J. M. and Bellon, P. (2001). Vacancy-assisted phase separation with asymmetric atomic mobility: Coarsening rates, precipitate composition, and morphology. *Physical Review B* 6318: art. no.-184114.

Rubin, G. and Khachaturyan, A. G. (1999). Three-dimensional model of precipitation of ordered intermetallics. *Acta Materialia* 47: 1995-2002.

Sagui, C., Orlikowski, D., Somoza, A. M. and Roland, C. (1998). Three-dimensional simulations of Ostwald ripening with elastic effects. *Physical Review E* 58: R4092-R4095.

Schweika, W. (1998). Disordered alloys : diffuse scattering and Monte Carlo simulations. Berlin ; New York: Springer.

Socrate, S. and Parks, D. M. (1993). Numerical Determination of the Elastic Driving Force for Directional Coarsening in Ni-Superalloys. *Acta Metallurgica Et Materialia* 41: 2185-2209.

Soisson, F. and Martin, G. (2000). Monte Carlo simulations of the decomposition of metastable solid solutions: Transient and steady-state nucleation kinetics. *Physical Review B* 62: 203-214.

Stadler, L.-M., Sepiol, B., Weinkamer, R., Hartmann, M., Fratzl, P., Kantelhardt, J. W., Zotone, F., Grübel, G. and Vogl, G. (2003). Long-term correlations distinguish coarsening mechanisms in alloys. *Physical Review B* 68: 180101R-1-4.

Sur, A., Lebowitz, J. L., Marro, J. and Kalos, M. H. (1977). Time Evolution of a Quenched Binary Alloy .4. Computer-Simulation of a 3-Dimensional Model System. *Physical Review B* 15: 3014-3026.

Thornton, K., Agren, J. and Voorhees, P. W. (2003). Modelling the evolution of phase boundaries in solids at the meso- and nanoscales. *Acta Materialia* 51: 5675-5710.

Thornton, K., Akaiwa, N. and Voorhees, P. W. (2001). Dynamics of late-stage phase separation in crystalline solids. *Phys Rev Lett* 86: 1259-1262.
Vaithyanathan, V. and Chen, L. Q. (2000). Coarsening kinetics of delta '-Al3Li precipitates: Phase-field simulation in 2D and 3D. *Scripta Materialia* 42: 967-973.
Vaithyanathan, V., Wolverton, C. and Chen, L. Q. (2002). Multiscale modeling of precipitate microstructure evolution. *Physical Review Letters* 88: -.
Vives, E. and Planes, A. (1993). Ordering Kinetics by Vacancies. *International Journal of Modern Physics C-Physics and Computers* 4: 701-720.
Voorhees, P. W. (1985). The Theory of Ostwald Ripening. *Journal of Statistical Physics* 38: 231-252.
Voorhees, P. W. (1992). Ostwald Ripening of 2-Phase Mixtures. *Annual Review of Materials Science* 22: 197-215.
Wagner, C. (1961). Theorie der Alterung von Niederschlägen durch Umlösen (Ostwald Reifung). *Zeitschrift für Elektrochemie* 65: 581-591.
Wagner, R., Kampmann, R. and Voorhees, P. W. (2001). Homogeneous second-phase precipitation. In Kostorz, G., Eds. *Phase Transformations in Materials,* Wiley-VCH. Chapter 5, 309-407.
Wang, Y., Chen, L. Q. and Khachaturyan, A. G. (1993). Kinetics of Strain-Induced Morphological Transformation in Cubic Alloys with a Miscibility Gap. *Acta Metallurgica Et Materialia* 41: 279-296.
Wang, Y., Chen, L. Q. and Khachaturyan, A. G. (1996). Modeling of dynamical evolution of micro/mesoscopic morphological patterns in coherent phase transformations. In Kirchner, H. O. and al., e., Eds. *Computer Simulation in Materials Science,* Kluwer Academic Publishers. 325-371.
Wang, Y. and Khachaturyan, A. G. (1997). Three-dimensional field model and computer modeling of martensitic transformations. *Acta Materialia* 45: 759-773.
Wang, Y. U., Jin, Y. M., Cuitino, A. M. and Khachaturyan, A. G. (2001). Nanoscale phase field microelasticity theory of dislocations: Model and 3D simulations. *Acta Materialia* 49: 1847-1857.
Wang, Y. U., Jin, Y. M. M. and Khachaturyan, A. G. (2003). Phase field microelasticity modeling of dislocation dynamics near free surface and in heteroepitaxial thin films. *Acta Materialia* 51: 4209-4223.
Weinkamer, R. and Fratzl, P. (2003). By which mechanism does coarsening in phase-separating alloys proceed? *Europhysics Letters* 61: 261-267.
Weinkamer, R., Fratzl, P., Gupta, H., Penrose, O. and Lebowitz, J. L. (2004). Using kinetic Monte Carlo simulations to study phase separation in alloys. *Phase Transitions* submitted:
Weinkamer, R., Fratzl, P., Sepiol, B. and Vogl, G. (1998). Monte Carlo simulation of diffusion in a B2-ordered model alloy. *Physical Review B* 58: 3082-3088.
Weinkamer, R., Fratzl, P., Sepiol, B. and Vogl, G. (1999). Monte Carlo simulations of Mossbauer spectra in diffusion investigations. *Physical Review B* 59: 8622-8625.
Weinkamer, R., Gupta, H., Fratzl, P. and Lebowitz, J. L. (2000). Dynamics of mesoscopic precipitate lattices in phase-separating alloys under external load. *Europhysics Letters* 52: 224-230.
Wolverton, C., Zunger, A. and Schonfeld, B. (1997). Invertible and non-invertible alloy Ising problems. *Solid State Communications* 101: 519-523.

Yaldram, K. and Binder, K. (1991). Monte-Carlo Simulation of Phase-Separation and Clustering in the Abv Model. *Journal of Statistical Physics* 62: 161-175.
Yaldram, K. and Binder, K. (1991). Spinodal Decomposition of a 2-Dimensional Model Alloy with Mobile Vacancies. *Acta Metallurgica Et Materialia* 39: 707-717.
Yaldram, K. and Binder, K. (1991). Unmixing of Binary-Alloys by a Vacancy Mechanism of Diffusion - a Computer-Simulation. *Zeitschrift Fur Physik B-Condensed Matter* 82: 405-418.
Zhu, J. Z., Liu, Z. K., Vaithyanathan, V. and Chen, L. Q. (2002). Linking phase-field model to CALPHAD: application to precipitate shape evolution in Ni-base alloys. *Scripta Materialia* 46: 401-406.

Utilization of the thermodynamic extremal principle for modelling in material science

J. Svoboda

Institute of Physics of Materials Academy of Sciences of the Czech Republic
Žižkova 22, CZ – 616 62 Brno, Czech Republic

Abstract

Modelling represents an important tool in the material science. It can simulate the technological steps as well as predict the material properties. A lot of technological steps are performed as well as several parts are operating at elevated temperatures. Under these conditions the thermodynamics of irreversible processes can describe successfully the processes occurring in the material. Each model represents a simplification of the reality and should be concentrated on the effects of the most interest. In many cases the system can be described by a set of the most characteristic parameters, the evolution of which we are interested in. In the classical way, the evolution of the parameters is obtained by solution of the phenomenological equations given by laws of irreversible thermodynamics. In an alternative way the laws of irreversible thermodynamics can be described by an extremal principle. Some examples of modelling in material science based on application of the extremal principle are presented in this chapter. It is demonstrated, that the application of the extremal principle represents a systematic way, how some models can effectively be developed.

List of symbols

A	parameter influencing the solute segregation at the interface
A_i	kinctic coefficient corresponding to the diffusing component i.
$\widetilde{B}_{ik}^{-1}$	matrix of generalized kinetic parameters
C	specific heat capacity or capacity of the condenser
c_i	concentration of the component i
c_i^*	concentration of the component i in the precipitate
$\bar{c}_i$	mean concentration of the component i in the matrix
c_{ki}	mean concentration of the component i in the precipitate k,
D_i	diffusion coefficient of the component i
E	additional potential term
$\vec{E}$	intensity of the electric field
f	correlation factor for the diffusive walk of atoms
g	Gibbs energy per one mole of site positions
G	total Gibbs energy
$\dot{G}$	rate of total Gibbs energy

$\dot{G}_{out}$	rate of Gibbs energy flowing out of the system
h	interface thickness
I	actual electric current in the circuit
$\vec{j}_e$	electric flux
$\vec{j}_q$	heat flux
$\vec{j}_i$	diffusive flux of the component i
L_{ik}	Onsager coefficients
M	interface mobility
M_b	mobility of the grain boundary
M_{eff}	effective interface mobility
N	number of moles of lattice positions in the representative volume
N_i	number of moles of component i in the representative volume
p	number of interstitial components
q	charge on the condenser
Q	total Gibbs energy dissipation
q_i	independent state parameters
$\dot{q}_i$	independent kinetic parameters
R	gas constant
r_{crit}	critical radius of the grain or of the precipitate
R_e	resistance of the resistor
r_k	effective radius of grain or precipitate k
s	number of substitutional components
S	total surface of the migrating interfaces
T	absolute temperature
U	voltage on the condenser
u_i	alternative independent kinetic parameters
U_k	fixed structure parameter of the precipitate k
u_{ki}	mean site fraction of component i in the precipitate k
v	velocity of the migrating grain boundary or of the phase interface
V	volume of the system
W	volume of the representative volume element
x_i	mole fraction of component i
$\vec{X}_i$	driving force corresponding to the flux $\vec{j}_i$
y_i	site fraction of component i
γ	interface Gibbs energy per unit area
γ_b	grain boundary Gibbs energy per unit area
γ_e	electric conductivity
Δ_k	thickness of the subregion
η, ξ	Lagrange multipliers
κ	grain boundary mean curvature
λ	heat conductivity

μ_i	chemical potential of component i
ρ_k	radius of precipitate k,
Φ	dissipative function
Ψ	total entropy production rate
Ω	volume corresponding to one mole of site positions
Ω_i	partial molar volume of the component i

1 Introduction

The increasing demands of the industry on the material properties raise the necessity to develop more and more sophisticated materials with complex microstructure. Modelling represents an important tool in material design and enables to simulate the material processing and properties. Numerous processes take place at elevated temperatures in metallic, ceramic or composite materials. Under such conditions the non-equilibrium thermodynamics can be applied to the systems to describe the processes occurring in the materials during processing and/or exploitation.

From the thermodynamic point of view the material exposed to elevated temperature, with or without an external loading, can be considered as a system not far from equilibrium being composed of a large number of equilibrium sub-systems which interact with each other (see e.g. Callen, 1960, Groot and Mazur, 1962 or Prigogine, 1967). Then it is possible to introduce the local values (corresponding to the sub-systems) of intensive state parameters like chemical composition, molar volume, temperature, pressure, chemical potentials of all components, lattice structure etc. and to describe the state of the system by 3-dimensional fields of the intensive state parameters. In order to treat the evolution of the system it is necessary to write down the evolution equations for the fields and to integrate them in time.

The thermodynamics of irreversible processes offers local phenomenological equations e.g. for heat conduction or diffusion. The equations express the linear relation between the heat flux or diffusive fluxes and the corresponding driving forces (gradients of temperature or of chemical potentials). If the equations are combined with the conservation laws the evolution equations in the form of partial differential equations are obtained. For a one-phase system the boundary conditions at the system surface and the starting conditions in the bulk of the system can be prescribed and the evolution equations can be integrated in time to obtain the system evolution.

The problem becomes dramatically more complicated if moving discontinuities are present in the multi-phase system. One example of such a moving discontinuity is a propagating phase interface occurring during phase transformations or during the precipitation process. To treat such a problem it is necessary first of all to introduce the equation for the interface motion relating the interface velocity and the driving force acting at the interface. In a second step it is necessary to write down the mass conservation laws for the migrating interface relating the interface velocity, the jumps across the interface of normal components of fluxes and the jumps in concentrations (see e.g. Fischer et al., 2003). Unfortunately, the number of jump conditions together with the equation for the interface motion is insufficient for the determination of the contact conditions at the migrating interface being necessary for the solution of heat conduction and/or diffusion in the bulk (Svoboda et al., 2003). That is why further contact conditions for the interface must be prescribed like continuity in the temperature, in the chemical potentials or in the chemical composition. These prescribed contact conditions are more or less artificial and they represent, together with the jump conditions for fluxes and relation for the interface velocity, a set of non-linear equations. This set must be solved for each integration step to provide the boundary conditions for the equations of heat conduction and/or diffusion in the bulk. One can easily imagine that even for a very simple system in which phase transformation occurs the simulation is extremely complicated.

The majority of structure materials exhibit rather complex microstructure. Thus modelling the evolution of the microstructure during processing (heat treatment) or exploitation at elevated temperatures (degradation) requires some reasonable simplifications still taking the complexity of the microstructure and all important phenomena into account. This is often impossible when applying and solving the standard local phenomenological equations of the thermodynamics of irreversible processes.

The thermodynamic extremal principle seems to be a very handy tool in modelling the above mentioned problems. Roughly speaking, the application of the principle has two big advantages:

1. You can choose the most suitable and representative state parameters of the system (these may be totally different from the 3-dimensional fields of the intensive state parameters) and the evolution equations for the state parameters can be derived in a direct way by means of application of the extremal principle.
2. There is no need to take care of the additional contact conditions at the migrating interface (like continuity of temperature, chemical potentials or chemical composition across the interface). The proper contact conditions are found by applying the principle, and the evolution equations ensure that they are automatically kept during the time integration of the evolution equations.

The aim of this chapter is to present the thermodynamic extremal principle and to demonstrate its application to several examples.

2 Thermodynamics of Non-Equilibrium Processes

The number of atoms involved in the usual specimens or parts is of the order of magnitude of about 10^{23}. To be able to simulate the kinetics of atoms (molecular dynamics) by means of quantum mechanics, the number of atoms cannot exceed a value of about 100. In the case of simulations by means of classical mechanics, the number of atoms cannot exceed about 1000. On the other hand thermodynamics and statistical physics are able to treat such large systems of atoms; there is, however, a loss of information about the individual motion of atoms. Very often this is no hindrance, because the motion of an individual atom cannot be measured and we are usually not interested in it. Thus thermodynamics in itself is already an excellent tool for the phenomenological treatment of large systems and it has an established position in material science.

Classical thermodynamics deals with equilibrium systems (see e.g. Callen, 1960). It introduces the extensive state parameters of the system e.g.

- volume
- mass
- internal energy

- entropy
- enthalpy
- Helmholtz energy
- Gibbs energy

and intensive state parameters of the system e.g.

- molar volume
- temperature
- pressure
- chemical composition
- chemical potentials
- lattice structure.

Strictly speaking the extensive and intensive parameters can be introduced only in equilibrium systems.

The main assumption of the non-equilibrium thermodynamics reads that a non-equilibrium system is composed of a large number of equilibrium sub-systems which interact with each other. This allows to assume that all intensive state parameters have a local character and can be characterized by 3-dimensional fields. Such an approach is not suitable for the description of processes like an explosion (systems very far from equilibrium); it is, however, an excellent means of describing the sluggish processes in solids.

2.1. Linear Phenomenological Laws

For slightly non-equilibrium systems, non-equilibrium thermodynamics assumes that the rates of state parameters or fluxes are proportional to corresponding driving forces. In isotropic systems they are expressed by the following phenomenological laws, i.e.

$$\vec{j}_e = \gamma_e \vec{E} \qquad \text{Ohm law} \tag{2.1}$$

$$\vec{j}_q = -\lambda \, gradT \qquad \text{Fourier law} \tag{2.2}$$

$$\vec{j}_i = -D_i \, gradc_i \qquad 1^{st} \text{ Fick law} \tag{2.3}$$

$$v = M_b \gamma_b \kappa \qquad \text{curvature driven grain boundary migration} \tag{2.4}$$

where $\vec{j}_e$, $\vec{j}_q$ and $\vec{j}_i$ are the electric flux, the heat flux and the diffusive flux of the component i, respectively, and $\vec{E}$, $-gradT$ and $-gradc_i$ are the respective driving forces - intensity of the electric field, gradient of temperature and gradient of concentration of the component i. γ_e, λ and D_i are the material kinetic coefficients - the electric conductivity, the heat conductivity and the diffusion coefficient of the component i. v is the velocity of the migrating grain boundary, M_b is the mobility of the grain boundary and the driving force for the grain boundary migration is given by the product of grain boundary energy per unit area γ_b and of grain boundary mean curvature κ. All driving forces and all rates and/or fluxes are zero for the state of thermodynamic equilibrium.

To be able to derive the evolution equations of the system the conservation laws must be utilized, which relate, among other things, the fluxes and rates of state parameters.

2.2. Conservation Laws and Evolution Equations

During the evolution of a system under specified conditions some quantities are conserved or must be in balance. The conservation laws are based on elementary physical laws like e.g. the mass conservation law or the first law of thermodynamics. Typical examples of conservation laws in thermodynamics are

$$C\dot{T} = -div\vec{j}_q \qquad \text{heat balance} \qquad (2.5)$$

$$\dot{c}_i = -div\vec{j}_i \qquad \text{mass conservation of component } i, \qquad (2.6)$$

where C is the specific heat capacity and the time derivative is denoted with the dot.

Combining phenomenological laws and conservation laws the partial differential equations for the system evolution can be obtained. Together with proper boundary conditions and starting conditions they can be solved by integration in time. The 2nd Fick law is a typical example of the evolution equation:

$$\dot{c}_i = div(D_i gradc_i) . \qquad (2.7)$$

3 Thermodynamic Extremal Principle

Lars Onsager (1931) has shown that the equations for the heat conduction can be replaced by an extremal principle from which the equations for the heat conduction can be derived.

Onsager introduces:

- Total entropy production Ψ – being a linear form of heat fluxes and of temperature gradients:

$$\Psi = -\int_V \frac{1}{T^2}\vec{j}_q gradT dV \qquad (3.1)$$

- Dissipative function Φ – being a quadratic form of heat fluxes and involving the material kinetic coefficient (heat conductivity λ):

$$2\Phi = \int_V \frac{1}{T^2\lambda}\vec{j}_q^2 dV \qquad (3.2)$$

Due to the balance of the total entropy in the system it is shown that

$$\Psi = 2\Phi . \qquad (3.3)$$

The principle formulated by Onsager asserts that the evolution in the system given by thermal fluxes corresponds to the **maximum** of the total entropy production Ψ (or of dissipative function Φ) constrained by relation (3.3). Furthermore, Onsager has shown that with regard to heat conduction, this principle is equivalent to the description by phenomenological equations and that these equations can be derived from the principle.

The principle can also be understood as a strengthened second law of thermodynamics. The second law of thermodynamics admits all evolution paths of the system with a positive total entropy production Ψ. The principle admits only one evolution path of the system which corresponds to the constrained maximum of the total entropy production Ψ. If the principle is handled correctly all linear phenomenological laws are a direct consequence of the principle.

This principle was nearly forgotten for 60 years after Onsager's publication. It was used or developed very sporadically, e.g. by Ziegler (1961,1963) or by Bazarov et al. (1989).

Already in 1931 Onsager (1931) derived from the principle **the principle of the least entropy production.** This principle is applicable only to stationary states with prescribed normal components of fluxes at the boundary of the system. The principle of the least entropy production asserts that the stationary case corresponds to the minimum of the entropy production with respect to variation of fluxes in the system interior fulfilling the given boundary conditions. This principle, dealing with the very special stationary case, is presented in almost every textbook of thermodynamics (e.g. Callen, 1960, Groot and Mazur, 1962 or Prigogine, 1967).

In general the extremal principles are applied for two reasons

- they help in the physical understanding of the problem
- they may offer an alternative mathematical formulation of the problem allowing effective solution.

In the case of the extremal thermodynamic principle the principle can also be understood as an effective tool in modelling i.e. it can offer sophisticated equations for the system evolution which cannot be derived from standard phenomenological equations.

3.1 Requirements on the Model

Problems of material science are often rather complex. Every model represents a certain simplification of the reality and in the ideal case it should have the following properties:

1) It must take into account all important processes connected with the modelled phenomenon.
2) It cannot deal with a particular process in much detail.
3) It cannot involve any physically or mathematically inconsistent assumptions.
4) It should provide the results within the usual accuracy of the input data.
5) It should offer only those output data that are of interest to us.
6) The solubility of equations of the model must be ensured.
7) The simulations utilizing the model should be performable on the computers within reasonable times.

The application of the thermodynamic extremal principle to the problems of material science may significantly help to fulfil the above-mentioned requirements on the model.

3.2 Thermodynamic Extremal Principle for a Closed System Under Constant Temperature and Pressure

In many practical applications the system can be considered (at least as a good approximation) as a closed system at constant temperature and pressure. In that case the characteristic potential of the system is its total Gibbs energy G. For a more detailed description of the theory in this paragraph see Svoboda and Turek (1991).

Let us consider a closed thermodynamic system in non-equilibrium under constant temperature and pressure. Let the state of the system be described by a set of independent state parameters q_i, $(i=1,...,n)$. The choice of the state parameters depends strictly on what interests us in the system and its evolution. The necessary condition for the choice of the state parameters is that the total Gibbs energy of the system, G, can be expressed as a function of the state parameters: $G=G(q_1,q_2,...,q_n)$.

During the system evolution the total Gibbs energy of the system dissipates by irreversible processes and thus the total Gibbs energy of the system G decreases. Let Q be the total Gibbs energy dissipation in the system and $\dot{q}_i \equiv \frac{dq_i}{dt}$, $(i=1,...,n)$ independent kinetic parameters of the system. Then we can perform the Taylor expansion of Q with respect to $\dot{q}_i$ as:

$$Q = P(q_1,q_2,...,q_n) + \sum_{i=1}^{n} H_i(q_1,q_2,...,q_n)\dot{q}_i + \sum_{i,k=1}^{n} B_{ik}(q_1,q_2,...,q_n)\dot{q}_i\dot{q}_k + ... \quad (3.4)$$

Due to second law of thermodynamics $Q > 0$ for any evolving system out of equilibrium and $Q = 0$ for the system in equilibrium. For the system in equilibrium $\dot{q}_i = 0$ $(i=1,...,n)$ and that is why P in Eq. (3.4) must be zero. For small values of $\dot{q}_i$ $(i=1,...,n)$ the second term dominates and due to the second law of thermodynamics it must be non-negative for arbitrary values of $\dot{q}_i$ $(i=1,...,n)$. This is possible only if $H_i = 0$ $(i=1,...,n)$. Thus the first non-zero term in Eq. (3.4) can be the third term and, due to the second law of thermodynamics, the matrix B_{ik} must be positive definite.

The determination of the matrix B_{ik} is perhaps one of the most complicated tasks. In cases where the evolution of the system is due to bulk diffusion and to migration of interfaces, the total Gibbs free energy dissipation Q is given by (Fischer et al., 2003)

$$Q = \int_S \frac{v^2}{M} dS + \sum_l \int_V \frac{RT}{c_l D_l} \vec{j}_l^{\,2} dV , \quad (3.5)$$

where S is the total surface of the migrating interfaces, v is the local velocity of the interface normal to the interface, M is the interface mobility, summation is over all components l, V is the volume of the system and R is the gas constant. Now some geometrical simplifications can be made and a proportionality relation between each kinetic parameter $\dot{q}_i$ on one side and interface velocities v and diffusive fluxes $\vec{j}_l$ on the other side can be assumed. Thus, the local values of v and $\vec{j}_l$ can be expressed as

$$v = v(q_1, q_2, \ldots, q_n, \dot{q}_1, \dot{q}_2, \ldots, \dot{q}_n) \tag{3.6}$$

$$\vec{j}_l = \vec{j}_l(q_1, q_2, \ldots, q_n, \dot{q}_1, \dot{q}_2, \ldots, \dot{q}_n) \tag{3.7}$$

where the relations (3.6) and (3.7) are linear forms in $\dot{q}_1, \dot{q}_2, \ldots, \dot{q}_n$. If the relations (3.6) and (3.7) are inserted in Eq. (3.5) and the integrals and summation is performed, equation (3.5) has the form

$$Q = \sum_{i,k=1}^{n} B_{ik}(q_1, q_{2,\ldots,} q_n)\dot{q}_i \dot{q}_k \tag{3.8}$$

and the B_{ik} coefficients become known.

If all dissipative processes in the system are taken into account, the balance equation for the total Gibbs energy

$$-\frac{dG}{dt} \equiv -\dot{G} = Q \tag{3.9}$$

must be fulfilled.

In comparison with the Onsager's definitions given by Eqs. (3.1) and (3.2) the functionals 2Φ and Ψ are given by:

$$\dot{G}/T = -\Psi \tag{3.10}$$

and

$$Q/T = 2\Phi\,. \tag{3.11}$$

Then Onsager's principle can be reformulated in terms of $\dot{G}$ and Q as:

Q is maximum with respect to kinetic parameters $\dot{q}_i$ under the constraint $\dot{G} + Q = 0$.

Mathematically the principle can be expressed as

$$\frac{\partial}{\partial \dot{q}_i}\left[Q + \eta\left(\dot{G} + Q\right)\right] = 0\,,\ (i=1,\ldots,n) \tag{3.12}$$

with

$$\dot{G} = \sum_{i=1}^{n} \frac{\partial G}{\partial q_i}\dot{q}_i\,, \qquad Q = \sum_{i=1}^{n}\sum_{k=1}^{n} B_{ik}\dot{q}_i \dot{q}_k \tag{3.13}$$

and η being the Lagrange multiplier. Inserting Eqs. (3.13) into Eq. (3.12) and performing the derivatives one gets

$$\sum_{k=1}^{n}(B_{ik}+B_{ki})\dot{q}_k+\eta\left(\frac{\partial G}{\partial q_i}+\sum_{k=1}^{n}(B_{ik}+B_{ki})\dot{q}_k\right)=0,\ (i=1,...,n). \tag{3.14}$$

Multiplication of Eqs. (3.14) by $\dot{q}_i$ and their summation over i gives:

$$\sum_{i=1}^{n}\sum_{k=1}^{n}(B_{ik}+B_{ki})\dot{q}_i\dot{q}_k+\eta\left(\sum_{i=1}^{n}\frac{\partial G}{\partial q_i}\dot{q}_i+\sum_{i=1}^{n}\sum_{k=1}^{n}(B_{ik}+B_{ki})\dot{q}_i\dot{q}_k\right)=0 \tag{3.15}$$

which can be rewritten as

$$2Q+\eta(\dot{G}+2Q)=0\ . \tag{3.16}$$

Comparing this equation with Eq. (3.9) we obtain the value of the Lagrange multiplier $\eta=-2$ and the leading equations (3.12) can be rewritten as:

$$-\frac{\partial G}{\partial q_i}=\frac{1}{2}\frac{\partial Q}{\partial \dot{q}_i},\ (i=1,...,n). \tag{3.17}$$

Using Eq. (3.13) one gets

$$\frac{1}{2}\frac{\partial Q}{\partial \dot{q}_i}=\sum_{k=1}^{n}\frac{B_{ik}+B_{ki}}{2}\dot{q}_k=\sum_{k=1}^{n}\widetilde{B}_{ik}\dot{q}_k\ ,\ (i=1,...,n), \tag{3.18}$$

where $\widetilde{B}_{ik}$ is the symmetric part of the matrix B_{ik}. Inserting equations (3.13) and (3.18) into Eqs. (3.17) we get a system of linear equations for $\dot{q}_i$ which can be inverted as:

$$\dot{q}_i=-\sum_{k=1}^{n}\widetilde{B}_{ik}^{-1}\frac{\partial G}{\partial q_k}\ ,\ (i=1,...,n). \tag{3.19}$$

The vector $-\frac{\partial G}{\partial q_k}$ can be understood as a vector of generalized driving forces and the matrix $\widetilde{B}_{ik}^{-1}$ as a matrix of generalized kinetic parameters.

From the treatment it is evident that only the symmetric part of B_{ik} can contribute to the kinetics of the system and that is why the well-known Onsager reciprocity relations can also be understood as a direct consequence of the extremal principle.

3.3 Further Modifications of the Extremal Thermodynamic Principle

1) In some applications it may be advantageous to calculate directly the rate of the total Gibbs energy $\dot{G}$ instead of G. Then using Eq. (3.13) one gets

$$\frac{\partial \dot{G}}{\partial \dot{q}_i} = \frac{\partial G}{\partial q_i} \tag{3.20}$$

and the set of the evolution equations has the form

$$-\frac{\partial \dot{G}}{\partial \dot{q}_i} = \frac{1}{2}\frac{\partial Q}{\partial \dot{q}_i}. \tag{3.21}$$

2) In some applications there may be a necessity to apply the principle to a system which is not closed. In that case the rate of Gibbs energy flowing out of the system, $\dot{G}_{out}$, must be evaluated as a linear form of kinetic parameters and the evolution equations assumes the form:

$$-\frac{\partial(\dot{G} + \dot{G}_{out})}{\partial \dot{q}_i} = \frac{1}{2}\frac{\partial Q}{\partial \dot{q}_i}. \tag{3.22}$$

3) In some cases we may have a higher number of independent kinetic parameters u_i $(i=1,\ldots,m,\ m>n)$ than independent state parameters q_i. In this case the independent kinetic parameters do not coincide with the rates of the state parameters. Diffusion in a 3-dimensional system is a typical example of such a case. The components of fluxes (vectors) may be chosen as independent kinetic parameters u_i the number of which exceeds three times the concentration rates (scalars) which can be chosen as the state parameters q_i. In this case the evolution equations can be written in the form:

$$-\frac{\partial \dot{G}}{\partial u_i} = \frac{1}{2}\frac{\partial Q}{\partial u_i}. \tag{3.23}$$

For diffusion the use of Eq. (3.23) leads to the conclusion that the components of fluxes not contributing to the change of the system get automatically zero. This condition reduces the number of degrees of freedom of the fluxes to the number of the concentration rates.

4) Some constraints may exist amongst state or kinetic parameters (the parameters are not independent). In that case two ways of treatment of the problem are possible:

- some parameters can be eliminated – the number of parameters decreases and the remaining parameters become independent;
- the constraints can be respected by means of the Lagrange multiplier method. This requires starting with the formulation of the principle in a way similar to Eq. (3.12) (with

a larger number of Lagrange parameters) and deriving the evolution equations by an analogous procedure. This procedure is demonstrated in the next paragraph.

4 Examples of Simple Models Derived from the Thermodynamic Extremal Principle

4.1 Discharging of a Condenser

The application of the principle can be demonstrated by a very simple example, i.e. discharging the condenser in a closed electric circuit with the resistor. The charge on the condenser q can be considered as the only state parameter and its rate $\dot{q}$ as the only kinetic parameter. Let U be the voltage on the condenser, R_e be the resistance of the resistor (constant), C be the capacity of the condenser (constant) and I be the actual electric current in the circuit. Furthermore it is known that the voltage U at the condenser is given by

$$U = q/C. \tag{4.1}$$

The energy of the condenser, which can be considered as the total Gibbs energy of the system, is given by

$$G = \frac{q^2}{2C} \tag{4.2}$$

and the relation between I and $\dot{q}$ is given by

$$\dot{q} = -I. \tag{4.3}$$

In the classical approach we can use Ohm's law

$$I = U/R_e \tag{4.4}$$

assuming more or less tacitly that the voltage on the condenser equals the voltage on the resistor (this is a certain kind of contact condition which must be taken into account in the classical approach). Putting equations (4.1), (4.3) and (4.4) together, we obtain the evolution equation for the circuit as:

$$\dot{q} = -\frac{1}{R_e C} q. \tag{4.5}$$

The Gibbs energy dissipation Q in the system equals the production of the Joule's heat at the resistor given by:

$$Q = R_e I^2 = R_e \dot{q}^2 . \tag{4.6}$$

Differentiation of Eq. (4.2) leads to

$$\dot{G} = \frac{q}{C}\dot{q} . \tag{4.7}$$

Now there are the following other ways how the kinetics of the system can be determined:

1. In the case of one kinetic variable $\dot{q}$ the use of the conservation law (3.9) is sufficient for the determination of the kinetics of the system. Putting equations (3.9), (4.6) and (4.7) together, their resolution with respect to $\dot{q}$ leads to Eq. (4.5).
2. The equations (4.2) and (4.6) can be combined with Eq. (3.17) and their resolution with respect to $\dot{q}$ leads to Eq. (4.5).
3. The equations (4.6) and (4.7) can be combined with Eq. (3.21) and their resolution with respect to $\dot{q}$ leads to Eq. (4.5).

It is worth noting that in all these treatments there is no need to assume that the voltage on the condenser equals the voltage on the resistor as in the classical approach. It might seem that this is not important; however, the importance of this property becomes clear when one applies the principle to more complicated cases.

4.2 Grain Growth in Polycrystals

The principle can be applied successfully to the problem of grain growth in a polycrystal driven by the decrease of the total energy of grain boundaries in the system. During the grain growth a certain grain size distribution exists. The large grains grow while the small ones shrink and disappear. We are usually interested in the time evolution of the size distribution of grains. Thus it is quite sufficient for the treatment if we approximate each grain k by a sphere of an effective radius r_k. Then the total Gibbs energy of the system is given by the energy of grain boundaries

$$G = \frac{1}{2}\gamma_b \sum_{k=1}^{n} 4\pi r_k^2 . \tag{4.8}$$

The total Gibbs energy dissipation due to grain boundary migration can be expressed according to Eq. (3.5) as

$$Q = \frac{1}{2}\frac{1}{M_b}\sum_{k=1}^{n} 4\pi r_k^2 \dot{r}_k^2 . \tag{4.9}$$

The factor ½ in both equations (4.8) and (4.9) is due to the fact that each grain boundary is common to two grains.

During the grain size evolution the volume V of the system remains the same; consequently, the total volume of grains must be constant:

$$\sum_{k=1}^{n} \frac{4\pi r_k^3}{3} = V \quad \text{or} \quad \sum_{k=1}^{n} r_k^2 \dot{r}_k = 0 \tag{4.10}$$

Now the evolution equations are derived in two ways:

1) by elimination of one dependent variable due to constraint (4.10)
2) by accounting the constraint (4.10) using the Lagrange multiplier method.

1) Combining Eqs. (4.8) and (4.10) we can eliminate the variable r_n as

$$G = 2\pi\gamma_b \left[\sum_{k=1}^{n-1} r_k^2 + \left(\frac{3V}{4\pi} - \sum_{k=1}^{n-1} r_k^3 \right)^{2/3} \right]. \tag{4.11}$$

Combining Eqs. (4.9) and (4.10) we can eliminate the variables r_n and $\dot{r}_n$ as

$$Q = \frac{2\pi}{M_b} \left[\sum_{k=1}^{n-1} r_k^2 \dot{r}_k^2 + \frac{\left(\sum_{k=1}^{n-1} r_k^2 \dot{r}_k \right)^2}{\left(\frac{3V}{4\pi} - \sum_{k=1}^{n-1} r_k^3 \right)^{2/3}} \right]. \tag{4.12}$$

The insertion of Eqs. (4.11) and (4.12) into Eq. (3.17) using Eqs. (4.10) leads after some algebraic manipulations to

$$2\gamma_b \left[\frac{1}{r_n} - \frac{1}{r_k} \right] = \frac{\dot{r}_k - \dot{r}_n}{M_b}, \; (k=1,\dots,n\text{-}1). \tag{4.13}$$

Equations (4.13) can be rewritten as

$$r_k^2 \dot{r}_k = r_k^2 \dot{r}_n + 2M_b\gamma_b \left[\frac{r_k^2}{r_n} - r_k \right], \; (k=1,\dots,n\text{-}1). \tag{4.14}$$

The summation of equations (4.14) from 1 to n-1 and comparison with Eq. (4.10) leads to this equation for $\dot{r}_n$:

$$-r_n^2 \dot{r}_n = \sum_{k=1}^{n-1} r_k^2 \dot{r}_n + 2M_b\gamma_b \left[\frac{\sum_{k=1}^{n-1} r_k^2}{r_n} - \sum_{k=1}^{n-1} r_k \right]. \tag{4.15}$$

with the solution

$$\dot{r}_n = -2M_b\gamma_b \left[\frac{1}{r_n} - \frac{\sum_{k=1}^{n} r_k}{\sum_{k=1}^{n} r_k^2} \right]. \tag{4.16}$$

Insertion of Eq. (4.16) into Eq. (4.13) enables to resolve $\dot{r}_k$, *(k=1,...,n-1)* and together with Eq. (4.16) the resultant evolution equations read:

$$\dot{r}_k = -2M_b\gamma_b \left[\frac{1}{r_k} - \frac{\sum_{l=1}^{n} r_l}{\sum_{l=1}^{n} r_l^2} \right], \ (k=1,...,n). \tag{4.17}$$

2) The application of the Lagrange multiplier method for accounting the constraint (4.10) seems to be a more elegant way. The requirement of the maximum of Q given by Eq. (4.9) constrained by relations (3.9) and (4.10) leads to

$$\frac{\partial}{\partial \dot{r}_k} \left[Q + \eta\left(\dot{G} + Q\right) + \xi \sum_{l=1}^{n} r_l^2 \dot{r}_l \right] = 0, \ (k=1,...,n) \tag{4.18}$$

with $\dot{G}$ given from Eq. (4.8) as

$$\dot{G} = 4\pi\gamma_b \sum_{k=1}^{n} r_k \dot{r}_k \tag{4.19}$$

and η and ξ being the Lagrange multipliers. The performance of the derivative in Eq. (4.18) offers:

$$\frac{4\pi}{M_b} r_k^2 \dot{r}_k + \eta \left(4\pi\gamma_b r_k + \frac{4\pi}{M_b} r_k^2 \dot{r}_k \right) + \xi r_k^2 = 0, \ (k=1,...,n). \tag{4.20}$$

Multiplication of Eqs. (4.20) by $\dot{r}_k$ and their summation gives

$$2Q + \eta(\dot{G} + 2Q) = 0 \tag{4.21}$$

and using $\dot{G} = -Q$ leads to $\eta = -2$. Then Eqs. (4.20) can be resolved with respect to $r_k^2 \dot{r}_k$ as

$$r_k^2 \dot{r}_k = -2\gamma_b M_b r_k + \frac{\xi M_b}{4\pi} r_k^2 \text{, } (k=1,...,n). \tag{4.22}$$

Summation of Eqs. (4.22) and its comparison with Eq. (4.10) enables to determine the value of ξ as

$$\xi = 8\pi\gamma_b \frac{\sum_{l=1}^{n} r_l}{\sum_{l=1}^{n} r_l^2} . \tag{4.23}$$

Inserting this value into Eq. (4.22) and resolving the equation with respect to $\dot{r}_k$ immediately gives Eqs. (4.17). Formally one can rewrite Eq. (4.17) into the form

$$\dot{r}_k = -2M_b\gamma_b \left[\frac{1}{r_k} - \frac{1}{r_{crit}} \right] \text{, } (k=1,...,n). \tag{4.24}$$

where

$$r_{crit} = \frac{\sum_{l=1}^{n} r_l^2}{\sum_{l=1}^{n} r_l} \tag{4.25}$$

represents the critical radius of the grain which neither grows nor shrinks.

The resulting evolution equations for the grain growth are identical to those derived by means of rather complicated considerations in classical works on grain growth (see e.g. Hillert, 1965). It can be seen that the application of the thermodynamic extremal principle to the problem of grain growth represents, from the physical point of view, an easy and straightforward way leading to the same results as those obtained by standard (maybe sometimes wearisome) mathematical procedures.

4.3 Precipitate Coarsening

The precipitation process consists of three stages: nucleation, growth and coarsening of precipitates. The precipitates nucleate from the supersaturated matrix by means of thermal fluctuations. During the nucleation period the precipitates reach the supercritical size and may spontaneously grow by diffusion and interface migration. At the end of the growth stage the number of precipitates equals the number of nucleated precipitates and the supersaturation of the matrix reaches a negligible value. The dominant Gibbs energy of the system which can be released by dissipative phenomena is stored in the interfaces. Analogously to the grain growth the total Gibbs energy of the system can decrease by the growth of large precipitates and the shrinkage and dissolution of small precipitates. This process is called coarsening or Oswald ripening.

Let r_k be the effective radius of the precipitate k, n be the number of precipitates in the system and m be the number of components in the system. Furthermore, we assume no chemical driving force and thus the total chemical Gibbs energy of the system can be set to zero. Then the total Gibbs energy of the system is given by

$$G = \gamma \sum_{k=1}^{n} 4\pi r_k^2 \, . \tag{4.26}$$

The total Gibbs energy dissipates by both interface migration and diffusion. Let c_i^* be the concentration of the component i in the precipitate and $\bar{c}_i$ the mean concentration of the component i in the matrix. We can assume that during the sluggish precipitate coarsening the concentration gradients in the matrix are negligible. During the change of the radius r_k of the precipitate k the radial diffusional fluxes of the component i in the matrix very near to the interface is given by

$$j_{ki}\big|_{r=r_k} \equiv j_{ki}^* = (\bar{c}_i - c_i^*)\dot{r}_k \, . \tag{4.27}$$

We can assume that during the coarsening the components must be transported over large distances compared to the precipitate radius (mathematically to infinity) and thus the diffusional flux at the distance $\rho\,(\rho > r_k)$ from the precipitate centre is given by:

$$j_{ki}\big|_{r=\rho} \equiv \frac{r_k^2}{\rho^2} j_{ki}^* \, . \tag{4.28}$$

Using Eq. (3.5) the total dissipation in the system due to diffusion is given by:

$$Q_{diff} = \sum_{k=1}^{n}\sum_{i=1}^{m}\int_{r_k}^{\infty} \frac{RT}{\bar{c}_i D_i} 4\pi\rho^2 j_{ki} dV = \sum_{k=1}^{n} 4\pi RT r_k^3 \dot{r}_k^2 \sum_{i=1}^{m} \frac{(c_i^* - \bar{c}_i)^2}{\bar{c}_i D_i} \, , \tag{4.29}$$

where Eqs. (4.25) and (4.26) are utilized in Eq. (4.27).

The total dissipation due to both diffusion and interface migration is given according to Eq. (3.5) as:

$$Q = \frac{1}{M}\sum_{k=1}^{n} 4\pi r_k^2 \dot{r}_k^2 + \sum_{k=1}^{n} 4\pi RT r_k^3 \dot{r}_k^2 \sum_{i=1}^{m} \frac{(c_i^* - \bar{c}_i)^2}{\bar{c}_i D_i} \, . \tag{4.30}$$

Similar to the case of grain growth we can assume that during coarsening the volume fraction and thus also the total volume of precipitates does not change. This can be expressed by the constraint

$$\sum_{k=1}^{n} r_k^2 \dot{r}_k = 0 . \tag{4.31}$$

To derive the evolution equations one can repeat one of the procedures used for the grain growth in the preceding paragraph. We can also introduce the effective interface mobility $\widetilde{M}_k$ as

$$\frac{1}{\widetilde{M}_k} = \frac{1}{M} + RTr_k \sum_{i=1}^{m} \frac{(c_i^* - \bar{c}_i)^2}{\bar{c}_i D_i} . \tag{4.32}$$

Then the equations (4.26) and (4.30) will formally be the same as Eqs. (4.8) and (4.9) with the same constraint (4.10) or (4.31). Both procedures lead again to the same result:

$$\dot{r}_k = -\frac{\gamma}{\frac{1}{M} + RTr_k \sum_{i=1}^{m} \frac{(c_i^* - \bar{c}_i)^2}{\bar{c}_i D_i}} \left(\frac{1}{r_k} - \frac{1}{r_{crit}} \right) \tag{4.33}$$

with

$$r_{crit} = \frac{\sum_{l=1}^{n} \left[\frac{1}{M} + RTr_l \sum_{i=1}^{m} \frac{\left(c_i^* - \bar{c}_i\right)^2}{\bar{c}_i D_i} \right]^{-1} r_l^2}{\sum_{l=1}^{n} \left[\frac{1}{M} + RTr_l \sum_{i=1}^{m} \frac{\left(c_i^* - \bar{c}_i\right)^2}{\bar{c}_i D_i} \right]^{-1} r_l} . \tag{4.34}$$

If the mobility M can be considered infinite (the case of diffusion controlled coarsening), the critical radius equals the mean precipitate radius

$$r_{crit} = \frac{1}{n} \sum_{l=1}^{n} r_l . \tag{4.35}$$

5 Diffusion in Multi-Component Crystalline Systems

5.1 Introduction

Diffusion in solids represents an industrially important phenomenon. Diffusion is utilized for the refinement of the microstructure of structural materials by homogenization and precipitation. Diffusion also plays an important role in powder metallurgy – during sintering the pores are eliminated from the powder compact by means of grain boundary and surface diffusion and/or bulk diffusion. Some negative aspects of diffusion are represented by dislocation and diffusional creep or by degradation of the materials by coarsening of precipitates or by nucleation and growth of intergranular cavities.

In simple crystalline alloys (random alloys) we usually distinguish two kinds of lattice positions – substitutional and interstitial ones, and three kinds of components:

1) substitutional components – Fe, Cr, Ni, Cu,...occupying the ordinary lattice positions
2) interstitial components – C, N, H,... occupying the spaces between ordinary lattice positions (interstitial positions)
3) lattice vacancies (vacancies) – represented by ordinary lattice positions occupied by no atom.

In principle also the interstitial vacancies can be introduced – they are, however, not used in our description.

During diffusion both substitutional and interstitial atomic components must be conserved. On the other hand vacancies can be generated and annihilated. Grain boundaries, free surfaces or jogs at dislocations may act as sources and sinks for vacancies. Generation and annihilation of vacancies is due to local deviation of vacancy concentration from the equilibrium one and is connected with the deformation of the specimen (local swelling due to generation of vacancies and local shrinkage due to annihilation of vacancies). Diffusion of components may cause significant deviations of vacancy concentration from the equilibrium one. In the case of diffusion driven by concentration gradients of atomic components the deformation of the specimen due to vacancy generation/annihilation process is known as Kirkendall effect. In the case of diffusion of vacancies connected with the vacancy generation/annihilation driven by the applied stress we talk about the diffusional creep.

Local chemical composition is usually characterized by mole fractions x_i given as a ratio of number of moles of the atomic component i divided by number of moles of all atomic components in the representative volume element. The mole fractions cannot, however, describe the concentration of vacancies. For this reason site fractions y_i are introduced by the relation

$$y_i = N_i / N \tag{5.1}$$

where N is the number of moles of lattice positions and N_i is the number of moles of component i (including vacancies) in the representative volume element.

Diffusion in crystals is due to the vacancy mechanism. The usual site fraction of vacancies is about 10^{-6}. During diffusion the substitutional atoms exchange positions with vacancies and thus the diffusion is conditioned by the existence of vacancies in the crystal. The number of interstitial positions is comparable with the number of lattice positions. The usual site fraction of interstitial components is of the order 10^{-2} and thus the site fraction of interstitial vacancies is near to 1. The atoms of interstitial components are practically always surrounded only by interstitial vacancies, they do not influence each other and they exhibit uncorrelated free jumping. To learn more about diffusion in solids we can recommend the textbooks by Shewmon (1989) or by Glicksman (2000). A more detailed description of the theory presented in this paragraph can be found in Svoboda et al. (2002) and Hartmann et al. (2003).

In the most general isotropic case diffusion in the multi-component system can be described by

$$\vec{j}_i = -\sum_{k=0}^{s+p} L_{ik} grad\mu_k \ , \ (i=0,...,s+p) \tag{5.2}$$

where L_{ik} are the Onsager coefficients represented by a symmetric positive definite matrix, i=0 corresponds to vacancies, i=1,...,s to substitutional components and i=s+1,...,s+p to interstitial components. The chemical potential μ_k is a function of site fractions y_i. Due to the vacancy mechanism of diffusion of substitutional components the constraint

$$\sum_{k=0}^{s} \vec{j}_i = \vec{0} \tag{5.3}$$

must be fulfilled.

To obtain the evolution equations for the system the diffusion laws must be completed by conservation laws. Let W be the volume of the representative volume element, Ω the volume corresponding to one mole of site positions and Ω_k the partial molar volume of the component k. The elementary relations

$$W = \sum_{k=0}^{s+p} N_k \Omega_k \ , \quad \Omega = \sum_{k=0}^{s+p} y_k \Omega_k \ , \quad \sum_{k=0}^{s} N_k = N \ , \quad \sum_{k=0}^{s} y_k = 1 \tag{5.4}$$

can be written down. Then the conservation laws for all components are given by

$$\dot{N}_k = -W \, div\vec{j}_k \ , k=1,...,s+p \tag{5.5}$$

and

$$\dot{N}_0 = \dot{N} - W \, div\vec{j}_0 \, , \tag{5.6}$$

where $\dot{N}$ represents the rate of number of site positions in the representative volume element being equal to the difference of generated and annihilated vacancies in the representative volume element during a time unit (for details see Svoboda et al., 2002). Then putting equations (5.1), (5.4), (5.5) and (5.6) together one obtains

$$\dot{y}_k = -\Omega \, div\vec{j}_k - y_k \frac{\dot{N}}{N} \ , k=1,...,s+p, \tag{5.7}$$

$$\dot{y}_0 = -\Omega \, div\vec{j}_0 + (1 - y_0) \frac{\dot{N}}{N} \ , \tag{5.8}$$

$$\dot{\Omega} = -\Omega \sum_{k=0}^{s+p} \Omega_k div\vec{j}_k + (\Omega_0 - \Omega) \frac{\dot{N}}{N} \tag{5.9}$$

and

$$\dot{W} = \dot{N}\Omega + N\dot{\Omega} = -W\sum_{k=0}^{s+p}\Omega_k div\vec{j}_k + \dot{N}\,\Omega_0 \,. \tag{5.10}$$

For no sources and sinks for vacancies in the representative volume one can expect $\dot{N} \equiv 0$, and equations (5.7) and (5.8) display their classical form. In the case of ideal sources and sinks for vacancies $y_0 = y_0^{eq}$ and $\dot{y}_0 \equiv 0$. Then using the Eq. (5.8) $\dot{N}$ is given by

$$\frac{\dot{N}}{N} = \frac{\Omega}{1-y_0^{eq}}\, div\vec{j}_0 \,. \tag{5.11}$$

5.2 Formulation of the Thermodynamic Extremal Principle for Diffusion

The total Gibbs energy in the system is given by (see Svoboda et al., 2002)

$$G = \int_V \frac{g}{\Omega} dV \,, \tag{5.12}$$

where

$$g = \sum_{k=0}^{s+p} y_k \mu_k \tag{5.13}$$

is the Gibbs energy per one mole of site positions

Using the relations $d\dot{V}/dV = \dot{W}/W$ and $W = N\Omega$ one can differentiate Eq. (5.12) in time

$$\dot{G} = \int_V \left[\frac{\dot{g}}{\Omega} + \frac{g}{\Omega}\left(\frac{\dot{W}}{W} - \frac{\dot{\Omega}}{\Omega}\right)\right] dV = \int_V \left[\frac{\dot{g}}{\Omega} + \frac{g}{\Omega}\frac{\dot{N}}{N}\right] dV \tag{5.14}$$

and using Eq. (5.13) and the Gibbs-Duhem relation

$$\sum_{k=0}^{s+p} y_k \dot{\mu}_k = 0 \tag{5.15}$$

equation (5.14) can be rewritten as

$$\dot{G} = \int_V \frac{1}{\Omega}\sum_{k=0}^{s+p}\mu_k\left(\dot{y}_k + \frac{\dot{N}}{N}y_k\right) dV = \int_V \left[\frac{\mu_0}{\Omega}\frac{\dot{N}}{N} - \sum_{k=0}^{s+p}\mu_k div\vec{j}_k\right] dV \,. \tag{5.16}$$

The Gauss theorem for the closed system can be applied to Eq. (5.16). For no sources and sinks for vacancies ($\dot{N} \equiv 0$) or for ideal sources and sinks for vacancies ($\mu_0 \equiv 0$) the Gauss theorem application leads to the same equation:

$$\dot{G} = \int_V \sum_{k=0}^{s+p} \vec{j}_k \cdot grad\mu_k \, dV . \tag{5.17}$$

The total dissipation in the system is given by the relation

$$Q = \int_V \sum_{k=0}^{s+p} \vec{j}_k \cdot \vec{X}_k dV , \tag{5.18}$$

where $\vec{X}_k$ is the driving force corresponding to the flux $\vec{j}_k$. Within the framework of linear thermodynamics and for an isotropic material (e. g. a material with cubic lattice) one can expect the relation:

$$\vec{j}_k = A_k \vec{X}_k \tag{5.19}$$

where A_k is a kinetic coefficient corresponding to the component k. Insertion of Eq. (5.19) into Eq. (5.18) yields:

$$Q = \int_V \sum_{k=0}^{s+p} \frac{\vec{j}_k^{\,2}}{A_k} dV \tag{5.20}$$

The thermodynamic extremal principle asserts that the evolution in the system corresponds to the maximum rate of Gibbs energy dissipation Q constrained by the condition $\dot{G} + Q = 0$. This implies that

$$\delta\left(\dot{G} + Q/2\right) = 0 . \tag{5.21}$$

Densities of $\dot{G}$ and of Q can be introduced as:

$$\dot{g} = \sum_{k=0}^{s+p} \vec{j}_k \cdot grad\mu_k \; , \quad q = \sum_{k=0}^{s+p} \frac{\vec{j}_k^{\,2}}{A_k} . \tag{5.22}$$

Then $\dot{g} + q/2$ must be maximum with respect to the local fluxes $\vec{j}_i$ in each point of the system.

5.3 Derivation of Diffusion Equations in Systems With Constrained Fluxes

The constraint to fluxes:

$$\vec{j}_0 = -\sum_{k=1}^{n} \vec{j}_k \tag{5.23}$$

can be used for the elimination of the flux $\vec{j}_0$ in Eqs. (5.22) as:

$$\dot{g} = \sum_{k=1}^{s+p} \vec{j}_k \cdot grad\mu_k - \sum_{k=1}^{s} \vec{j}_k \cdot grad\mu_0 \ , \ q = \sum_{k=1}^{s+p} \frac{\vec{j}_k^{\,2}}{A_k} + \frac{\left(\sum_{k=1}^{s} \vec{j}_k\right)^2}{A_0} \ . \tag{5.24}$$

The requirement of the extreme of $\dot{g} + q/2$ with respect to the independent fluxes can be expressed by:

$$\frac{\partial}{\partial \vec{j}_i}(\dot{g} + q/2) = \vec{0} \ , \ (i=1,...,s+p). \tag{5.25}$$

Insertion of Eqs. (5.24) into Eqs. (5.25), performance of the derivatives and resolution of the set of linear equations together with the constraint (5.23) with respect to fluxes leads to (see Svoboda et al., 2002 for details)

$$\vec{j}_i = A_i \left[-grad\mu_i + \frac{\sum_{k=0}^{s} A_k grad\mu_k}{\sum_{k=0}^{s} A_k} \right], \ (i=0,...,s) \tag{5.26}$$

and

$$\vec{j}_i = -A_i grad\mu_i \ , \ (i = s+1,...,s+p). \tag{5.27}$$

The Onsager coefficients from Eq. (5.2) are then given by

$$L_{ik} = A_i \delta_{ik} - \frac{A_i A_k}{\sum_{l=0}^{s} A_l} \ , (i=0,...,s \text{ and } k=0,...,s) \tag{5.28}$$

and

$$L_{ik} = A_i \delta_{ik} \ , \ (i=s+1,...,s+p \text{ or } k= s+1,...,s+p). \tag{5.29}$$

5.4 Determination of the Kinetic Coefficients A_i

Let us consider the tracer diffusion experiment when only the component k and its radioisotope i are in non-equilibrium and all other components are in equilibrium. Some manipulations of Eqs. (5.26) lead to

$$\vec{j}_i = -A_i(\, grad\, \mu_i - grad\, \mu_0) - \frac{A_i}{A_0}\vec{j}_0 = -A_i grad\, \mu_i\,, \tag{5.30}$$

where the second equality in Eq. (5.30) is due to the fact that the system is chemically in equilibrium and thus no gradient of chemical potential of vacancies and no vacancy flux occurs in the system. The diffusion law for the dilute tracer is given by:

$$\vec{j}_i = -\frac{D_i}{\Omega} grad\, y_i\,, \tag{5.31}$$

where D_i is the tracer diffusion coefficient for the equilibrium site fraction of vacancies y_0^{eq}. The chemical potential for the dilute tracer can be calculated as

$$\mu_i = \mu_i^0 + RT\, ln\, y_i\,, \tag{5.32}$$

μ_i^0 represents a constant. Comparison of Eqs. (5.30) to (5.32) yields:

$$A_i = \frac{y_i D_i}{\Omega RT}, \quad (i=1,\dots,s). \tag{5.33}$$

To express the fact that the diffusion kinetics is proportional to the concentration of vacancies, equation (5.33) can be modified as

$$A_i = \frac{y_0 y_i D_i}{y_0^{eq}\, \Omega RT}, \quad (i=1,\dots,s). \tag{5.34}$$

For interstitial atoms the situation is much easier: comparison of Eqs. (5.27), (5.31) and (5.32) yields directly

$$A_i = \frac{y_i D_i}{\Omega RT}, \quad (i=s+1,\dots,s+p). \tag{5.35}$$

The diffusion mechanism of interstitial atoms is not connected with the existence of lattice vacancies, and thus there is no dependence of A_i on y_0.

The determination of A_0 is not trivial. The problem of diffusive motion of vacancies and the consequences on diffusive law have been studied intensively by many researchers and the works are not easy to be understood (see e.g. Manning, 1971, Lidiard, 1986 or Moleko et al., 1989). We have to limit ourselves to the presentation of the final formula

$$A_0 = -\frac{1}{1-f}\sum_{i=1}^{s} A_i \,, \tag{5.36}$$

where f is the correlation factor for the diffusive walk of atoms (in general $0<f<1$). It is worth noting that A_0 is negative, for $0<f<1$, however, Q given by Eq. (5.20) is always non-negative. For more details see Svoboda et al. (2002) and Hartmann et al. (2003).

5.5 Concluding Remark

In this chapter it was demonstrated that the diffusion law given e.g. by Eqs. (5.25) and (5.26) is independent of the existence of sources and sinks for vacancies. On the other hand the sources and sinks for vacancies influence the mass conservation laws with consequences on the evolution equations for the chemical composition and molar volume, and on the equations describing local swelling and shrinkage (see Eqs. (5.7) to (5.10)).

6 Solute Segregation and Drag at Migrating Interfaces

6.1 Introduction

A non-coherent phase interface can be imagined as a thin film of the typical thickness of a few inter-atomic distances. The film enables the transition from one crystal with a certain lattice structure and orientation to a neighbouring one with another lattice structure and orientation. The degree of disorder in the interface is much higher than in crystals, which enables segregation of solutes in the interface. If the interface migrates, the solutes are dragged by the migrating interface (they have to move by diffusion together with the interface), which can significantly decrease the interface mobility. For high interface velocities the solutes cannot keep up with the migrating interface, they detach from the interface, and the interface mobility increases.

The classical approach to the problem makes the following assumptions (see e.g. Cahn, 1962 or Lücke and Stüwe, 1971):

- The solute segregation in the interface is caused by an additional potential term.
- The interface velocity v is taken as an independent variable.
- The diffusion equation of the solute is formulated in the frame of reference connected with the migrating interface.
- The diffusion equation is solved under steady-state conditions in the region involving the interface and the adjacent semi-infinite grains.
- Zero diffusive fluxes are assumed at both ends of the infinite region.

In the lattice fixed frame of reference (actual configuration) the evolution of the concentration profile is given by the 2nd Fick law:

$$\dot{c} = \frac{\partial}{\partial x}\left[D\frac{\partial c}{\partial x}\right]. \tag{6.1}$$

We can introduce an additional potential term E causing a correction of the solubility of the solute in the interface and in the parent phase β. In the product phase α, E=0 is assumed. Then the evolution equation for the profile in the lattice fixed frame of reference reads:

$$\dot{c} = \frac{\partial}{\partial x}\left[D\frac{\partial c}{\partial x} + \frac{cD}{RT}\frac{\partial E}{\partial x}\right]. \tag{6.2}$$

Let the interface move with the velocity v; our intention is to solve the evolution equation in the frame of reference connected with the interface. Then the evolution equation must be brought into the form:

$$\frac{\partial c}{\partial t} = \frac{\partial}{\partial x}\left[D\frac{\partial c}{\partial x} + \frac{cD}{RT}\frac{\partial E}{\partial x} + vc\right]. \tag{6.3}$$

Now we can search after stationary solutions in the moving frame of reference ($\partial c/\partial t = 0$) which are given by the differential equation:

$$0 = \frac{\partial}{\partial x}\left[D\frac{\partial c}{\partial x} + \frac{cD}{RT}\frac{\partial E}{\partial x} + vc\right]. \tag{6.4}$$

This equation is solved with the boundary conditions $\left.\frac{\partial c}{\partial x}\right|_{x=\pm\infty} = 0$, which offers steady state solutions for massive transformations with $c(-\infty) = c(\infty)$.

The application of this concept to diffusive transformations occurring under transient conditions is impossible. That is why we propose a new approach: to search the steady state solution only in the interface with prescribed boundary conditions at the interface surface. This approach is then compatible with transient conditions occurring during diffusive phase transformations. The detailed description of the model is published by Svoboda et al. (2002a)

6.2 The Model

Let us assume a binary system of the solvent A and of the interstitial solute B (e.g. Fe – C system). Let the interface be represented by a layer of a thickness h, in which the thermodynamic quantities like the chemical potential μ_B or the mole fraction x of the solute change continuously from the grain α to the grain β. For the ideal solution the chemical potential of the solute can be expressed by:

$$\mu_B = \mu_B^0 + RT\,ln\,x\,, \tag{6.5}$$

where μ_B^0 and the mole fraction of the solute B, x, depend only on the coordinate y normal to and moving with the interface. We can choose an explicit expression for μ_B^0 (see Fig. 6.1):

$$\mu_B^0 = \frac{y}{h}\mu_B^\beta + \frac{h-y}{h}\mu_B^\alpha - RT\,ln\left[\frac{y}{h}x^\beta + \frac{h-y}{h}x^\alpha\right] + A\left[\left(y-\frac{h}{2}\right)^2 - \frac{h^2}{4}\right]. \tag{6.6}$$

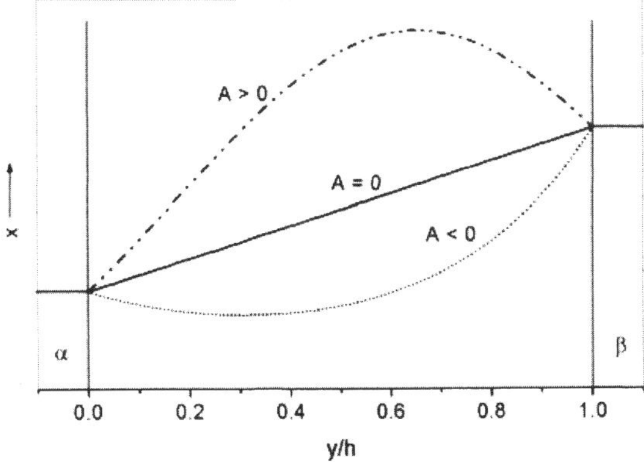

Figure 6.1. Schematic plot of the mole fraction profile in the interface in thermodynamic equilibrium for different parameters A.

This choice of μ_B^0 makes it possible that in the equilibrium ($\mu_B = const$) for A=0 the mole fraction x of the solute B changes linearly with the coordinate y across the interface, for A>0 the solute segregates at the interface and for A<0 the solute is expelled from the interface (see Fig. 6.1.). Moreover, equation (6.6) ensures due to the continuity in the mole fraction x also the continuity of the chemical potential μ_B and vice versa.

Now we assume that the interface migrates with a velocity v and a steady state is established in the migrating interface. From the mass conservation law we infer that

$$j = j^\alpha + \frac{x - x^\alpha}{\Omega}v\,, \tag{6.7}$$

where j is the diffusive flux of the solute, j^α is the diffusive flux at the left hand side surface of the interface and x^α is the mole fraction of the solute at the same place. The diffusive flux j is given by the diffusion equation as:

$$j = -\frac{xD}{\Omega RT}\frac{d\mu_B}{dy}. \tag{6.8}$$

Putting equations (6.7) and (6.8) together one obtains:

$$\frac{dx}{dy} = -\frac{j^\alpha \Omega}{D} - \frac{v}{D}(x - x^\alpha) - \frac{x}{RT}\left(\frac{\mu_B^\beta - \mu_B^\alpha}{h} + 2Ay - Ah\right) + x\frac{x^\beta - x^\alpha}{(x^\beta - x^\alpha)y + x^\alpha h}. \tag{6.9}$$

This is a differential equation for the mole fraction profile in the interface which has a unique solution for given values of v, D and j^α and the boundary condition $x(0) = x^\alpha$. We assume that $\mu_B^\alpha = \mu_B^\alpha(x^\alpha)$ and $\mu_B^\beta = \mu_B^\beta(x^\beta)$ are known functions.

6.3 Application of the Model to the Fe-C System

For calculations we choose the Fe-C system, where the phase α represents ferrite and the β phase represents austenite. The mole fraction profiles $x(y)$ are plotted in Fig. 6.2. for different interface velocities v assuming $j^\alpha = 0$ and $A=\kappa RT/h^2$.

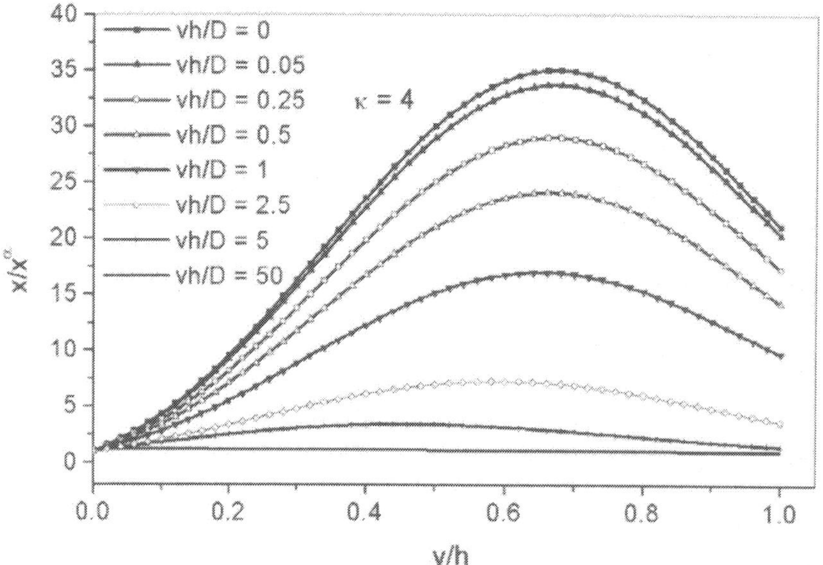

Figure 6.2. Solute mole fraction profiles for different interface velocities v assuming j^α=0 and $A=\kappa RT/h^2$.

Using the dependences $\mu_B^\alpha = \mu_B^\alpha(x^\alpha)$ and $\mu_B^\beta = \mu_B^\beta(x^\beta)$ and Eqs. (6.5) and (6.6) one can calculate also the profiles of the chemical potential μ_B, which are plotted in Fig. 6.3.

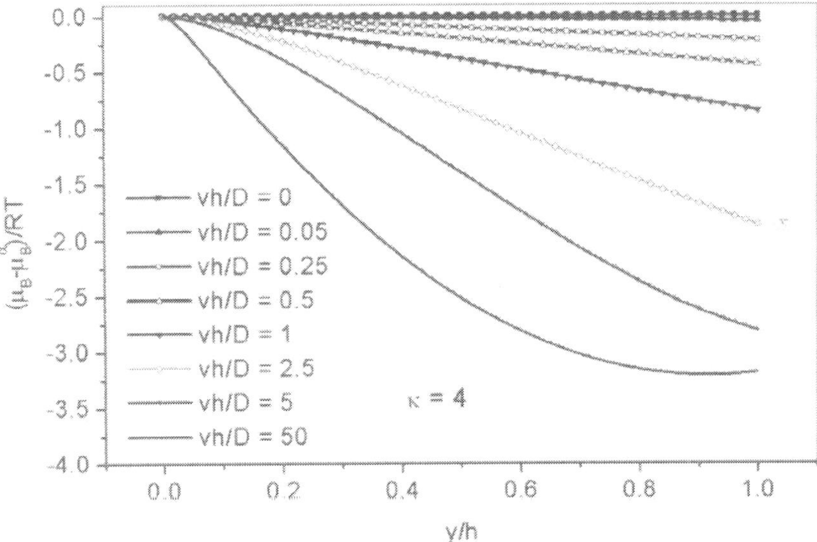

Figure 6.3. Profiles of chemical potentials of the solute for different interface velocities v assuming j^α=0 and $A=\kappa RT/h^2$.

From the mole fraction profile the diffusion flux j can be calculated using Eq. (6.7). Then the rate of Gibbs energy dissipation Q due to diffusion in the interface is given by:

$$Q = \int_0^h \frac{\Omega RT}{xD} j^2 dy \tag{6.10}$$

and the drag force necessary to move the solute in the interface is given by

$$F = Q/v \tag{6.11}$$

The drag force in dependence on the interface velocity v for different values of κ is plotted in Fig. 6.4. Note that the drag force assumes maximum for $vh/D \approx 1$.

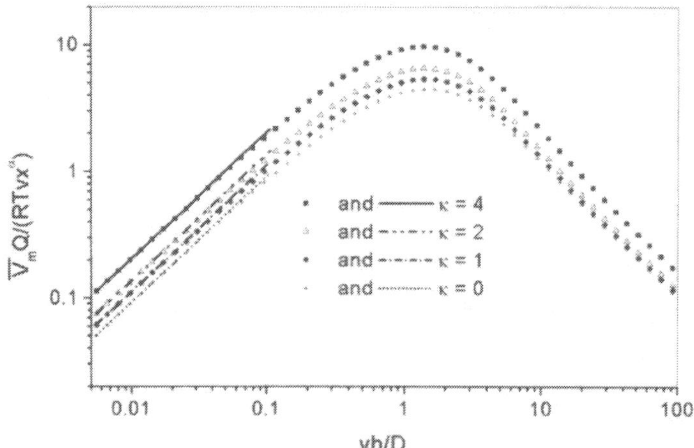

Figure 6.4. The drag force in dependence on the interface velocity v for different values of κ assuming $j^{\alpha}=0$ and $A=\kappa RT/h^2$. $\overline{V_m} \equiv \Omega$.

Let M_{eff} be the effective interface mobility including the trans-interface diffusion and the drag of the solute in the interface. If M denotes the interface mobility due to rearrangement of atoms in the interface without drag and trans-interface diffusion, then the total dissipation rate in the interface is given by:

$$Q_{tot} = Mv^2 + Q\,. \tag{6.12}$$

The same dissipation must be obtained due to migration of the interface with the effective interface mobility M_{eff}. Thus,

$$Q_{tot} = M_{eff}\, v^2\,. \tag{6.13}$$

Comparing Eqs. (6.12) and (6.13) one gets

$$M_{eff} = \left[\frac{1}{M} + \frac{Q}{v^2}\right]^{-1}. \tag{6.14}$$

Another important characteristic feature of the migrating interface is the difference of the chemical potential $\Delta\mu_B = \mu_B^{\alpha} - \mu_B^{\beta}$ across the interface. This difference is given by

$$\Delta\mu_B = \int_0^h \frac{\Omega RT}{xD} j\,dy\,. \tag{6.15}$$

The difference of the chemical potential $\Delta\mu_B$ in dependence on the interface velocity v for different values of κ is plotted in Fig. 6.5.

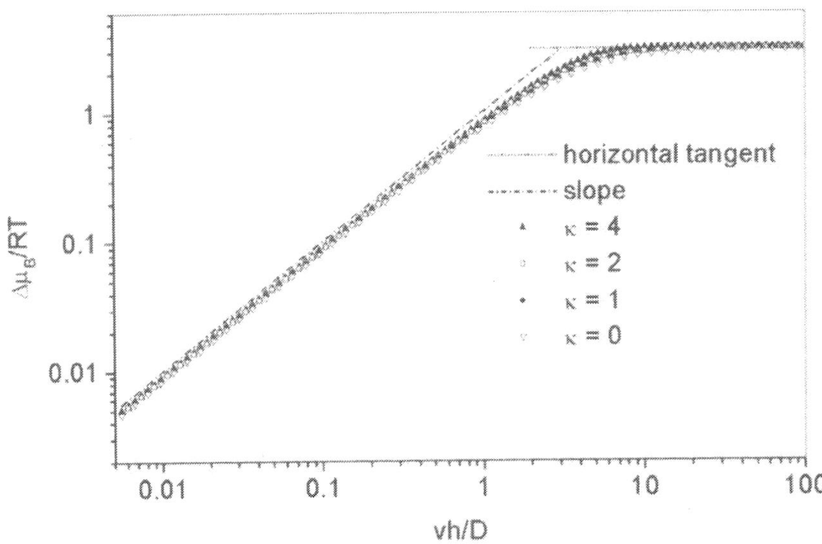

Figure 6.5. The difference of the chemical potential $\Delta\mu_B$ in dependence on the interface velocity v for different values of κ assuming $j^\alpha=0$ and $A=\kappa RT/h^2$.

To account for the solute drag and the trans-interface diffusion in the interface it is necessary to change the standard treatment in two items:

1) Replacing the interface mobility by the effective interface mobility M_{eff} given by Eq. (6.14)
2) Replacing the standard contact conditions $\mu_B^\alpha - \mu_B^\beta = 0$ by $\mu_B^\alpha - \mu_B^\beta = \Delta\mu_B$, where $\Delta\mu_B$ is given by Eq. (6.15).

6.4 Concluding Remarks

- The present concept enables to couple the steady-state solute diffusion in the migrating interface with transient diffusion in adjacent grains by proper boundary conditions. This makes it possible to treat the solute drag during the diffusional phase transformations.
- Based on the Gibbs energy dissipation due to solute diffusion in the interface the effective interface mobility is calculated. Furthermore, the jump in the chemical potential of the solute across the interface is determined. The solute trans-interface diffusion and drag in the interface can be efficiently taken into consideration by the effective interface mobility and by the jump in the chemical potential across the interface.

7 Application of the Thermodynamic Extremal Principle to the Diffusional Phase Transformations

7.1 Introduction

To simulate diffusional phase transformations it is necessary to solve the coupled problem of bulk diffusion and interface migration. The solution of diffusion equations makes it necessary to know the boundary conditions at the surface of the system and the contact conditions at the migrating interface. Usually it is assumed that the system is closed and no fluxes exist on the surface of the system. One class of contact conditions at the migrating interface stems from the mass conservation law. It relates the jumps in the fluxes across the interface, the jumps in concentrations and the interface velocity. Unfortunately, these contact conditions are not sufficient, and additional conditions are required to solve the coupled problem. Further contact conditions are usually added, e.g. the ortho-equilibrium conditions prescribing the continuity in chemical potentials of all components across the migrating interface or the para-equilibrium conditions prescribing the continuity in chemical potentials of interstitial components and the continuity in site fractions for substitutional components. Both conditions differ drastically, and they are selected according to rather qualitative physical arguments. Moreover, the ortho-equilibrium contact conditions imply no driving force at the migrating interface, which means a deliberate assumption of an infinite interface mobility.

Furthermore, fulfilling any type of contact conditions at the migrating interface for the multi-component systems makes it necessary to solve a set of complex non-linear equations before each time integration step. This may be a difficult task which may cause some solution instabilities.

To overcome such difficulties one can engage the thermodynamic extremal principle. In this chapter a model for diffusional phase transformations in substitutional alloys is derived from the thermodynamic extremal principle. The approach needs no additional contact conditions at the migrating interface and admits a finite value of the interface mobility. The detailed description of the treatment can be found in Svoboda et al. (2004).

7.2 System Description

Let us assume a closed one-dimensional multi-component system with s substitutional components i, $i=1,\dots,s$. We assume two grains of different phases α and β separated by a migrating incoherent sharp interface (see Fig. 7.1.).

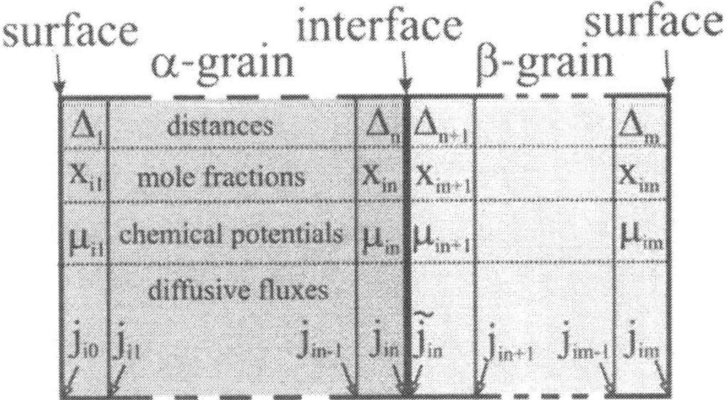

Figure 7.1. Schematic figure of the transforming system introducing physical quantities

Starting from the left (see Fig. 7.1), the first grain α is divided into n subregions of the thickness Δ_k, $k=1,\ldots,n$, with the mean mole fractions x_{ik} of substitutional components, $i=1,\ldots,s$, in each subregion. The position of the centre of a subregion is defined by its coordinate z_k. The second grain β is divided into m-n regions of the thickness Δ_k, $k=n+1,\ldots,m$, with the mean mole fractions x_{ik}. To each region mean chemical potentials $\mu_{ik}=\mu_{ik}(x_{1k}, x_{2k},\ldots,x_{s-1k})$ can be assigned. The interface is located between the subregions n and $n+1$. The thicknesses Δ_k, $k=1,\ldots,n-1,n+2,\ldots,m$, are supposed to be fixed. The atomic mobilities in individual subregions are described by the tracer diffusion coefficients D_{ik} .

The physical quantities are constrained by the conservation laws as:

$$N_i = \frac{1}{\Omega}\sum_{k=1}^{m} \Delta_k x_{ik} \text{ , } (i=1,\ldots,s), \tag{7.1}$$

where N_i is the fixed number of moles of the component i in the system (we assume a unit cross section of the system). Furthermore

$$\sum_{i=1}^{s} x_{ik} = 1 \text{ , } (k=1,\ldots,m). \tag{7.2}$$

To model the interface migration we assume that Δ_n and Δ_{n+1} are not constants and

$$\Delta_n + \Delta_{n+1} = \Delta \text{ ,} \tag{7.3}$$

Δ is fixed. Then we can introduce the interface velocity v as

$$v = \dot{\Delta}_n = -\dot{\Delta}_{n+1} \text{ .} \tag{7.4}$$

Due to the constraints (7.1) to (7.3) the state of the system is uniquely described by independent state parameters x_{ik} $(k=1,\ldots,m-1)$, $(i=1,\ldots,s-1)$ and Δ_n. Analogously, the kinetics of the system is given by independent kinetic parameters (rates of independent state parameters) $\dot{x}_{ik}$ $(k=1,\ldots,m-1)$, $(i=1,\ldots,\ s-1)$ and $\dot{\Delta}_n$. The rates of independent state parameters are related to the diffusive fluxes as:

$$\Delta_k \dot{x}_{ik} = \Omega(j_{ik-1} - j_{ik}),\ (i=1,\ldots,s-1;\ k=1,\ldots,n,\ n+2,\ldots,m), \tag{7.5}$$

$$(\Delta - \Delta_n)\dot{x}_{in+1} = \Omega(\tilde{j}_{in} - j_{in+1}) = \Omega(j_{in} - j_{in+1}) - (x_{in} - x_{in+1})\dot{\Delta}_n,\ (i=1,\ldots,s-1), \tag{7.6}$$

with $j_{i0}=0$ and $j_{im}=0$ $(i=1,\ldots,s-1)$ for a closed system. Then, in an alternative (and mathematically more convenient) way, the independent kinetic parameters can be chosen as j_{ik}, $(k=1,\ldots,m-1)$, $(i=1,\ldots,\ s-1)$ and $\dot{\Delta}_n$.

7.3 The Total Gibbs Energy of the System and Its Rate

The total Gibbs energy of the system can be calculated as:

$$G = \frac{1}{\Omega}\sum_{i=1}^{s}\sum_{k=1}^{m}\Delta_k x_{ik}\mu_{ik}\ . \tag{7.7}$$

Using the relations (7.1) to (7.6) and using the Gibbs-Duhem equation one can express

$$\frac{\partial \dot{G}}{\partial j_{ik}} = \overline{\mu}_{ik+1} - \overline{\mu}_{ik}\ ,\ (k=1,\ldots,m-1),\ (i=1,\ldots,\ s-1) \tag{7.8}$$

with $\overline{\mu}_{ik} = \mu_{ik} - \mu_{sk}$ and

$$\frac{\partial \dot{G}}{\partial \dot{\Delta}_n} = \frac{1}{\Omega}\sum_{i=1}^{s} x_{in}(\mu_{in} - \mu_{in+1})\ . \tag{7.9}$$

7.4 The Total Gibbs Energy Dissipation in the System

The total Gibbs energy dissipation in the system is given by the general expression (3.5). If the integrals are replaced by sums one obtains:

$$Q = \frac{1}{M}\dot{\Delta}_n^2 + \sum_{i=1}^{s}\left[\sum_{k=1}^{n} U_{ik}(j_{ik-1}^2 + j_{ik}^2) + \sum_{k=n+2}^{m} U_{ik}(j_{ik-1}^2 + j_{ik}^2) + U_{in+1}(\tilde{j}_{in}^2 + j_{in+1}^2) \right]\ , \tag{7.10}$$

with

$$U_{ik} = \frac{RT\Omega\Delta_k}{2x_{ik}D_{ik}}, \; (k=1,...,m\text{-}1), \; (i=1,..., \; s\text{-}1). \tag{7.11}$$

Using $j_{i0}=0$ and $j_{im}=0$ $(i=1,...,s\text{-}1)$, relations (7.6) and

$$j_{sk} = -\sum_{i=1}^{s-1} j_{ik}, \; (k=1,...,m\text{-}1), \tag{7.12}$$

one can express Q by means of independent fluxes j_{ik}, $(k=1,...,m\text{-}1)$, $(i=1,...,s\text{-}1)$ and $\dot{\Delta}_n$ and perform the derivatives with respect to the independent kinetic parameters:

$$\frac{\partial Q}{\partial j_{ik}} = 2(U_{ik} + U_{ik+1})j_{ik} - 2(U_{sk} + U_{sk+1})j_{sk}, \; (i=1,...,s\text{-}1; \; k=1,...,n\text{-}1, \; n+1,...,m\text{-}1), \tag{7.13}$$

$$\frac{\partial Q}{\partial j_{in}} = 2(U_{in} + U_{in+1})j_{in} - 2(U_{sn} + U_{sn+1})j_{sn} - RT\left[\frac{x_{in} - x_{in+1}}{x_{in+1}D_{in+1}} - \frac{x_{sn} - x_{sn+1}}{x_{sn+1}D_{sn+1}}\right](\Delta - \Delta_n)\dot{\Delta}_n,$$

$$(i=1,..., \; s\text{-}1) \tag{7.14}$$

and

$$\frac{\partial Q}{\partial \dot{\Delta}_n} = \frac{2\dot{\Delta}_n}{M} + \frac{RT}{\Omega}\sum_{\ell=1}^{s}\frac{(x_{in} - x_{in+1})(\Delta - \Delta_n)((x_{in} - x_{in+1})\dot{\Delta}_n - \Omega j_{in})}{x_{in+1}D_{in+1}}. \tag{7.15}$$

7.5 Evolution Equations Based on the Application of the Thermodynamic Extremal Principle

From the thermodynamic extremal principle it follows that the evolution of the system is given by the set of linear equations

$$\frac{\partial \dot{G}}{\partial \dot{q}_l} = -\frac{1}{2}\frac{\partial Q}{\partial \dot{q}_l}, \; l = 1,...,(s-1)(m-1)+1, \tag{7.16}$$

where $\dot{q}_l$ represent the independent fluxes and the interface velocity $\dot{\Delta}_n$. The utilization of Eqs. (7.8), (7.12), (7.13) and (7.16) yields the fluxes in the bulk:

$$j_{ik} = -B_{ik}\left(\mu_{ik+1} - \mu_{ik} - \frac{\sum_{l=1}^{s} B_{lk}(\mu_{lk+1} - \mu_{lk})}{\sum_{l=1}^{s} B_{lk}}\right), \; (i=1,...,s; \; k=1,...,n\text{-}1, \; n+1,...,m\text{-}1) \tag{7.17}$$

with $B_{ik} = [U_{ik} + U_{ik+1}]^{-1}$.

To determine the fluxes at the interface and the interface velocity the equations (7.9), (7.12), (7.14) to (7.16) must be used, with the solution:

$$\dot{\Delta}_n = \frac{\dfrac{1}{\Omega}\sum_{i=1}^{s} x_{in}F_i - \sum_{i=1}^{s} C_i B_{in}\left[F_i - \dfrac{\sum_{\ell=1}^{s} B_{\ell n}F_\ell}{\sum_{\ell=1}^{s} B_{\ell n}}\right]}{\dfrac{1}{M} + P - \sum_{i=1}^{s} C_i B_{in}\left[C_i - \dfrac{\sum_{\ell=1}^{s} B_{\ell n}C_\ell}{\sum_{\ell=1}^{s} B_{\ell n}}\right]}, \tag{7.18}$$

and

$$j_{in} = -B_{in}\left(F_i - C_i\dot{\Delta}_n - \frac{\sum_{\ell=1}^{s} B_{\ell n}(F_\ell - C_\ell\dot{\Delta}_n)}{\sum_{\ell=1}^{s} B_{\ell n}}\right), \; (i=1,\dots,s). \tag{7.19}$$

with

$$C_i = \frac{RT}{2}\frac{x_{in} - x_{in+1}}{x_{in+1}D_{in+1}}(\Delta - \Delta_n), \; (i=1,\dots,s), \tag{7.20}$$

$$P = \frac{RT}{2\Omega}\sum_{\ell=1}^{s}\frac{(x_{in} - x_{in+1})^2(\Delta - \Delta_n)}{x_{in+1}D_{in+1}}, \tag{7.21}$$

$$F_i = \mu_{in+1} - \mu_{in}, \; (i=1,\dots,s). \tag{7.22}$$

7.6 Discussion

The presented evolution equations describe the diffusion in the bulk of both grains and the coupling of diffusion and interface migration at the interface. The system is considered to be closed and this implies the boundary conditions at the system surface represented by zero fluxes. The conservation laws for the bulk and for the migrating interface are used to relate the fluxes and the rates of state parameters. This is sufficient for the expression of the rate of the total Gibbs energy $\dot{G}$ and the total dissipation Q as functions of independent kinetic parameters and for the derivation of evolution equations by means of the application of the principle. The rates of change of all state parameters depend on the actual values of chemical potentials in the regions. The evolution of the chemical potentials is given by the evolution of the mole fractions in the individual regions. The evolution of mole fractions is given by Eqs. (7.5) and (7.6). The set of equations is complete and gives the unique description of the system evolution.

In recent treatments, e.g. DICTRA, some additional contact conditions must be prescribed and solved. It is necessary to point out that no additional contact conditions at the migrating interface (like e.g. ortho-equilibrium conditions) are required in our case. This is an analogy to the example described in Paragraph 4.1 where it was shown, that the treatment based on the application of the principle can do without the assumption that the voltage on the condenser equals the voltage on the resistor, as in the classical approach. Thus the contact conditions can be understood as conditions which are not required by the principle but which, all the same, follow from the principle and are automatically met by equations derived from the principle.

It can be shown that in the limit case $\Delta_k \to 0$ the differential equations for bulk diffusion with constrained fluxes are reproduced. Moreover, it can be shown that the standard expression for the driving force acting at the interface is reproduced and the jumps in chemical potentials across the interface are the same for all substitutional components. This can be interpreted in such a way that during the lattice rearrangement process at the migrating interface all components dissipate the same Gibbs energy per rearranged mole. For the infinite interface mobility ($M \to \infty$) the contact conditions turn into ortho-equilibrium conditions used e.g. in DICTRA software.

7.7 Examples

Let us assume the $\gamma \rightarrow \alpha$ phase transformation in the Fe-rich part of the Fe-Cr-Ni system. The chemical potentials in both phases as functions of mole fractions can be calculated in a standard way using e.g. THERMOCALC; the kinetic parameters like diffusion coefficients and interface mobility are given in Lee and Oh (1996) and by Krielaart (1995). The initial chemical composition in both phases was set to be identical: x_{Fe}=0.98, x_{Cr}=0.001 and x_{Ni}=0.019. The details of the numerical method are presented in Svoboda et al. (2003). The total length of the system is chosen to be 10^{-5}m.

The transformation kinetics for different temperatures is given in Fig. 7.2. The transformation for temperatures higher or equal than 1053K is a diffusional one (Fig. 7.2a)), while the transformation for temperatures lower or equal than 1043K has all the features of the massive transformation (Fig. 7.2b)).

During the massive transformation very thin spikes are formed within a short time in front of the migrating interface. The spikes in the Cr and Ni mole fraction profile constituted during the first 0.01s of phase transformation are presented in Fig. 7.3. In comparison to Cr, the thickness of the Ni spike is much smaller due to its much lower diffusion coefficient (by one order of magnitude). It can be shown that the diffusional phase transformation is controlled by the diffusion. On the other hand the massive transformation is controlled by a rearrangement process in the migrating interface characterized by the interface mobility M (for details see Svoboda et al. 2004).

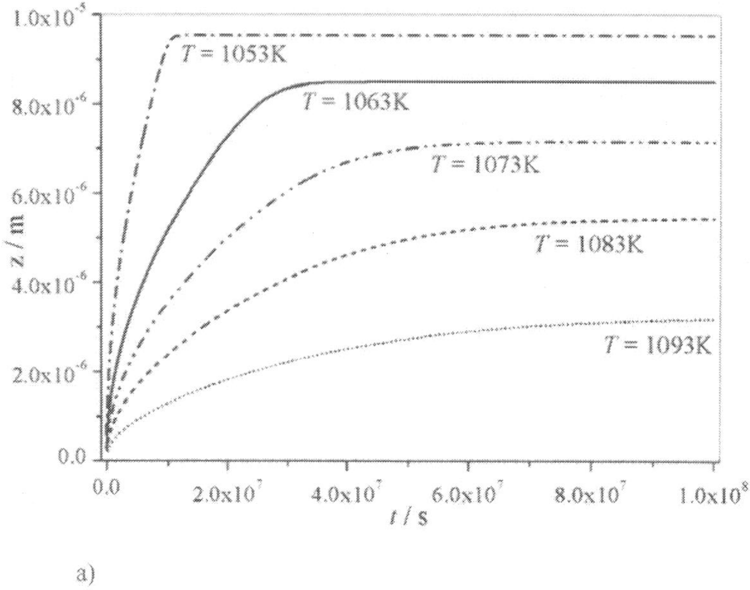

a)

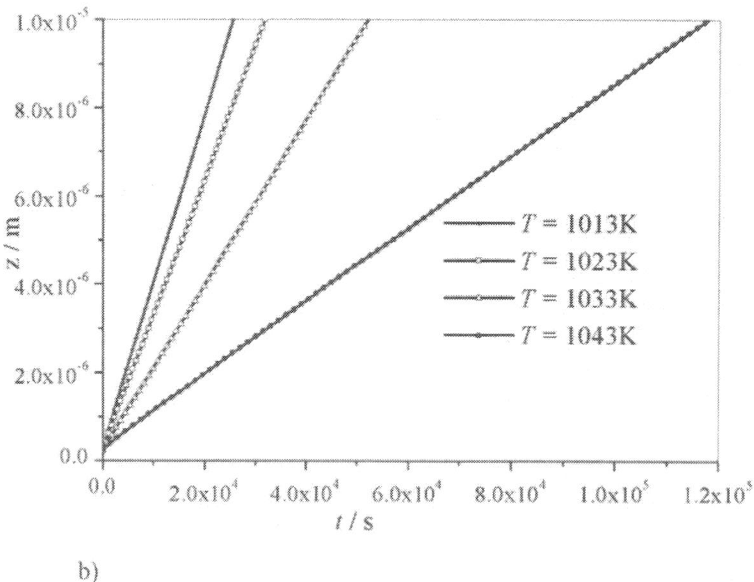

b)

Figure 7.2. Transformation kinetics a) diffusional transformation at temperatures 1053K-1093K, b) massive transformation at temperatures 1013K-1043K.

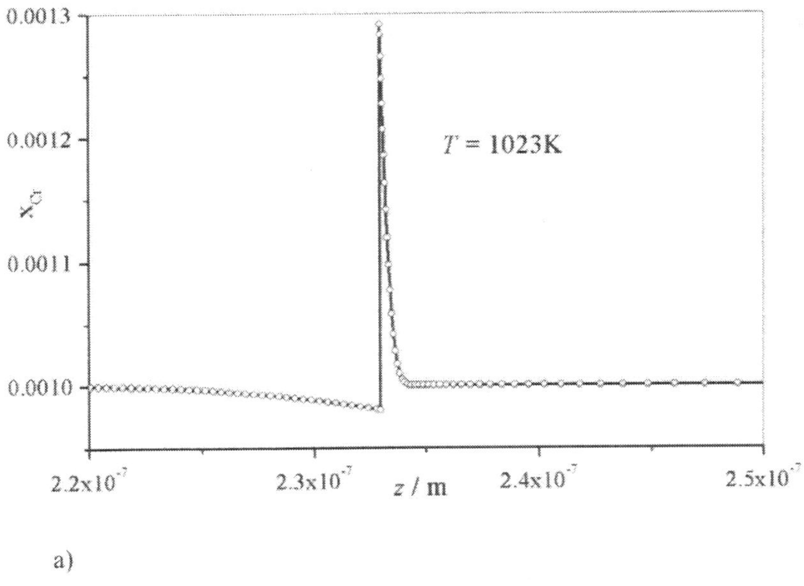

a)

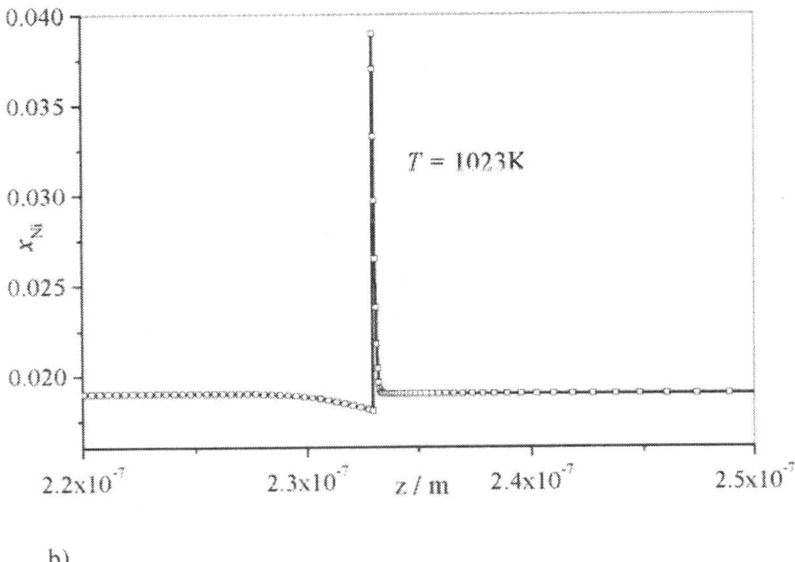

b)

Figure 7.3. Spikes in a) Cr and b) Ni mole fractions formed during 0.01s of transformation time at a temperature of 1023K. Symbols correspond to nodal points (centres of regions) used in the calculations.

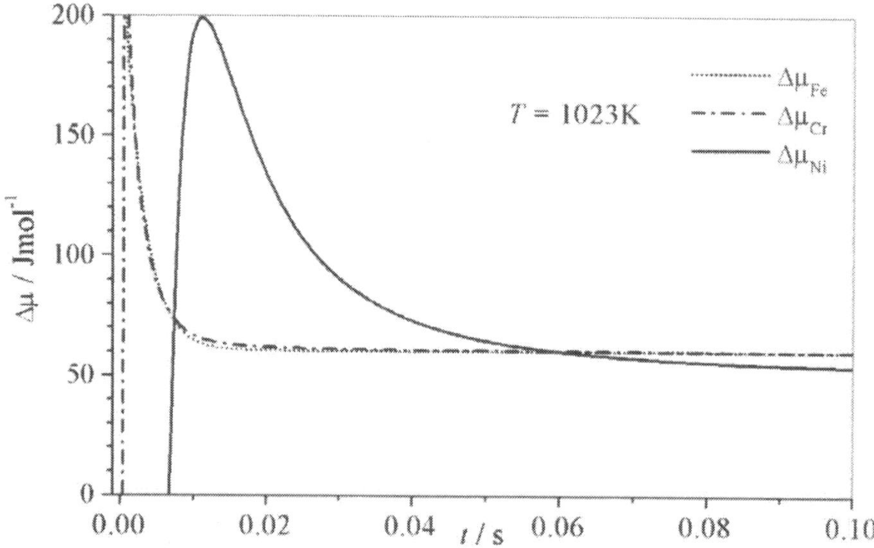

Figure 7.4. Time evolution of the jumps of chemical potentials at the interface for all components. After about 0.05s the differences are nearly the same.

The analysis of the model presented in paragraph 7.6. leads to the conclusion that the jumps of the chemical potentials across the interface must be the same for all components. The jumps in chemical potentials given by the initial chemical composition are 236.5 Jmol^{-1} for Fe, -2122.7 Jmol^{-1} for Cr and -6236,9 Jmol^{-1} for Ni. As can be seen in Fig. 7.4. the jumps approach each other very quickly (within less than 0.1s). For finer nets of nodal points than that presented in Fig. 7.3. the agreement in jumps after the transition time gets even better.

In figures 7.5. and 7.6. the mole fraction profiles for different times are presented for diffusional transformation and massive transformation, respectively. Unlike the massive transformation the diffusional transformation stops before the whole specimen is transformed. The massive transformation can be characterized by the steady state mole fraction profiles and by the constant transformation rate.

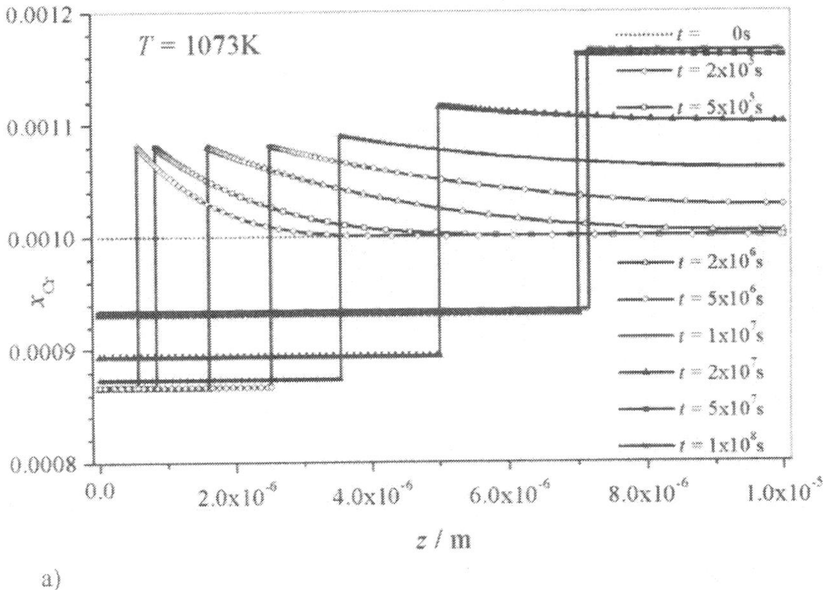

a)

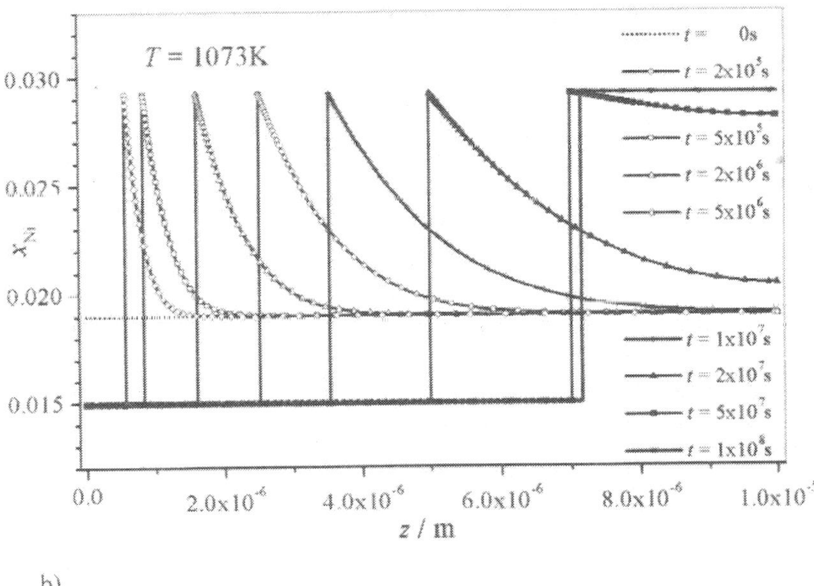

b)

Figure 7.5. Mole fraction profiles of a) Cr and b) Ni for different times during diffusional transformation

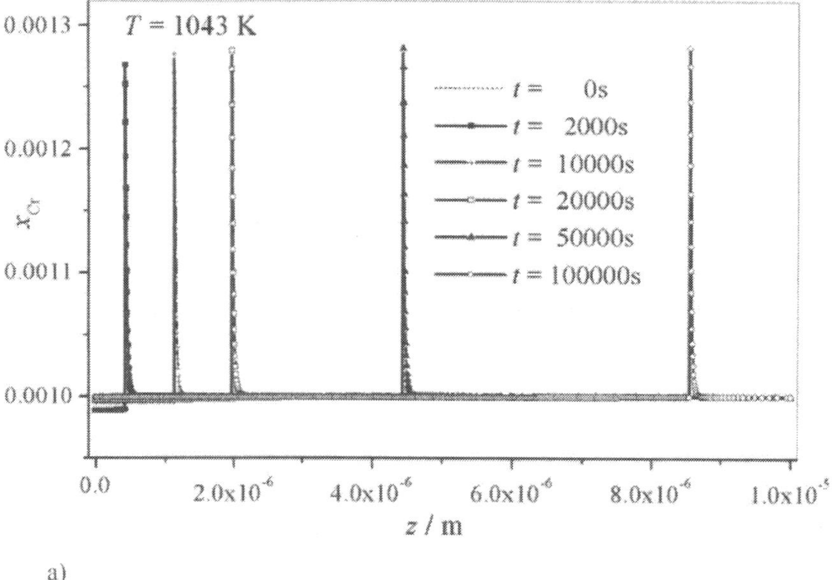

a)

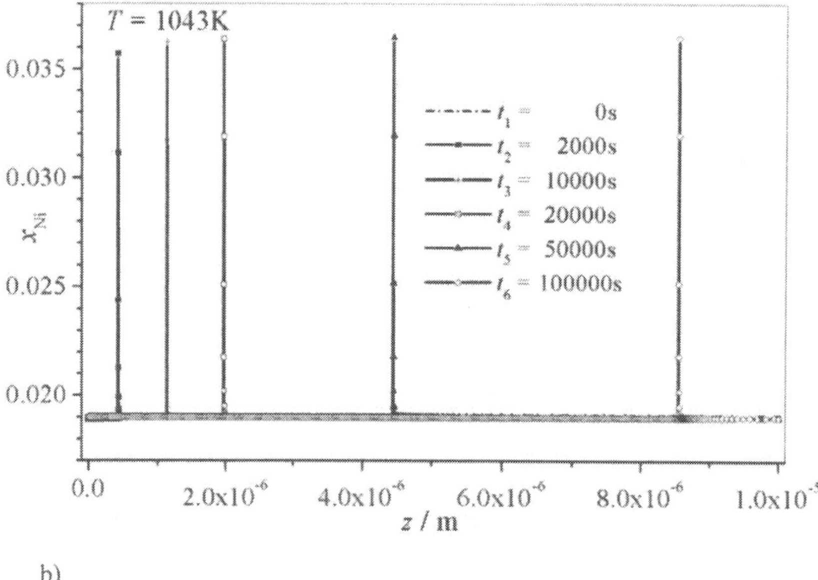

b)

Figure 7.6. Mole fraction profiles of a) Cr and b) Ni for different times during massive transformation

7.8 Concluding Remarks

Based on the application of the thermodynamic extremal principle a sharp-interface model for diffusional and massive transformation in substitutional alloys is developed. The model has the following properties:

- the model enables to assume a finite value of the interface mobility
- standard equations for diffusion and interface migration are reproduced
- no contact conditions for the chemical potentials at the migrating interface are required
- the same jumps of chemical potentials across the interface for all components are kept automatically by equations derived from the principle

8 Modelling of Kinetics in Multi-Component Multi-Phase Multi-Particle Systems

8.1 Introduction

Precipitates play an exclusive role in microstructure refinement of structure materials like steels. On the other hand coarsening of precipitates during long term exploitation of steels under elevated temperature degrades the precipitate microstructure. Thus the modelling of precipitation process is of high relevance.

There occur three stages of evolution of precipitate microstructure which may considerably overlap.

1) Nucleation stage
2) Growth stage
3) Coarsening stage

Moreover, the chemical composition of the precipitates may also change during all stages. Steels are multi-component systems in which precipitates of different phases with different size distribution may coexist. This represents concurrent conditions leading to very complex evolution of the system during which some fast nucleating and growing phases can in later stages dissolve due to nucleation and growth of more stable phases. Modelling of the precipitation process in its whole complexity requires finding the appropriate tool. The proper model can be derived simply by application of the extremal thermodynamic principle.

Let us assume m precipitates embedded in the matrix. Let each precipitate be characterized by the type of its lattice (correspondence to a phase), by its size, and by its mean chemical composition. The system consisting of the matrix and of the precipitates is considered to be closed. Then the actual mean composition of the matrix is given by the initial state of the system (e.g. the chemical composition of the supersaturated matrix without precipitates) and by the actual size and chemical composition of all precipitates. A detailed description of the model and of results of simulation can be found in Svoboda et al. (2004a) and Kozeschnik et al. (2004).

8.2 System Description

The system consists of matrix and precipitates. Let s be the fixed number of substitutional components and p the fixed number of interstitial components in the system. $n \equiv s + p$. Let N_i $(i=1,...,n)$ be the fixed number of moles of component i in the system, m be the number of precipitates in the system, ρ_k $(k=1,...,m)$ be the radius of precipitate k, u_{ki} $(k=1,...,m,\ i=1,...,n)$ be the mean site fraction of component i in the precipitate k and U_k be the fixed structure parameter of the precipitate k (e.g. in case $M_{23}C_6$, U_k=6/23). Then trivial relations follow from the definition of the site fractions

$$\sum_{i=1}^{s} u_{ki} = 1\,,\ \sum_{i=s+1}^{n} u_{ki} = U_k\,,\ (k = 1....m)\,. \tag{8.1}$$

We assume that the partial molar volumes of all substitutional components are the same $\Omega_i = \Omega$ $(i=1,...,s)$ and for interstitial components they are zero - $\Omega_i = 0$ $(i=s+1,...,n)$. Then we can introduce the mean concentration of the component i in the precipitate k, c_{ki}

$$c_{ki} = u_{ki} / \Omega\ \ ,(k=1,...,m,\ i=1,...,n). \tag{8.2}$$

The number of moles of the component i in the matrix N_{0i} is given by

$$N_{0i} = N_i - \sum_{k=1}^{m} \frac{4\pi\rho_k^3 c_{ki}}{3}\,,\ (i=1,...,n), \tag{8.3}$$

the set of which also characterizes the mean chemical composition of the matrix.

The state of the system is uniquely described by independent state parameters m, ρ_k $(k=1,...,m)$ and c_{ki} $(k=1,...,m,\ i=2,...,s,\ s+2,...,n)$. The time evolution of these parameters describes the evolution of the whole system.

8.3 The Total Gibbs Energy of the System

The total Gibbs energy of the system is given by

$$G = \sum_{i=1}^{n} N_{0i}\mu_{0i} + \sum_{k=1}^{m} \frac{4\pi\rho_k^3}{3}\left(\lambda_k + \sum_{i=1}^{n} c_{ki}\mu_{ki}\right) + \sum_{k=1}^{m} 4\pi\rho_k^2\gamma_k\,, \tag{8.4}$$

where μ_{0i} is the chemical potential of the component i in the matrix, μ_{ki} is the chemical potential of the component i in the precipitate k, λ_k accounts for the elastic and plastic strain around the precipitate k and γ_k is the interface energy density corresponding to the precipitate k.

Using equations (8.1) to (8.3) the partial derivatives of G with respect to independent state parameters (m is assumed to be fixed for the moment) read

$$\frac{\partial G}{\partial \rho_k} = 8\pi\rho_k\gamma_k + 4\pi\rho_k^2\left[\lambda_k + \frac{1}{\Omega}\{\mu_{k1} - \mu_{01} + U_k(\mu_{ks+1} - \mu_{0s+1})\} + \right.$$
$$\left. + \sum_{i=2}^{s} c_{ki}(\mu_{ki} - \mu_{0i} - \mu_{k1} + \mu_{01}) + \sum_{i=s+2}^{n} c_{ki}(\mu_{ki} - \mu_{0i} - \mu_{ks+1} + \mu_{0s+1})\right], (k=1,...,m), \quad (8.5)$$

$$\frac{\partial G}{\partial c_{ki}} = \frac{4\pi\rho_k^3}{3}(\mu_{ki} - \mu_{0i} - \mu_{k1} + \mu_{01}), \; (k=1,...,m, \; i=2,...,s), \quad (8.6)$$

$$\frac{\partial G}{\partial c_{ki}} = \frac{4\pi\rho_k^3}{3}(\mu_{ki} - \mu_{0i} - \mu_{ks+1} + \mu_{0s+1}), \; (k=1,...,m, \; i=s+2,...,n). \quad (8.7)$$

8.4 The Total Gibbs Energy Dissipation in the System

We assume that three processes contribute to the total Gibbs energy dissipation in the system:

1) Migration of interfaces described by interface mobilities M_k $(k=1,...,m)$.
2) Diffusion in the precipitates described by diffusion coefficients D_{ki}, $(k=1,...,m,\ i=1,...,n)$.
3) Diffusion in the matrix described by diffusion coefficients D_{0i}, $(i=1,...,n)$.

Dissipation due to interface migration is given by (see Eq. (3.5))

$$Q_1 = \sum_{k=1}^{m} \frac{4\pi\rho_k^2\dot{\rho}_k^2}{M_k}. \quad (8.8)$$

Dissipation due to diffusion in the precipitates can be calculated in the following way. We assume that in precipitates all components are deposited or collected uniformly. Then the radial diffusive flux is given by:

$$j_{ki} = -\frac{r\dot{c}_{ki}}{3}, \quad (0 \le r \le \rho_k). \quad (8.9)$$

Using Eq. (3.5) one can calculate the dissipation rate due to diffusion in precipitates as:

$$Q_2 = \sum_{k=1}^{m}\sum_{i=1}^{n}\int_0^{\rho_k} \frac{RT}{c_{ki}D_{ki}} 4\pi r^2 j_{ki}^2 dr = \sum_{k=1}^{m}\sum_{i=1}^{n} \frac{4\pi RT\rho_k^5\dot{c}_{ki}^2}{15c_{ki}D_{ki}}, \quad (8.10)$$

where $\dot{c}_{ik}$ are constrained (see Eqs. (8.1) and (8.2)) with

$$\sum_{i=1}^{s}\dot{c}_{ki} = 0 \text{ and } \sum_{i=s+1}^{n}\dot{c}_{ki} = 0. \quad (8.11)$$

Dissipation due to diffusion in the matrix is calculated under similar assumptions as those in precipitates. It is assumed that fluxes corresponding to each precipitate are radial, the diffusion zone is large and the deposition rate in the whole diffusion zone is constant. Then the rate of dissipation due to diffusive fluxes in the matrix is given by

$$Q_3 = \sum_{k=1}^{m} \sum_{i=1}^{n} \int_{\rho_k}^{Z} \frac{RT}{c_{0i} D_{0i}} 4\pi r^2 J_{ki}^2 dr \approx \sum_{k=1}^{m} \sum_{i=1}^{n} \frac{4\pi RT \rho_k^3 (\dot{\rho}_k (c_{ki} - c_{0i}) + \rho_k \dot{c}_{ki}/3)^2}{c_{0i} D_{0i}} \tag{8.12}$$

where c_{0i} is the concentration of the component i in the matrix, J_{ki} is the radial flux of the component i around the precipitate k and Z is the radius of the diffusion zone.

The total dissipation rate in the system is then given by

$$Q = Q_1 + Q_2 + Q_3 . \tag{8.13}$$

The partial derivatives of Q with respect to the independent kinetic parameters (the constraints (8.11) must be taken into account) are given by:

$$\frac{\partial Q}{\partial \dot{\rho}_k} = 8\pi \rho_k^2 \left[\frac{1}{M_k} + RT\rho_k \sum_{i=1}^{n} \frac{(c_{ki} - c_{0i})^2}{c_{0i} D_{0i}} \right] \dot{\rho}_k +$$

$$+ \frac{8\pi RT \rho_k^4}{3} \left[\sum_{i=2}^{s} \left(\frac{c_{ki} - c_{0i}}{c_{0i} D_{0i}} - \frac{c_{k1} - c_{01}}{c_{01} D_{01}} \right) \dot{c}_{ki} + \sum_{i=s+2}^{n} \left(\frac{c_{ki} - c_{0i}}{c_{0i} D_{0i}} - \frac{c_{ks+1} - c_{0s+1}}{c_{0s+1} D_{0s+1}} \right) \dot{c}_{ki} \right],$$

$$(k=1,\ldots,m), \tag{8.14}$$

$$\frac{\partial Q}{\partial \dot{c}_{ki}} = \frac{8\pi RT \rho_k^4}{3} \left(\frac{c_{ki} - c_{0i}}{c_{0i} D_{0i}} - \frac{c_{k1} - c_{01}}{c_{01} D_{01}} \right) \dot{\rho}_k +$$

$$+ \frac{8\pi RT \rho_k^5}{15} \left[\left(\frac{1}{c_{ki} D_{ki}} + \frac{5}{3 c_{0i} D_{0i}} \right) \dot{c}_{ki} + \sum_{j=2}^{s} \left(\frac{1}{c_{k1} D_{k1}} + \frac{5}{3 c_{01} D_{01}} \right) \dot{c}_{kj} \right],$$

$$(k=1,\ldots,m,\ i=2,\ldots,s), \tag{8.15}$$

$$\frac{\partial Q}{\partial \dot{c}_{ki}} = \frac{8\pi RT \rho_k^4}{3} \left(\frac{c_{ki} - c_{0i}}{c_{0i} D_{0i}} - \frac{c_{ks+1} - c_{0s+1}}{c_{0s+1} D_{0s+1}} \right) \dot{\rho}_k +$$

$$+ \frac{8\pi RT \rho_k^5}{15} \left[\left(\frac{1}{c_{ki} D_{ki}} + \frac{5}{3 c_{0i} D_{0i}} \right) \dot{c}_{ki} + \sum_{j=s+2}^{n} \left(\frac{1}{c_{ks+1} D_{ks+1}} + \frac{5}{3 c_{0s+1} D_{0s+1}} \right) \dot{c}_{kj} \right],$$

$$(k=1,\ldots,m,\ i=s+2,\ldots,n). \tag{8.16}$$

8.5 Evolution Equations

The evolution equations are found by application of the extremal thermodynamic principle; they have the form:

$$\frac{\partial G}{\partial \rho_k} = -\frac{1}{2}\frac{\partial Q}{\partial \dot{\rho}_k},\ (k=1,\ldots,m) \tag{8.17}$$

and

$$\frac{\partial G}{\partial c_{ki}} = -\frac{1}{2}\frac{\partial Q}{\partial \dot{c}_{ki}}\ (k=1,\ldots,m,\ i=2,\ldots,s \text{ or } i=s+2,\ldots,n). \tag{8.18}$$

The respective partial derivatives are given by Eqs. (8.5) to (8.7) and (8.14) to (8.16). For each precipitate with a given k one can make a substitution: $w_1 \equiv \dot{\rho}_k$, $w_i \equiv \dot{c}_{ki}$ $(i=2,\ldots,s)$ and $w_{i-1} \equiv \dot{c}_{ki}$ $(i=s+2,\ldots,n)$ and the evolution equations take the form:

$$\sum_{j=1}^{n-1} B_{ij} w_j = F_i\ ,\ (i=1,\ldots,n\text{-}1). \tag{8.19}$$

Equations (8.19) represent a set of linear equations of the dimension n-1. The quantities B_{ij} and F_i are given by comparison of Eqs. (8.5) to (8.7) and (8.14) to (8.19). B_{ij} represents a symmetric positive definite matrix of generalized kinetic coefficients and F_i represents a vector of generalized driving forces.

The total number of precipitates can change due to nucleation of new precipitates with the rate $\dot{m}_{nuc}$ and by dissolution of instable precipitates given by the rate $\dot{m}_{diss}$. The evolution is given by:

$$\dot{m} = \dot{m}_{nuc} - \dot{m}_{diss}\ , \tag{8.20}$$

where $\dot{m}_{nuc}$ is supplied by a proper nucleation model and $\dot{m}_{diss}$ is supplied by the number of shrinking precipitates the radius of which drops under a certain limit during a time unit. After each time integration step the newly nucleated precipitates are added to the system and the dissolved ones are eliminated.

8.6 Results of Simulations

The model was implemented in the MatCalc computer software by Dr. Kozeschnik. The software has direct access to existing thermodynamic databases which offer the chemical potentials and diffusion coefficients of all components.

First of all it is necessary to compare the model with existing detailed models in a simple example. That is why the growth kinetics of a cementite precipitate in a ferrite matrix was calculated by means of finite elements and by the present model for the same initial conditions. The growth kinetics for 700°C is compared in Fig. 8.1.

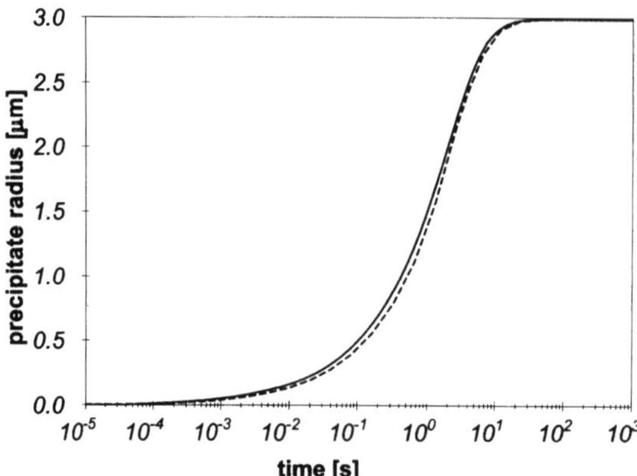

Figure. 8.1. Development of the cementite precipitate radius with time. Solid line: present model; dashed line: discrete model.

From the figure it can be seen that for many applications the difference in kinetics is acceptable and within the usual errors of determination of diffusional coefficients.

In the next example we demonstrate the model in its whole complexity. We assume continuous nucleation, growth and coarsening of precipitate phases MC, $M_{23}C_6$, M_7C_3, M_3C (cementite), M_2C and M_6C in the Fe-Mn-Si-Cr-Mo-Ti-C system with 0.18% C, 10% Cr, 0.5% Mn, 0.3% Si, 1.5% Mo, 0.5% Ti in wt%. The radii of precipitates of each precipitate phase are represented by 100 size classes. Two simulations, A and B, are performed differing by interface energy densities γ_k and by nucleation sites. The parameters are summarized in Tabs. 8.1. and 8.2.

Table 8.1. Interfacial energies and nucleation sites used in the simulation A

	γ *[J/m²]*	*Nucleation sites*
MC	0.35	bulk
M_3C	0.03	dislocations
M_7C_3	0.17	(sub) grain boundaries
$M_{23}C_6$	0.16	(sub) grain boundaries
M_2C	0.24	bulk
M_6C	0.20	(sub) grain boundaries

Table 8.2. Interfacial energies and nucleation sites used in the simulation B

	γ *[J/m²]*	*Nucleation sites*
MC	0.08	bulk
M_3C	0.10	dislocations
M_7C_3	0.12	(sub) grain boundaries
$M_{23}C_6$	0.14	(sub) grain boundaries
M_2C	0.05	dislocations
M_6C	0.25	(sub) grain boundaries

The evolution of the system for both sets of parameters A and B is depicted in Fig. 8.2. The precipitation sequences differ dramatically due to different nucleation conditions given by parameter sets A and B. On the other hand, the final precipitate structure is practically the same given by stable MC and $M_{23}C_6$ carbides. The chemical composition of all carbides in early stage of precipitation is summarized in Tab. 8.3. As can be seen, the carbides MC and $M_{23}C_6$ involve different metallic elements. For this reason, they are not in a concurrent position and they represent a stable coexistence of carbides.

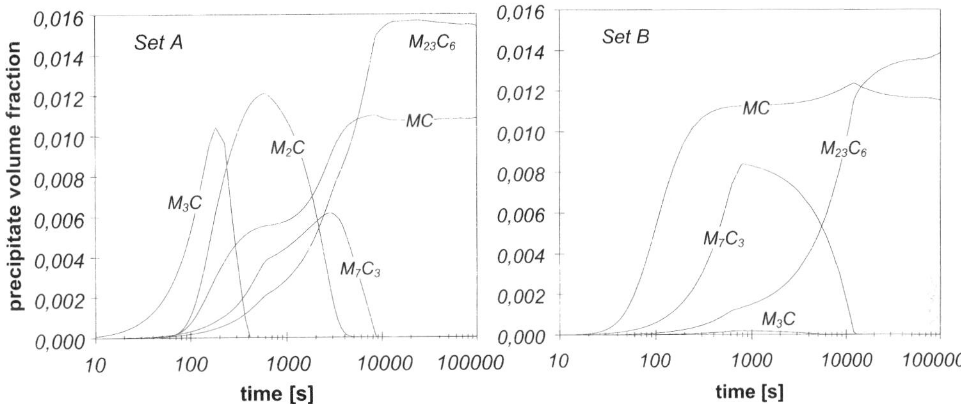

Figure 8.2. Precipitation sequences in a Fe-Mn-Si-Cr-Mo-Ti-C system at 600 °C

Table 8.3. Mean chemical composition (mole fraction) of the particles at t=100s from simulation with input data set A.

	C	Cr	Mn	Mo	Si	Ti	Fe
MC	0.479	0.0036	<<	0.0003	<<	0.5171	<<
M_3C	0.250	0.0896	0.0047	0.0077	<<	-	0.6479
M_7C_3	0.300	0.6600	0.0048	0.0050	-	-	0.0299
$M_{23}C_6$	0.207	0.6032	0.0003	0.1003	-	-	0.0896
M_2C	0.333	0.0041	0.0006	0.3881	<<	0.2370	0.0001
M_6C	0.143	<<	-	0.5449	-	-	0.3312

8.7 Concluding Remarks

- Based on the application of the thermodynamic extremal principle a model for the evolution of multi-particle, multi-component, multi-phase systems has been developed. The model is suitable for modelling of precipitation in all nucleation, growth and coarsening stages. The model also includes the evolution of chemical composition of precipitates.
- Software implementation makes use of existing thermodynamic and kinetic databases.

- Due to the complexity of the model and its generality, a wide range of applications of the model to simulations of precipitation process in real steels can be expected.

9 Summary

The aim of this chapter is to demonstrate the wide applicability of the thermodynamic extremal principle. The classical treatment of irreversible processes in linear thermodynamics represents a rather complicated system of relations amongst thermodynamic quantities (see e.g. Callen, 1960, Groot and Mazur, 1962 or Prigogine, 1967). The relations are given by phenomenological equations expressing linear dependences of local kinetic variables of the system (e.g. fluxes) on the local state parameters of the system (thermodynamic forces).

The determination of the time evolution of the state of a thermodynamic system seems to be one of the major problems in material science. For this purpose the phenomenological equations complemented by conservation laws and respective boundary conditions can be used to provide the detailed description of the system kinetics.

Most thermodynamic systems in material science are distinguished by a considerable degree of complexity. However, only a limited number of characteristic parameters can usually be used for the description of the system and only a limited number of characteristic parameters are of our interest. This limitation implies a lower or higher idealization of the system. The number of state parameters chosen for the system description depends on the requirements regarding the accuracy of the description and on the complexity of the system. Then the standard task is to determine the equations for the time evolution of these parameters serving as equations of motion of the system.

The classical way of determining the kinetics of the system is represented by the solution of phenomenological equations with proper boundary and contact conditions. By setting some assumptions on system geometry and boundary and contact conditions a certain degree of idealization of the system can be achieved. The time evolution of the chosen state parameters can then be obtained from the solutions of the phenomenological equations. In general the solutions of the phenomenological equations bear more information than is necessary for the determination of the time evolution of the state parameters of our interest. Often the necessary idealization must drastically reduce the complexity of the problem and the evolution of the state parameters of our interest cannot be obtained at all. This problem can often be overcome by applying the thermodynamic extremal principle, a useful tool for deriving in a direct way the evolution equations of state parameters. Moreover, the constraints of the local or global character can easily be taken into consideration. Then by a proper choice of the number and of the physical meaning of the state parameters one can find a compromise between the requirements on accuracy and complexity of the system description.

The extremal formulation of a thermodynamic problem was presented first by Onsager (1931), who showed that the equations for heat conduction in a closed system can be derived from the requirement of the maximum of a functional having a close relation to the entropy production in the system. During the next sixty years only few works dealt with the formulation of the extremal principle describing the thermodynamics of irreversible processes, see e.g. Ziegler (1961,1963) or Bazarov et al. (1989). These authors showed that the extremal principle concentrates all substantial knowledge on thermodynamics of irreversible processes, but they did not show any advantages of the extremal formulation for the solution of the problems. That is probably why the formulation of the extremal principle did not appear in classical textbooks of thermodynamics of irreversible processes (see e.g. Callen, 1960, Groot and Mazur, 1962 or Prigogine, 1967). However, recent works (Fischer et al., 2003, Svoboda and Turek, 1991 or Svoboda et al., 2002, 2004,2004a) have demonstrated that the extremal principle seems to be a handy tool for the solution of problems of thermodynamics of irreversible processes, which are probably unsolvable or solvable with great complications in the classical way.

Acknowledgement

The authors are indebted to Materials Center Leoben for financial support within the scope of the strategic projects SP9 and SP11. The work on the project, which was also supported by Grant Agency AS CR (project No. K-1010104) and by Kontakt – Austrian-Czech Scientific Cooperation (project No. A-15/03), is performed within the framework of the Research Plan of AS CR (No. Z-204 1904).

References

Bazarov I. P., Gevorkyan E. V., Nikolaev P. N. (1989) *Non-Equilibrium Thermodynamics and Physical Kinetics*. Moscow University Press, Moscow (in Russian).

Callen H. B. (1960). *Thermodynamics*. Wiley, New York.

Fischer F. D., Simha N. K., Svoboda J. (2003). Kinetics of diffusional phase transformation in multicomponent elastic-plastic materials. *ASME J Eng Mat & Technology* 125:266-276.

Fischer F. D., Svoboda J., Fratzl P. A thermodynamic approach to grain growth and coarsening. *Philos. Mag.* 83:1075-1093.

Glicksman M. E. (2000). *Diffusion in Solids*. New York: John Wiley & Sons, Inc.

Groot S. R., Mazur P. (1962). *Non-Equilibrium Thermodynamics*. North-Holland, Amsterdam.

Hartmann M. A., Weinkammer R., Fratzl P., Svoboda J., Fischer F. D. (2003). Onsager's coefficients and diffusion laws – a study based on Monte-Carlo simulations, to be published.

Hillert M. (1965). On the theory of normal and abnormal grain growth. *Acta metall.* 13:227-238.

Kozeschnik E., Svoboda J., Fratzl P., Fischer F. D. (2004). Modelling of kinetics in multi-component multi-phase multi-particle systems II. – Numerical solution and application, *Mater. Sci. Eng. A*, in press.

Krielaart G. (1995). *Primary ferrite formation from supersaturated austenite*, Ph. D. thesis, TU-Delft.

Lee B. J., Oh K. H. (1996). Numerical treatment of the moving interface in diffusional reactions. *Z. Metallkd.* 87:195-204.

Lidiard A. B. (1986). A note on Manning's relations for concentrated multicomponent alloys. *Acta metall.* 34:1487-1490.

Lücke, K., Stüwe, H. P. (1971). On the theory of impurity controlled grain boundary motion. *Acta metall.* 19:1087-1099.

Manning J. R. (1971). Correlation factors for diffusion in nondilute alloys *Phys. Rew.* B4:1111-1121.

Moleko L. K., Allnatt A. R., Allnatt E. L. (1989). A self-consistent theory of matter transport in a random lattice gas and some simulation results. *Philos. Mag.* A59:141-160.

Onsager L. (1931). Reciprocal relations in irreversible processes. I. *Phys. Rev.* 37:405-426.

Prigogine I. (1967). *Introduction to Thermodynamics of Irreversible Processes*, 3rd edn. Interscience, New York.

Shewmon P.. (1989). *Diffusion in Solids.* The Minerals, Metals & Materials Society, Warrendale.

Svoboda J., Turek I. (1991). On diffusion-controlled evolution of closed solid-state thermodynamic systems at constant temperature and pressure. *Philos. Mag.* B64:749-759.

Svoboda J., Fischer F. D., Fratzl P., Kroupa A. (2002). Diffusion in multi-component systems with no or dense sources and sinks for vacancies. *Acta mater.* 50:1369-1381.

Svoboda J., Fischer F. D., Gamsjäger E. (2002a). Influence of solute segregation and drag on properties of migrating interfaces. *Acta mater.* 50:967-977.

Svoboda J., Gamsjäger E., Fischer F. D., Fratzl P. (2004). Application of the thermodynamic extremal principle to the diffusional phase transformations, *Acta mater.* 52:959-967.

Svoboda J., Fischer F. D., Fratzl P., Kozeschnik E. (2004a). Modelling of kinetics in multi-component multi-phase multi-particle systems I. – Theory. *Mater. Sci. Eng. A*, in press.

Ziegler H. (1961). Zwei Extremalprinzien der irreversiblen Thermodynamik. *Ing.-Arch.* 30:410-416.

Ziegler H. (1963). in I.N. Sneddon and R. Hill eds., *Progress in Solid Mechanics,* Vol. 4, North-Holland, Amsterdam. 93-193.

Thermodynamics and Kinetics of Phase and Twin Boundaries

Franz Dieter Fischer[1 *)] and Narendra K. Simha[2]

[1] Institute of Mechanics, Montanuniversität Leoben, Leoben, Austria and
Erich Schmid Institute for Materials Science, Austrian Academy of Sciences, Leoben, Austria
[2] Department of Mechanical Engineering, University of Miami, Coral Gables, USA

Abstract. This chapter describes a framework for studying the propagation of singular interfaces such as phase and twin boundaries in multicomponent bodies. The balance equations for mass, energy and entropy at an interface between two phases are examined in detail. The chemical and mechanical contributions to the total thermodynamic driving force are identified for both ideal and real interfaces. Transformation kinetics, based on these thermodynamic driving forces, is formulated for the martensitic and diffusional transformations. The framework is illustrated by two examples, the first estimates the thickness of a deformation twin, and the second describes the growth of an allotriomorphic ferrite film at the expense of the austenitic phase.

List of Notations

Note: Symbols, which are used only locally in the text and which are explained there, are omitted in this list.

c_i concentration of the i-th component, $[c_i] = \text{mo}\ell/\text{m}^3$

$\underline{\underline{\underline{\underline{C}}}}$ 4th order elasticity tensor

d_{chem} dissipation due to the chemical process

D_{tot} the total energy dissipated by interface motion in time period $[t_3, t_f]$

e_{chem} specific internal chemical energy, $[e_{\text{chem}}] = \text{J}/\text{m}^3$

$\underline{e}_\Gamma$ surface growth term due to energy flow of component i, $[\underline{e}_\Gamma] = \text{J}/(\text{m}^2\text{s})$

E Young's modulus

*) F.D. Fischer expresses his heart-felt thanks for the intensive discussions, the help and the contributions from (in alphabetic order) Prof. P. Fratzl, former Professor of Metal Physics and Director of the Erich Schmid Institute until September 2003, now with Max-Planck-Institut für Kolloid- und Grenzflächenforschung, Potsdam, Germany; Prof. V. Levitas, Department of Mechanical Engineering, Texas Technical University, Lubbock, USA; and Dr. J. Svoboda, Institute of Physics of Materials, Academy of Sciences of the Czech Republic, Brno. Furthermore, the fundings of Christian Dopper Laboratory of Functionally Oriented Materials Design, Leoben/Vienna, Austria, Materials Centre Leoben, and the FWF (Project number: P15251) are kindly appreciated.

F thermodynamic force, $[F] = Nm/m^3$, labels "chem", "in,chem", "va,chem", "mech", "c"
g_{chem} specific chemical energy, $[g_{chem}] = J/mo\ell$
h thickness of a twin band
$\underline{i}_\Gamma$ surface growth term due to mass flow of component i, $[\underline{i}_\Gamma] = mo\ell/(m^2 s)$
$\underline{\underline{I}}$ unity tensor
$\underline{j}_i$ mass flux of the i-th component, $[\underline{j}_i] = mo\ell/(m^2 s)$; label "v" for vacancy
J_i total inflow of chemical energy into a region S
K strain energy factor for a twin band, $[K] = N/m^2$
ℓ length of a twin band
$\underline{m}$ normal vector to the interface Γ
M mobility, M_0 its reference value
$\underline{\underline{\underline{\underline{M}}}}$ compliance tensor
$\underline{n}$ normal vector, labels "S", "R"
N_i number of moles of the i-th component
$\underline{\underline{p}}$ twin eigenstrain orientation matrix
$\underline{\dot{q}}$ heat flux vector
$\dot{Q}$ total inflow of heat into region S
Q_M activation energy for interface motion
R body with surface ∂R
R_f radius of a ferritic spherical inclusion, its increment ΔR_f
R_g gas constant, 8.3144 $J/(mo\ell\ K)$
R_{ref} radius of a cylinder or a sphere
s_ℓ sliding component along Γ, $\overset{\circ}{s}_\ell$ its specific time rate
S body inside R with surface $\partial S_1 + \partial S_2$; interface Γ splits S into two parts $S_1 + S_2 = S$, its increment ΔS
t time, label "s" for start, label "f" for finish, symbol "•" for a total time derivative
$\underline{t}$ traction vector on ∂R, $[\underline{t}] = N/m^2$
$\underline{t}_\Gamma$ unit tangent vector to Γ
$\underline{u}$ displacement vector, $\underline{\dot{u}}$ velocity vector
U' specific elastic strain energy $[U'] = Nm/m^3$
$\underline{v}$ interface velocity vector, $[\underline{v}]$=m/s
V_i partial molar volume, $[V_i] = m^3/mo\ell$, labels "su" and "in" instead of "i"
V_T volume of the twin band

W	work, label "ex" for external work, "el" for elastic work, "pl" for plastic work, Γ for total interface (surface) energy, ΔW_{ex}, $\Delta W_{e\ell}$, $\Delta W_{p\ell}$ their increments, $\dot{W}$ the rate of W, W' its specific value, label " τ " for work increments only due to $\underline{\underline{\tau}}$
X_i	mole fraction of the i-th component, $[X_i]=1$, labels "eq", "int"
α	factor 1 or $(1-\nu)$
γ_T	twinning shear eigenstrain
Γ	moving interface inside R intersecting $S, \partial S$
δ	transformation volume strain
ΔS	volume change of a region from S_1 at t_s to S_1 at t_f
$\underline{\underline{\varepsilon}}$	strain tensor, labels "el", "pl", "*"
η	entropy density, labels "b", "Γ", $[\eta]=J/(m^3K)$
θ	temperature, $[\theta]=K$
κ	curvature of interface Γ
κ_p	plastic modulus
μ	shear modulus, $[\mu]=N/m^2$
μ_i	chemical potential of the i-th component, $[\mu_i]=J/mo\ell$
ν	Poisson's ratio
σ	surface energy, $[\sigma]=J/m^2$
σ_H	mean value of normal stress components
$\underline{\sigma}^{int}$	stress state due to internal sources (e.g. eigenstrain)
σ_y	yield stress of austenite
$\underline{\underline{\Sigma}}$	stress state corresponding to $\underline{t}$ on ∂R
τ	shear stress component, label "$_T$" for resolved shear stress, label "$_{cr,T}$" for critical resolved shear stress
$\underline{\underline{\tau}}$	additional internal stress state due the development of ΔS
φ_{chem}	specific free chemical energy, $[\varphi_{chem}]=J/m^3$
ω	intrinsic normal velocity of the interface Γ, $[\omega]=m/s$

1 Introduction

Think of a polycrystalline material e.g. steel consisting of iron as the main component and several other substitutional additional metals as nickel, manganese etc. as well as other interstitial elements like carbon. We look at the microstructure of the material (on the µm-scale) and assume that we observe two phases, which are separated by an interface. Think again of steel and let us distinguish between ferrite or martensite with a body-centered lattice (a bcc-lattice) and austenite with a face-centered lattice (an fcc-lattice). The atoms are arranged in different ways in the two lattices, suggesting the existence of two different phases. One can easily imagine that due to the different arrangement of the atoms, the atomic binding energy, expressed by the internal energy of the material, differs between the two phases. Therefore, we contemplate two phases separated by an interface. This topic is, by no means, an antiquated one as the recent review by Thornton, Ågren and Voorhees (2003) shows.

The goal of this work is to explain within the context of continuum mechanics and thermodynamics how such an interface Γ moves through the body R or, in other words, how the new, product phase develops from the old, parent phase. We consider both an ideal interface and a real interface. The ideal interface is a sharp interface in the mathematical sense. It does not have any internal structure and cannot carry any material with it and does not act as a source or sink of vacancies. Contrarily, a real interface is regarded as a layer of the thickness of some atomic distances, supposed to be an ideal source and sink of vacancies. Atoms of all components can be "caught" or "ejected" by the interface and carried with it – here we refer to the so-called solute drag, Svoboda, Fischer and Gamsjäger (2002) and the chapter by Svoboda in this book.

2 Problem Definition

2.1 The Actual Configuration

We consider a body in the actual or current configuration R with the surface ∂R and its normal $\underline{n}_R$. The body contains an elastic-plastic, heat conducting material consisting of n components. We observe a motion of the interface Γ and simplify our consideration to a body R, which originally, at the time $t = 0$, consists only of the parent phase with the label "2". At the time t it is separated by Γ into a product phase "1" (which develops at the expense of the parent phase "2") and the remaining parent phase "2", see Fig. 1.

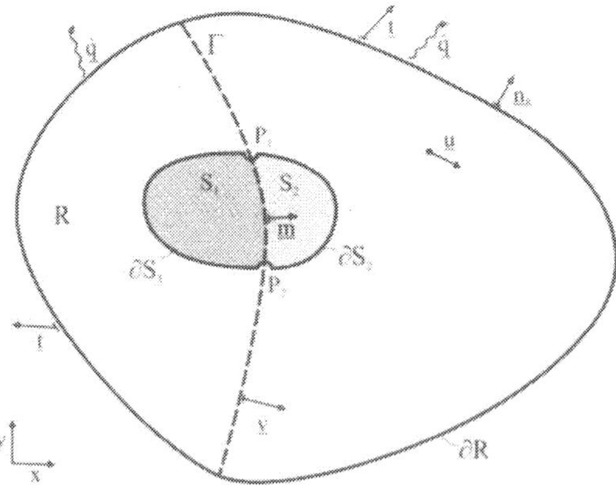

Figure 1. The body R with a subregion S=S1+S2 crossed by an interface Γ moving with the velocity $\underline{v}$.

The body is loaded by external tractions $\underline{t}$, $[t] = N/m^2$, on its surface ∂R. For the sake of simplicity body forces are ignored. The tractions $\underline{t}$ lead to a deformation state described by the displacement vector $\underline{u}$. The interface Γ moves with a velocity [$\underline{v}$]; note that the interface moves with respect to the material points and hence $\underline{v}$ is not a material velocity. We can then define an intrinsic velocity ω, $\omega = (\underline{v} - \dot{\underline{u}}) \cdot \underline{m}$, on both sides of the interface with $\underline{m}$ being the normal to the interface and $\dot{\underline{u}}$ being the material velocity as the time derivative of $\underline{u}$. Generally the intrinsic velocity ω differs on both sides of Γ.

2.2 Mathematical Prerequisites

We look at a region S inside R with the surface ∂S and an outer normal $\underline{n}_s$, moving with the material velocity $\dot{\underline{u}}$. The region S is crossed by the interface Γ, separating the material into two phases with different compositions and lattices in regions S_1 and S_2 ($S_1+S_2=S$). Similarly, the moving interface Γ divides the surface ∂S into two parts, ∂S_1 and ∂S_2. If we consider the volumes S_1 and S_2, we denote the side of interface Γ belonging to S_1 and S_2, as Γ_1 and Γ_2, respectively.

The limit of a quantity a as it approaches the interface from S_1 is a_1 and from S_2 it is a_2. The jump at the interface is denoted by a double bracket e.g. $[\![a]\!] = a_2 - a_1$; note a_2 is the limit from the old or parent phase while a_1 is the corresponding value from the new product phase. We use the symbol $\langle a \rangle$ for the average along Γ, $\langle a \rangle = (a_1 + a_2)/2$. We note the important identities

$$[\![a \cdot b]\!] = \langle a \rangle \cdot [\![b]\!] + [\![a]\!] \cdot \langle b \rangle, \tag{1}$$

$$\langle a \rangle \langle b \rangle - \langle a\, b \rangle = -[\![a]\!] \cdot [\![b]\!]/4. \tag{2}$$

We use the following two relations, which are required for regions containing a propagating singular interface (see Simha and Bhattacharya, 1998): The first one is the Divergence Theorem for a vector field $\underline{b}$ specifically adapted for the actual configuration

$$\int_{\partial S_1 + \partial S_2} \underline{b} \cdot \underline{n}\, dA = \int_{S_1 + S_2} \operatorname{div} \underline{b}\, dV + \int_{\Gamma \cap S} [\![\underline{b} \cdot \underline{m}]\!]\, dA, \tag{3}$$

where $\Gamma \cap S$ denotes the part of the interface Γ which lies inside of S. The second one is the Transport Theorem yielding for a specific scalar quantity ψ and its total value Ψ, $\Psi = \int_V \psi\, dV$,

$$\frac{d}{dt}\Psi = \dot{\Psi} = \int\limits_{S_1+S_2} \left(\frac{\partial \psi}{\partial t} + \operatorname{div}(\psi \underline{\dot{u}}) \right) dV - \int\limits_{\Gamma \cap S} [[\psi(\underline{v} - \underline{\dot{u}}) \cdot \underline{m}]] dA . \tag{4}$$

If the volume S_1 and S_2 collapse on to the interface Γ, then only the surface integrals in (3) and (4) will survive. Relation (4) can be further simplified for incompressible materials or volume preserving deformations ($\operatorname{div}\underline{\dot{u}} = 0$), with the jump $[[\omega]]$ is set to zero to obtain

$$\frac{d}{dt}\Psi = \dot{\Psi} = \int\limits_{S_1+S_2} \dot{\psi}\, dV - \int\limits_{\Gamma \cap S} [[\psi]] \omega dA . \tag{5}$$

Here, the Lagrangian or material time derivative $\dot{\psi} = \partial\psi/\partial t + \underline{\dot{u}} \cdot \operatorname{grad}\psi$ has been introduced. This relation was used in the derivations outlined in Fischer, Simha and Svoboda (2003).

2.3 Kinematics

We work within a small strain setting, so the strain tensor $\underline{\underline{\varepsilon}}$, $\varepsilon_{ij} = \frac{1}{2}\left(\partial u_i/\partial x_j + \partial u_j/\partial x_i\right)$, x_1, x_2, x_3 being the Cartesian coordinates, can be decomposed into

$$\underline{\underline{\varepsilon}} = \underline{\underline{\varepsilon}}^{el} + \underline{\underline{\varepsilon}}^{pl} + \underline{\underline{\varepsilon}}^{*} . \tag{6}$$

$\underline{\underline{\varepsilon}}^{el}$ is the elastic contribution to $\underline{\underline{\varepsilon}}$, described by Hooke's law, see below. $\underline{\underline{\varepsilon}}^{pl}$ is the plastic contribution to $\underline{\underline{\varepsilon}}$; here we refer to the literature. $\underline{\underline{\varepsilon}}^{*}$ is the so-called transformation eigenstrain which includes in our case also the difference of the thermal strains. The eigenstrain $\underline{\underline{\varepsilon}}^{*}$ describes the deformation state of a stress free body after its phase transformation in relation to its parent phase. Typically, diffusive transformations are accompanied by a volume change δ yielding $\underline{\underline{\varepsilon}}^{*} = \frac{\delta}{3}\underline{\underline{I}}$, $\underline{\underline{I}}$ unity tensor. The martensitic transformation is accompanied by a significant shear strain and a volume change (which is nearly zero in the case of shape memory alloys), see e.g. the chapter by Berveiller et al. in this book. Twinning can also be counted as such a type of transformation e.g. with a shear strain $\gamma_T = 1/\sqrt{2}$ in the case of Ti-Al and no volume change. Of course, in the case of twinning the lattice is not changed, but significantly rotated, for details see Christian and Mahajan (1995). The total strain $\underline{\underline{\varepsilon}}$ is assumed to be continuous within the individual phases. However, at the interface Γ a jump of $\underline{\underline{\varepsilon}}$ appears. Note, that for a coherent interface the jump $[[\underline{u}]]$ is zero with the consequence

$$[[\underline{\dot{u}}]] = -[[\operatorname{grad}\underline{u}]] \cdot \underline{v} , \tag{7}$$

$$[\![\mathrm{grad}\underline{u}]\!] = [\![\mathrm{grad}\underline{u}]\!] \cdot (\underline{m} \otimes \underline{m}).$$

The first relation shows that the jumps in the material velocity and the displacement gradients are not independent; instead they are related through the interface velocity $\underline{v}$ (Simha and Bhattacharya, 1998). The second is the usual Hadamard compatibility condition.

Generally, we consider a coherent interface. However, in the case of a diffusive transformation process at higher temperatures, the interface may be a "sliding one" allowing an opposite motion on both sides of the interface; then only the relation $[\![\underline{u} \cdot \underline{m}]\!] = 0$ is valid, for details see and Fischer, Simha and Svoboda (2003).

2.4 Mechanical Energy Terms

The total stress state $\underline{\underline{\sigma}}$ at a certain time $t = t_s$ is represented by the sum $\underline{\underline{\sigma}} = \underline{\underline{\Sigma}} + \underline{\underline{\sigma}}^{int}$. The contribution $\underline{\underline{\Sigma}}$ corresponds to the tractions $\underline{t}$ as $\underline{t} = \underline{\underline{\Sigma}} \cdot \underline{n}_R$. The internal stress state $\underline{\sigma}^{int}$ is caused by the evolution of the microstructure, e.g. by the accommodation of the transformation strain $\underline{\underline{\varepsilon}}^*$, see e.g. Fischer and Oberaigner (2000), (2001). From the time t_s on we keep $\underline{t}$ and $\underline{\underline{\Sigma}}$ constant and observe the evolution of the product phase until t_f, whereby $\underline{\sigma}^{int}$ increases by an additional stress tensor $\underline{\underline{\tau}}(t)$, $t_s \le t \le t_f$, yielding $\underline{\underline{\sigma}} = \underline{\underline{\Sigma}} + \underline{\underline{\sigma}}^{int}(t_s) + \underline{\underline{\tau}}(t)$.

The stress state $\underline{\sigma}$ is related to $\underline{\underline{\varepsilon}}^{e\ell}$ by Hooke's law with $\underline{\underline{\underline{C}}} = C_{ijkl}$ being the fourth order elasticity tensor, $\underline{\underline{\sigma}} = \underline{\underline{\underline{C}}} \cdot \underline{\underline{\varepsilon}}^{e\ell}$, or $\underline{\underline{\underline{M}}} = M_{ijkl}$ being the fourth order compliance tensor, $\underline{\underline{\varepsilon}}^{e\ell} = \underline{\underline{\underline{M}}} \cdot \underline{\underline{\sigma}}$. The specific elastic strain energy U' and its rate $\dot{U}'$ follow as

$$U' = \frac{1}{2} \underline{\underline{\sigma}} : \underline{\underline{\varepsilon}}^{e\ell}, \quad \dot{U}' = \underline{\underline{\sigma}} : \dot{\underline{\underline{\varepsilon}}}^{e\ell}. \tag{8}$$

The accommodation may also produce plastic straining yielding a specific plastic work rate

$$\dot{W}'_{p\ell} = \underline{\underline{\sigma}} : \dot{\underline{\underline{\varepsilon}}}^{p\ell}. \tag{9}$$

2.5 Chemical Energy Terms

We assume that R is a multicomponent body consisting of n components. For the i-th component let X_i denote the mole fraction, V_i the partial molar volume and c_i the concentration. The mole fraction is dimensionless ($[X_i] = 1$),the dimension for the partial molar volume is $[V_i] = m^3/mo\ell$ and for the concentration it is $[c_i] = mo\ell/m^3$. Let the molar volume of the multicomponent material be denoted by V_m. If N_i is the number of moles of the i-th component in a region $S \subset R$, then the relations between these quantities are

$$V_m = \sum_{i=1}^{n} X_i V_i \,, \quad c_i = X_i / V_m \,, \quad N_i = \int_R (X_i / V_m) dV \,, \quad \sum_{i=1}^{n} X_i = 1 \,. \tag{10}$$

Both substitutional and interstitial components are considered. The first m components, $i = 1, \ldots m$, are assumed to be interstitial components, the rest, $i = m+1 \ldots n$, are substitutional components. As already discussed the individual atoms are situated in a lattice; the substitutional atoms are located only in the lattice positions whereas the interstitial atoms can also be located in-between the lattice positions. Some lattice positions are not occupied and these are denoted as vacancies.

Each component i may move in the lattice and this is described by a mass flux $\underline{j}_i$, $[\underline{j}_i] = \text{mo}\ell/(\text{m}^2\text{s})$. The interstitials may squeeze through the lattice, but the substitutionals can only move by exchanging their positions with vacancies. Therefore, we introduce in addition a flux of the vacancies $\underline{j}_v$, see e.g. Svoboda, Fischer, Fratzl et al. (2002). The simplest balance equation is that due to the fluxes of the substitutional components, namely

$$\sum_{i=m+1}^{n} \underline{j}_i = -\underline{j}_v \,. \tag{11}$$

The chemical potential of the i-th component is $\mu_i, [\mu_i] = \text{J/mo}\ell$. μ_i is generally a function of the mole fractions of all components and the temperature θ, $\mu_i = \mu_i\,(X_1, \ldots X_n; \theta\,)$. The pressure may also influence μ_i. However, a constant pressure can be assumed in many solid state phase transformations with respect to the chemical potentials. For example, the chemical potentials of the components of both phases during austenite-to-ferrite phase transformation are not significantely altered for a pressure being below the GPa-range. The specific energy associated with the i-th component is the product $X_i\mu_i, [X_i\mu_i] = \text{J/mo}\ell$. The chemical energy of the composition is expressed either per mole of material as g_{chem}, $[g_{chem}] = \text{J/mo}\ell$, or per unit volume as $\varphi_{chem}, [\varphi_{chem}] = \text{J/m}^3$, and these are given by

$$g_{chem} = \sum_{i=1}^{n} X_i \mu_i \,, \quad \varphi_{chem} = g_{chem} / V_m = \sum_{i=1}^{n} c_i \mu_i \,. \tag{12}$$

Lastly, we introduce the specific internal energy e_{chem}, $[e_{chem}]=\text{J/m}^3$. This describes the energy on the continuum level and depends on the composition and the entropy η, $e_{chem}=e_{chem}(X_1, \ldots X_n; \eta\,)$, see also the next section.

2.6 Entropy Terms

Let the heat influx vector to the body be $-\underline{\dot{q}}$, $[\underline{\dot{q}}] = J/(m^2 s)$; note that $\underline{\dot{q}} \cdot \underline{n}_R$ is an outflow of heat. Finally, we define the bulk entropy density η, $[\eta] = J/(m^3 K)$ and the bulk entropy production rate $\dot{\eta}_b$, $[\dot{\eta}_b] = J/(m^3 Ks)$. At the interface we may have a different entropy production rate $\dot{\eta}_\Gamma$, $[\dot{\eta}_\Gamma] = J/(m^2 Ks)$, due to additional dissipation mechanisms like lattice rearrangement. Neither a mass nor an energy influx from or outflux to the outside is assumed at the interface.

3 Dissipation due to Multicomponent Diffusion

3.1 General Comments

We split the dissipation into two contributions, the first one due to the chemical process (multicomponent diffusion and a moving phase interface) and the second one due to the mechanical process (deformation process due to the tractions $\underline{t}$ on ∂R and the accommodation process due to the transformation strain). We assume that both dissipation contributions are non-negative, which is a stronger condition than requiring that the total dissipation be non-negative. This assumption gives the impression that two independent processes are considered. However, this is not the case, since the chemical and mechanical processes are coupled by the material velocity $\underline{\dot{u}}$ which contributes to the intrinsic normal velocity ω. The material velocity $\underline{\dot{u}}$ is a consequence of the fluxes of the components (the "chemical" deformation) as well as the deformation process controlled externally by the tractions $\underline{t}$.

3.2 The Ideal Interface

3.2.1 The Mass Balance

In the context of multicomponent diffusion the balance of mass can be written as follows. For each component i and for each subregion $S \subset R$ the balance of mass requires that

$$\frac{d}{dt}\int_S c_i dV = -\int_{\partial S_1 + \partial S_2} \underline{j}_i \cdot \underline{n}_S \, dA \, . \tag{13}$$

In other words, the rate of change of component i inside the region S must be equal to the flux of the same component over the boundary. We apply the transport theorem (4) to the integral on the left to obtain,

$$\frac{d}{dt}\int_{S_1+S_2} c_i dV = \int_{S_1+S_2} \left(\frac{\partial c_i}{\partial t} + \mathrm{div}(c_i \underline{\dot{u}})\right) dV - \int_{\Gamma \cap S} [[c_i \omega]] \, dA \, . \tag{14}$$

Applying the divergence theorem (3) to the right hand side of (13) leads to

$$-\int_{\partial S_1+\partial S_2} \underline{j}_i \cdot \underline{n} dA = -\int_{S_1+S_2} \operatorname{div} \underline{j}_i \, dV - \int_{\Gamma \cap S} [\![\underline{j}_i \cdot \underline{m}]\!] dA . \qquad (15)$$

Equating (14) and (15) yields, since the mass balance is valid for any subregion $S \subset R$,

$$\frac{\partial c_i}{\partial t} = -\operatorname{div}\left(\underline{j}_i + c_i \underline{\dot{u}}\right), \qquad (16)$$

$$[\![c_i \omega - \underline{j}_i \cdot \underline{m}]\!] = 0 . \qquad (17)$$

Relation (16) is the mass conservation relation in the bulk and is known as Fick's first law in the diffusion theory. What is often ignored is the fact that the total flux consists of two contributions, the diffusive flux $\underline{j}_i$ and the convective flux $c_i \underline{\dot{u}}$. Equation (17) is the mass balance at an ideal interface and can be found in the literature, see e.g. Lehner (1990), Gurtin (1993), Cermelli and Gurtin (1994), Morland and Gray (1995).

3.2.2 The Energy Balance

We consider a situation where there is mass exchange by diffusion in the presence of thermal fields. Then the balance of energy can be written as follows. For each subregion $S \subset R$

$$\frac{d}{dt} \int_S e_{chem} dV = -\int_{\partial S} \underline{\dot{q}} \cdot \underline{n}_S \, dA - \sum_{i=1}^{n} \int_{\partial S} \mu_i \underline{j}_i \cdot \underline{n}_S \, dA . \qquad (18)$$

The inflow of chemical energy due to the flux $-\underline{j}_i \cdot \underline{n}_S$ of the component i at the boundary ∂S can be expressed with the chemical potential μ_i as $-\mu_i \cdot \underline{j}_i \cdot \underline{n}_S$. Thus, the total chemical energy inflow is obtained by summing over all the components and this is the second integral in equation (18). The first integral on the right in (18) is the heat influx into the region S from the rest of the body. This is the first principle of thermodynamics applied to the chemical setting and states that the rate of change of internal energy is equal to the heat as well as the total chemical energy influx.

Applying the divergence theorem (3), to the chemical energy influx and the heat influx in equation (18) leads to

$$-\int_{\partial S_1+\partial S_2} \left(\sum_{i=1}^{n} \mu_i \underline{j}_i + \underline{\dot{q}} \right) \cdot n_s \, dA = -\int_{S_1+S_2} \operatorname{div}\left(\sum_{i=1}^{n} \mu_i \underline{j}_i + \underline{\dot{q}} \right) dV - \int_{\Gamma \cap S} \left[\!\!\left[\sum_{i=1}^{n} \mu_i \underline{j}_i + \underline{\dot{q}} \right]\!\!\right] \cdot \underline{m} \, dA . \qquad (19)$$

On the other hand, the change of the total internal energy can be written by applying the transport theorem (4) as

$$\frac{d}{dt}\int_{S_1+S_2} e_{chem}\,dV = \int_{S_1+S_2}\left(\frac{\partial e_{chem}}{\partial t} + \mathrm{div}\left(e_{chem}\,\dot{\underline{u}}\right)\right)dV - \int_{\Gamma\cap S}[[e_{chem}\omega]]\,dA\,. \tag{20}$$

Consequently, equating (19) and (20), the local forms of the first principle of thermodynamics become

$$\frac{\partial e_{chem}}{\partial t} = -\mathrm{div}\left(e_{chem}\dot{\underline{u}} + \dot{\underline{q}} + \sum_{i=1}^{n}\mu_i \underline{j}_i\right), \tag{21}$$

$$\left[\left[e_{chem}\left(\underline{v} - \dot{\underline{u}}\right) - \dot{\underline{q}} - \sum_{\mu=1}^{n}\mu_i \underline{j}_i\right]\right]\cdot\underline{m} = 0\,. \tag{22}$$

3.2.3 The Entropy Balance

Diffusion is the dissipative mechanism in the bulk. At the interface, energy is dissipated due to the lattice rearrangement, some interface diffusion and the motion of the interface. In addition also solute segregation and drag may result in energy dissipation connected with the moving interface. We refer the reader to the section 5 of this chapter dealing with kinetics. Entropy only satisfies an imbalance relation. We convert this to a balance law, by introducing entropy production terms for the bulk $\dot{\eta}_b$ and for the interface $\dot{\eta}_\Gamma$ as outlined in section 2.6.

The first task is to formulate the balance of entropy. For every subregion $S \subset R$

$$\frac{d}{dt}\int_S \eta_{chem}dV = \int_{\partial S} -\frac{\dot{\underline{q}}\cdot\underline{n}_S}{\theta}\,dA + \int_S \dot{\eta}_b dV + \int_{\Gamma\cap S}\dot{\eta}_\Gamma dA\,. \tag{23}$$

Then we apply the transport theorem (4) to the left side of (23) and the divergence theorem (3) to the first integral on the right side of (23) and obtain for a material point in the bulk

$$\frac{\partial\eta_{chem}}{\partial t} + \mathrm{div}\left(\eta\,\dot{\underline{u}} - \dot{\underline{q}}/\theta\right) = \dot{\eta}_b\,, \tag{24}$$

and at a point on the interface

$$\left[\left[\eta_{chem}\left(\underline{v} - \dot{u}\right) - \dot{\underline{q}}/\theta\right]\right]\cdot\underline{m} = -\dot{\eta}_\Gamma\,. \tag{25}$$

3.2.4 The Dissipation at the Interface

A further reasonable assumption is to consider a continuous temperature across the interface, $[\![\theta]\!] = 0$. Now insert (25) after multiplication with θ into (22), keeping in mind the Legendre transformation $[\![\eta_{chem}(\underline{v} - \dot{u}) - \underline{\dot{q}}/\theta]\!] \cdot \underline{m} = -\dot{\eta}_\Gamma$. This step allows for the elimination of η and $\underline{\dot{q}}$ in (24), yielding finally the dissipation d_{chem}

$$d_{chem} = \left[\!\!\left[\varphi_{chem}\omega - \sum_{i=1}^{n} \mu_i \underline{j}_i \cdot \underline{m} \right]\!\!\right] = \theta\dot{\eta}_\Gamma > 0 . \tag{26}$$

The label "chem" for the dissipation d and the Helmholtz free energy density φ is to keep in mind that we deal with a chemical process accompanied by heat flux. The dissipation relation (26) can be found at several places in the open literature, see e.g. the references for (17) in section 3.2.1 and Gurtin and Voorhees (1993), Fried and Gurtin (1999), Cocks, Gill and Pan (1999), Maugin (1999) and also earlier Grinfield (1991) in a specific form. Here, we have tried to show a consistent derivation applying the two mathematical theorems (3), (4) together with the standard balance laws. The jump conditions are a direct outcome of this procedure. We modify now (26) by inserting (12) in the form of $\varphi_{chem} = \sum_{i=1}^{n} c_i \mu_i$, applying (1) on the jump of φ_{chem} as a sum of products c_i and μ_i, and keeping in mind the mass balance (17), yielding finally

$$d_{chem} = \sum_{i=1}^{n} [\![\mu_i]\!] \left\langle c_i \omega - \underline{j}_i \cdot \underline{m} \right\rangle . \tag{27}$$

The term $\left\langle c_i \omega - \underline{j}_i \cdot \underline{m} \right\rangle$ can be substituted by $c_i \omega - \underline{j}_i \cdot \underline{m}$ from either the left side or the right side of the interface, since it must be equal on both sides due to the mass balance.

Therefore, we can also write

$$d_{chem} = \sum_{i=1}^{n} [\![\mu_i]\!] \left(c_i \omega - \underline{j}_i \cdot \underline{m} \right). \tag{28}$$

3.2.5 The Chemical Driving Force

We discuss now relation (28) for d_{chem}. Since we have not restricted ourselves on any kind of transformation, we study first the thermally-driven martensitic transformation. In this case no fluxes do exist; $\underline{j}_i \equiv \underline{0}$, $i = 1,....n$. If we assume the same intrinsic velocity ω on both sides of the interface, or, in other words, ω is taken to be the interface normal velocity, then

$$d^{\alpha'}_{chem} = \left(\sum_{i=1}^{n} [\![\mu_i]\!] c_i \right) \omega = F^{\alpha'}_{chem} \omega . \tag{29}$$

The label "α'" is applied to refer to martensite (in the case of steel the α' phase). The c_i are equal on both sides of the interface. The chemical driving force $F^{\alpha'}_{chem}$ depends on the jumps of the chemical potentials of all components at the interface.

Investigating the diffusional phase transformation with fluxes $\underline{j}_i \neq \underline{0}$ operative, one may assume no resistance to the "squeezing through" of the interstitials through the interface, which means that the chemical potential of the interstitial atoms is the same on the two sides of the interface $[\![\mu_i]\!] = 0, i = 1,....m$. It is also reasonable to attach the identical resistance to the motion of the substitutionals in the lattice allowing for $[\![\mu_i]\!] = [\![\mu_n]\!], i = m+1,....n$. This assumption on the jumps of the chemical potentials, as discussed in Fischer, Simha and Svoboda (2003) and derived in Svoboda, Gamsjäger, Fischer et al. (2004), does not anticipate thermodynamic equilibrium. At the interface, equilibrium would enforce different equilibrium concentrations $c_{i,eq}$ at both sides of the interface depending on the composition and the lattice. In summary, there is no contribution from the interstitial atoms, and the jump in the chemical potential of all substitutionals is the same. Then, taking ω to be continuous, (28) can be rewritten as

$$d_{chem} = \left([\![\mu_n]\!] \sum_{i=m+1}^{n} c_i \right) \omega - [\![\mu_n]\!] \sum_{i=m+1}^{n} \underline{j}_i \cdot \underline{m} .$$

Using relation (11) yields

$$d_{chem} = \left([\![\mu_n]\!] \sum_{i=m+1}^{n} c_i \right) \omega + [\![\mu_n]\!] \underline{j}_v \cdot \underline{m} . \tag{30}$$

Thus, there are two contributions to the driving force – one from the change in chemical energy of the substitutional atoms on the two sides of the interface and the second from the diffusion of vacancies through the interface. We can now follow a common procedure in physical chemistry assuming the same molar V_{in} for all the interstitial elements, $V_i = V_{in}, i = 1,....m$, and the same molar volume V_{su} for all the substitutional elements,

$V_i = V_{su}$, $i = m+1,....n$, yielding $V_m = \left(V_{in} \sum_{i=1}^{m} X_i + V_{su} \sum_{i=m+1}^{n} X_i \right)$. Often V_{in} is neglected in relation to V_{su} yielding finally $V_m = V_{su} \sum_{i=m+1}^{n} X_i$ or $1/V_{su} = \sum_{i=m+1}^{n} c_i$. If we use the last relation we can reformulate (30) to

$$d_{chem} = ([\![\mu_n]\!]/V_{su})\omega + [\![\mu_n]\!]\underline{j}_v \cdot \underline{m}. \tag{31}$$

The first term on the right side of d_{chem}, (31), is the product of the thermodynamic force $F_{in,chem} = [\![\mu_n]\!]/V_{su}$ and the intrinsic velocity of the interface ω. The interface chemical driving force $F_{in,chem}$ is the thermodynamic force that drives the interface overcoming the energy necessary for the lattice rearrangement and the friction due to motion. The second term on the right side of d_{chem} is the product of the thermodynamic force $F_{va,chem} = [\![\mu_n]\!]$ and the normal component of the flux of the vacancies $\underline{j}_v \cdot \underline{m}$. Obviously $F_{va,chem}$ is the thermodynamic force with respect to the presence of vacancies in the interface. A detailed explanation is given in the next section.

We explain the interface driving force $F_{in,chem}$ in (30) by a simple example for a material with two components A and B with B being the lattice forming element. The label 1 refers to the product phase, 2 to the parent phase. We draw $\varphi_{chem} = \mu_A c_A + \mu_B c_B$ for both phases in Fig. 2 and set $V_m = "1"$ for simplicity, then from (10) $c_A + c_B = 1$ $(0 \le c_A \le 1)$ and $\varphi_{chem} = (\mu_A - \mu_B)c_A + \mu_B$. Since the chemical potentials also depend on concentration, φ_{chem} is not linear in c_A, but instead has a nonlinear dependence. The concentration c_A has different values on the two sides of the interface and we have two curves for φ_{chem} corresponding to each of these values. If we draw the tangent to one of these curves for a fixed value c_A, the equation for φ_{chem} teaches us that the tangent intersects the line $c_A = 0$ at $\varphi_{chem} = \mu_B$ and the line $\varphi_{chem} = \mu_A$ at $c_A = 1$. From this "tangent argument" it follows immediately that in the case of thermodynamic equilibrium, $\mu_A^1 = \mu_A^2, \mu_B^1 = \mu_B^2$, the equilibrium concentrations $c_{A,eq}^1, c_{A,eq}^2$ are the concentrations where the common tangent touches the two φ_{chem} curves. The process starts with an initial concentration c_A^2, and the interface moves as long as $c_{A,eq}^2$ is maintained. Then the driving force on the interface is nearly exhausted, and the interface continues moving due to the bulk diffusion, generally taking place on both sides of the interface. The velocity follows directly from (17) with c_i being now the equilibrium concentrations $c_{A,eq}^1, c_{A,eq}^2$. In the case of an interstitial atom A the condition $[\![\mu_A]\!] = 0$ enforces, due to the "tangent argument", to draw a tangent from A_i to the curve φ_{chem}^1, allowing to find $c_{A,i}^1$. The distance between the two tangents represents at $c_A = 0$ the potential jump $[\![\mu_B]\!]$ and at $c^1_{A,i}$ just the chemical driving force $F_{in,chem}$, denominated as F_i, since $1/V_{su}$ obtains now the value $c^1_{B,i}$ due to (10) with $V_m = 1$ and $V_{in} = 0$.

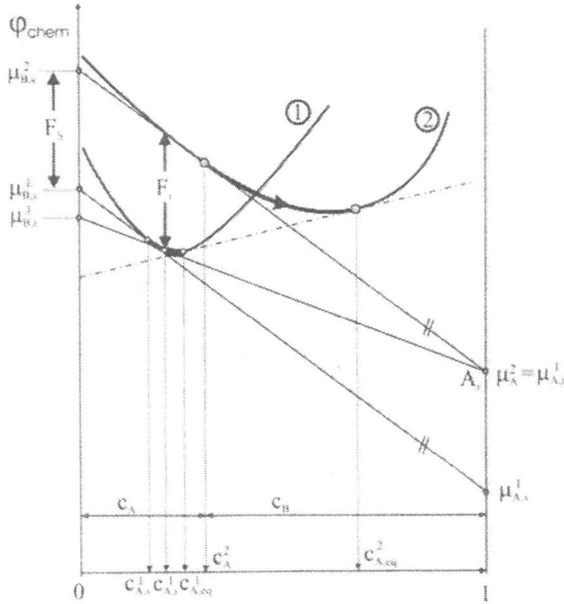

Figure 2. The thermodynamic forces on the interface in the case of interstitial and substitutional atomic motion, F_i and F_s, respectively.

In the case of a substitutional atom A the condition $[\![\mu_A]\!] = [\![\mu_B]\!]$ enforces, due to the "tangent argument", to draw a parallel tangent on the curve φ^1_{chem} ensuring that the a.m. jumps are identical. Again the tangent ordinate $c^1_{A,s}$ can be found, and at the line $c_A = 0$ the quantity $[\![\mu_B]\!]$ represents now the chemical driving force $F_{in,chem}$, denominated as F_s, since $1/V_{su}$ is now 1. As can be seen the chemical driving force differs for both cases.

3.3 Balance Relations and Driving Forces Assuming a Real Interface

From the mathematical point of view we introduce an interface source or sink as a flux: the term $\underline{i}_\Gamma$ is introduced as the source or sink due to mass flow and $\underline{e}_\Gamma$ as that due to the energy flow. The statement of the balance of mass is modified as follows: For every subregion $S \subset R$ and for each component i

$$\frac{d}{dt}\int_S c_i dV + \int_{\Gamma\cap S} \underline{i}_{\Gamma,i} \cdot \underline{m} dA = -\int_{\partial S_1+\partial S_2} \underline{j}_i \cdot \underline{n}_S \, dA .$$

Localizing this, we get the bulk balance relation to be the same as (16), but at the interface the interfacial flux appears as an additional term

$$\underline{i}_{\Gamma,i} \cdot \underline{m} + \left[\!\left[c_i \omega - \underline{j}_i \cdot \underline{m} \right]\!\right] = \underline{0} . \tag{32}$$

Similarly the balance of energy is modified as follows: For every subregion $S \subset R$

$$\frac{d}{dt}\int_S e_{chem} dV + \int_{\Gamma\cap S} e_\Gamma dA = -\int_{\partial S} \underline{\dot{q}} \cdot \underline{n}_S \, dA - \sum_{i=1}^{n} \int_{\partial S} \mu_i \underline{j_i} \cdot \underline{n}_S \, dA .$$

Localizing this, we get the bulk balance relation to be the same as (21), but at the interface the interfacial energy flux appears as an additional term

$$\underline{e}_{\Gamma,i} \cdot \underline{m} + \left[\!\!\left[e_{chem}\left(\underline{v} - \underline{\dot{u}}\right) - \underline{\dot{q}} - \sum_{\mu=1}^{n} \mu_i \underline{j}_i \right]\!\!\right] \cdot \underline{m} = 0 . \tag{33}$$

We now propose appropriate expressions for the interfacial mass and energy flux. A positive jump $\left[\!\left[\underline{j}_v \right]\!\right]$ of the vacancy-flux corresponds to a negative source term $\underline{i}_{\Gamma,i}$, since an increase of vacancies at the interface means automatically a loss of substitutional sites in the bulk phases adjacent to the interface. Since interstitial atoms may be located between the substitutional atoms, a jump $\left[\!\left[\underline{j}_v \right]\!\right] \neq \underline{0}$ affects both kinds of atoms. For $\underline{i}_{\Gamma,i}$ we suggest

$$\underline{i}_{\Gamma,i} = -\left[\!\left[\underline{j}_v \right]\!\right] \langle c_i \rangle \Big/ \sum_{i=m+1}^{n} \langle c_i \rangle .$$

Applying $\langle c_i \rangle$ can be considered as approximating the concentration profile of the component i in the interface Γ by its average value. Then equation (32) corresponds to an allocation of the atoms caught or ejected at the interface for each component i being proportional to its average concentration.

In an analogous way we suggest for $\underline{e}_{\Gamma,i}$ the expression

$$\underline{e}_{\Gamma,i} = -[\![\underline{j}_v]\!]\langle\mu_i c_i\rangle \Big/ \sum_{i=m+1}^{n}\langle c_i\rangle .$$

The denominator in the equation for $\underline{i}_{\Gamma,i}$ takes care that the sum of the $\underline{i}_{\Gamma,i}$ with respect to the substitutional elements in the interface guarantees that it be equal to $-[\![\underline{j}_v]\!]$. The jump conditions for the mass (32) and energy (33) now become with the suggested relations for $\underline{i}_{\Gamma,i}$, $\underline{e}_{\Gamma,i}$

$$\left[\!\!\left[\underline{j}_v\cdot\underline{m}\langle c_i\rangle \Big/ \sum_{i=m+1}^{n}\langle c_i\rangle - \left(c_i\omega - \underline{j}_i\cdot\underline{m}\right)\right]\!\!\right] = 0 , \tag{34}$$

$$\left[\!\!\left[\sum_{i=1}^{n}\underline{j}_v\cdot\langle\mu_i c_i\rangle \Big/ \sum_{i=m+1}^{n}\langle c_i\rangle - \left(e(\underline{v}-\dot{\underline{u}}) - \dot{\underline{q}} - \sum_{i=1}^{n}\mu_i\underline{j}_i\right)\right]\!\!\right]\cdot\underline{m} = 0 . \tag{35}$$

First we reformulate $\sum_{i=1}^{n}\langle\mu_i c_i\rangle$ with (2) to $\sum_{i=1}^{n}\left(\langle\mu_i\rangle\langle c_i\rangle + \frac{1}{4}[\![\mu_i]\!]\cdot[\![c_i]\!]\right)$. With the same arguments about $[\![\mu_i]\!], i=1....n$, from section 3.2.5 and replacing $\sum_{i=m+1}^{n} c_i$ by $1/V_{su}$ and, therefore, $\sum_{i=m+1}^{n}[\![c_i]\!]=0$, the product sum $\sum_{i=1}^{n}\langle\mu_i c_i\rangle$ can be replaced by the sum of the products $\sum_{i=1}^{n}\langle\mu_i\rangle\langle c_i\rangle$. If we insert again the entropy balance from section 3.2.3 in (35) we obtain

$$d_{chem} = \left[\!\!\left[\varphi_{chem}\omega - \sum_{i=1}^{n}\mu_i\underline{j}_i\cdot\underline{m} - \underline{j}_v\cdot\underline{m}\langle c_i\rangle \Big/ \sum_{i=m+1}^{n} c_i\right]\!\!\right] - \theta\dot{\eta}_\Gamma > 0 . \tag{36}$$

The same procedure as the one leading from (26) to (27) with mass balance (34) instead of (17) yields finally

$$d_{chem} = \left([\![\mu_n]\!]\sum_{i=m+1}^{n}\langle c_i\rangle\right)\omega + [\![\mu_n]\!]\langle\underline{j}_v\cdot\underline{m}\rangle . \tag{37}$$

The flux that accounts for the motion of vacancies through the interface as well as the generation of new or the annihilation of old lattice positions in the interface is now $\langle\underline{j}_v\cdot\underline{m}\rangle$, since the jump $[\![\underline{j}_v]\!] \neq \underline{0}$. Although the situation is much more complicated at the interface compared to the case with an ideal interface, the thermodynamic forces and fluxes are surprisingly simple. In fact, the forces are the same; only the flux due to vacancy motion is generalized to the average $\langle \underline{j}_v \cdot \underline{m} \rangle$. This means that the mathematical idealization of a sharp interface captures the correct terms for the thermodynamic driving forces and dissipation although it is not capable of explicitly accounting for vacancy flux, etc. in the interface.

Although many results of this chapter can be directly related to those in Hillert and Rettenmayr (2003), we note that they do not account for the role of vacancies. Moreover, we think that the approach followed here comes nearer to the real physical process in an interface.

Finally, we would like to mention cases where some of the above assumptions may be violated. For instance in the Fe-Mn-C system, the concentration profile of Mn degenerates to a spike type curve in the parent phase at high interface velocities. If the width of this spike becomes smaller than that of the real interface, the assumption on the equality of $[\![\mu_i]\!]$, $[\![\mu_i]\!] = [\![\mu_n]\!]$, $i = m+1, \ldots\ldots n$, breaks down. The corresponding fluxes $\underline{j}_i$ become $\underline{0}$, and we have, according to (11) and (34), no jump in the concentrations c_i, $i = m+1, \ldots\ldots n$. Such a state is referred to as paraequilibrium (Svoboda, Gamsjäger, Fischer et al., 2004). The chemical driving force becomes $F_{chem} = \sum_{i=m+1}^{n} [\![\mu_i]\!] c_i$.

4 Dissipation due to a Mechanical Deformation Process

4.1 General Comments

First we justify the word "mechanical" in the title of section 4. By a "mechanical" deformation process we mean the strain and stress state in a body R generated by the external tractions $\underline{t}$ on ∂R. Local variations of the density due to diffusion are assigned to a "chemical" deformation process and neglected at the moment.

Secondly, we have to note that the understanding of the thermodynamic force acting on an interface in an deformation process has been a topic of research for more than 50 years. Here we refer to the classical paper by Eshelby and corresponding references in the chapters by Kienzler and Berveiller in this booklet. Especially chapter by R. Kienzler is devoted to this task. Derivations within the context of the continuum mechanics have been presented by several authors, see e.g. Levitas (1999, 2000) or Fischer and Reisner (1998). A very recent investigation including a sliding interface was published by Fischer, Svoboda and Simha (2003). We will repeat the main results of their derivations in the next section.

4.2 The Stress Power

We use all the definitions and relations for mechanical terms as outlined in the sections 2.3, 2.4. In addition we neglect inertia terms, and this yields the continuity of the traction vector $[\![\underline{\underline{\sigma}} \cdot \underline{\tilde{n}}]\!] = \underline{0}$ at any area element with the normal $\underline{\tilde{n}}$. The principle of virtual power, applied for the sub-bodies S_1 and S_2 and then added together, yields

$$\int_{S_1+S_2} \underline{\underline{\sigma}} : \dot{\underline{\underline{\varepsilon}}} dV = \int_{\partial S} (\underline{\underline{\sigma}} \cdot \underline{n}_s) \cdot \underline{\dot{u}} \, dA - \int_{\Gamma} [\![(\underline{\underline{\sigma}} \cdot \underline{m}) \cdot \underline{\dot{u}}]\!] dA . \tag{38}$$

Inserting the decomposition of strains (6) and applying (1) for the expression in the integral along Γ, taking into account $[\![\underline{\underline{\sigma}} \cdot \underline{m}]\!] = \underline{0}$, yields

$$\int_{S_1+S_2} \underline{\underline{\sigma}} : \dot{\underline{\underline{\varepsilon}}}^{e\ell} dV + \int_{S_1+S_2} \left(\underline{\underline{\sigma}} : \dot{\underline{\underline{\varepsilon}}}^{p\ell} + \underline{\underline{\sigma}} : \dot{\underline{\underline{\varepsilon}}}^{*}\right) dV = \int_{\partial S} (\underline{\underline{\sigma}} \cdot \underline{n}_s) \cdot \underline{\dot{u}} \, dA - \int_{\Gamma} \langle \underline{\underline{\sigma}} \cdot \underline{m} \rangle \cdot [\![\underline{v}]\!] dA . \tag{39}$$

With (8) we introduce the total strain energy $U = \int_S U' dV$ and its rate $\dot{U}$ with (5) as

$$\dot{U} = \int_{S_1+S_2} \underline{\underline{\sigma}} : \dot{\underline{\underline{\varepsilon}}}^{e\ell} dV - \int_{\Gamma \cap S} [\![U']\!] \omega \, dA .$$

Furthermore, we introduce the total plastic work rate $\dot{W}_{p\ell} = \int_{S_1+S_2} \underline{\underline{\sigma}} : \dot{\underline{\underline{\varepsilon}}}^{p\ell}$ and keep in mind that $\dot{\underline{\underline{\varepsilon}}}^{*} = \underline{0}$ (- which simply means that $\underline{\underline{\varepsilon}}^{*}$ is only an outcome of crystallography). Applying the relations below yields

$$\int_{\Gamma} \left([\![U']\!] \omega + \langle \underline{\underline{\sigma}} \cdot \underline{m} \rangle \cdot [\![\underline{\dot{u}}]\!]\right) dA = \int_{\partial S} (\underline{\underline{\sigma}} \cdot \underline{n}_s) \cdot \underline{\dot{u}} \, dA - \dot{U} - \dot{W}_{p\ell} .$$

We can introduce the Hadamard relations (7) $[\![\underline{\dot{u}}]\!] = -[[\text{grad} u]]$ in $\langle \underline{\underline{\sigma}} \cdot \underline{m} \rangle [\![\underline{\dot{u}}]\!]$. Next consider the following identity for an arbitrary vector $\underline{v}$ (in index notation):

$$\langle \sigma_{ij} m_j \rangle [\![\partial u_i / \partial x_k]\!] v_k = \langle \sigma_{ij} m_j \rangle [\![\partial u_i / \partial x_\ell]\!] m_\ell m_k v_k = \langle \sigma_{ij} \rangle [\![\partial u_i / \partial x_\ell]\!] m_\ell m_j m_k v_k =$$
$$\langle \sigma_{ij} \rangle [\![\partial u_i / \partial x_j]\!] m_k v_k .$$

The symmetry $\sigma_{ij} = \sigma_{ji}$ allows us to write $\langle \sigma_{ij} \rangle \frac{1}{2} \left[\!\left[\frac{\partial u_i}{\partial x_j} + \frac{\partial u_j}{\partial x_i} \right]\!\right] m_k v_k = \langle \underline{\underline{\sigma}} \rangle : [\![\underline{\underline{\varepsilon}}]\!] \underline{v} \cdot \underline{m} \cdot$.

For $\underline{v} \cdot \underline{m}$ we insert ω. Finally we have

$$\int_\Gamma \left([\![U']\!] - \langle \underline{\underline{\sigma}} \rangle : [\![\underline{\underline{\varepsilon}}]\!] \right) \omega \, dA = \int_{\partial S} \left(\underline{\underline{\sigma}} \cdot \underline{n}_s \right) \cdot \dot{\underline{u}} \, dA - \dot{U} - \dot{W}_{p\ell} \, . \tag{40}$$

4.3 The Mechanical Driving Force

The quantity in brackets in the integral on the left side of (40) is the prominent Eshelby term, being the projection of the Eshelby tensor to the normal $\underline{m}$. Obviously the mechanical driving force F_{mech} in an interface point is

$$F_{mech} = [\![U']\!] - \langle \underline{\underline{\sigma}} \rangle : [\![\underline{\underline{\varepsilon}}]\!] . \tag{41}$$

One can understand F_{mech} more easily if one considers the same elastic properties of both phases. Then $[\![U']\!]$ can be written with (4) and (8)

$$[\![U']\!] = \frac{1}{2} [\![\underline{\underline{\sigma}} : \underline{\underline{\varepsilon}}^{e\ell}]\!] = \frac{1}{2} \langle \underline{\underline{\sigma}} \rangle : [\![\underline{\underline{\varepsilon}}^{e\ell}]\!] + \frac{1}{2} [\![\underline{\underline{\sigma}}]\!] : \langle \underline{\underline{\varepsilon}}^{e\ell} \rangle .$$

Since $\underline{\underline{\sigma}} = \underline{\underline{\underline{\underline{C}}}} \cdot \underline{\underline{\varepsilon}}^{e\ell}$ and $[\![\underline{\underline{\sigma}}]\!] = \underline{\underline{\underline{\underline{C}}}} \cdot [\![\underline{\underline{\varepsilon}}^{e\ell}]\!]$, $\langle \underline{\underline{\sigma}} \rangle = \underline{\underline{\underline{\underline{C}}}} \cdot \langle \underline{\underline{\varepsilon}}^{e\ell} \rangle$, the quantity $[\![U']\!]$ follows immediately as $\langle \underline{\underline{\sigma}} \rangle : [\![\underline{\underline{\varepsilon}}^{e\ell}]\!]$. Inserting this relation and (6) in (41) yields

$$F_{mech} = - \langle \underline{\underline{\sigma}} \rangle : [\![\underline{\underline{\varepsilon}}^{p\ell} + \underline{\underline{\varepsilon}}^*]\!] .$$

Since $\underline{\underline{\varepsilon}}^{p\ell}$ is usually continuous and $[\![\underline{\underline{\varepsilon}}^*]\!] = - \underline{\underline{\varepsilon}}^*$, F_{mech} follows as

$$F_{mech} = - \langle \underline{\underline{\sigma}} \rangle : [\![\underline{\underline{\varepsilon}}^*]\!] = \langle \underline{\underline{\sigma}} \rangle : \underline{\underline{\varepsilon}}^* . \tag{42}$$

In the case of $\underline{\underline{\varepsilon}}^* = \frac{\delta}{3} \underline{\underline{I}}$ the quantity F_{mech} is $\langle \sigma_H \rangle \delta$ for $\sigma_H = (\sigma_{11} + \sigma_{22} + \sigma_3)/3$.

Equations (40) and (42) are also used to derive an energy balance for the mechanical terms when the interface has moved over a certain subregion, say ΔS, in the time interval $[t_s, t_f]$

(with $(t_f - t_s)$ being a short time period). For this reason we define three work increments, namely

$$\int_{t_s}^{t_f} \dot{U}\,dt = U(t_f) - U(t_s) = \Delta W_{e\ell}, \quad \int_{t_s}^{t_f} \dot{W}_{p\ell}\,dt = \Delta W_{p\ell}, \quad \int_{t_s}^{t_f}\int_{\partial S} \left(\underline{\underline{\sigma}} \cdot \underline{n}_s\right) \cdot \underline{u}\,dA dt = \Delta W_{ex} .$$

The last term can be considered as the increment of total mechanical work provided from the surrounding body $R - S$ on S. Then relation (42) takes the form

$$\int_{ts}^{t_f}\int_{\Gamma} F_{mech}\,\omega dA dt = \Delta W_{ex} - \left(\Delta W_{e\ell} + \Delta W_{p\ell}\right). \tag{43}$$

Relation (43) shows clearly that the part of the work contribution from ΔW_{ex} that is not spent to generate strain energy and plastic work is used for the motion of the interface Γ within the time interval $[t_s, t_f]$. Relation (43) offers the further advantage, that the quantities $\Delta W_{e\ell}$, $\Delta W_{p\ell}$ are provided by continuum mechanics software like ABAQUS (http://www.hks.com). Also the term ΔW_{ex} can usually be calculated easily in an incremental procedure. This is specifically the case if S coincides with R. Then one has to calculate only the work done by the tractions on ∂R in time interval $[t_s, t_f]$.

The right side of (43) can also be modified by introducing two further incremental work terms

$$\Delta W_{e\ell}^{\tau} = \frac{1}{2}\int_S \underline{\underline{\tau}} : \underline{\underline{\underline{\underline{C}}}} : \underline{\underline{\tau}}\,dV \Big|_{t_f}, \quad \Delta W_{p\ell}^{\tau} = \int_{t_s}^{t_f}\int_S \underline{\underline{\tau}} : \dot{\underline{\underline{\varepsilon}}}^{p\ell}\,dV dt .$$

The first integral $\Delta W_{e\ell}^{\tau}$ represents the strain energy only due to the accommodation of $\underline{\underline{\varepsilon}}^*$ developed in ΔS in $[t_s, t_f]$ if the body was originally stress-free. The second integral $\Delta W_{p\ell}^{\tau}$ represents the contribution to the plastic work due to the additional eigenstrain $\underline{\underline{\tau}}$, see also sect. 2.4. Both quantities can be estimated in separate calculations. Berveiller et al. give values for $\Delta W_{e\ell}^{\tau}$ especially for martensitic plates, approximated as oblate spheroids using Eshelby's concept for an ellipsoidal inclusion, see e.g. the chapter of Berveiller in this book. Estimations for $\Delta W_{e\ell}^{\tau}$ and $\Delta W_{p\ell}^{\tau}$ are also reported studying the martensitic transformation in plate-type inclusions using the finite element method, see e.g. Reisner, Fischer, Wen et al. (1999). Specific values of $\Delta W_{e\ell}^{\tau}$ and $\Delta W_{p\ell}^{\tau}$ (which means work per transformed volume ΔS in $[t_s, t_f]$) are in the order of 1 to 10 Nmm/mm^3. Usually $\Delta W_{p\ell}^{\tau}$ is higher than $\Delta W_{e\ell}^{\tau}$. Detailed studies for a growing sphere with a transformation volume change can be taken from Fischer and Oberaigner (2000), (2001). Formulae for $\Delta W_{e\ell}^{\tau}$ applicable for spheroids in an elastic matrix can be taken from Böhm, Fischer and Reisner (1997). Relation (43) can then be replaced by

$$\int_{t_s}^{t_f}\int_{\Gamma} F_{mech}\omega dAdt = \int_{\Delta S}\left(\underline{\underline{\Sigma}} + \underline{\underline{\sigma}}^{int}\right):\underline{\underline{\varepsilon}}^{*}dV - \Delta W_{e\ell}^{\tau} - \Delta W_{p\ell}^{\tau}\,. \tag{44}$$

The detailed derivation can be taken from Fischer and Reisner (1998) and Fischer, Simha and Svoboda (2003). Keep in mind, that $\underline{\underline{\sigma}}^{int}$ is the internal stress state at the start time t_s. $\Delta W_{e\ell}^{\tau}$ and $\Delta W_{p\ell}^{\tau}$ are positive quantities.

The situation is more complicated if we deal with a noncoherent (i.e. a sliding) interface. Here we refer the reader to Simha and Bhattacharya (1998) with respect to the geometrical framework and to Fischer, Simha and Svoboda (2003), with respect to a detailed derivation. We limit ourselves to reporting the result of the latter paper concluding with

$$F_{mech}^{i} = [\![U']\!] - \left\langle\underline{\underline{\sigma}}\right\rangle : [\![\underline{\underline{\varepsilon}}]\!] + \left\langle\underline{\underline{\sigma}}\cdot\underline{m}\right\rangle\cdot\underline{t}_{\Gamma}\,\overset{\circ}{s}_{\ell}\,. \tag{45}$$

The label "i" added to F_{mech} stands for "incoherent". $\overset{\circ}{s}_{\ell}$ is a special time derivative of the sliding motion, $[\![u]\!] = s_{\ell}t_{\Gamma}$. The tangent to Γ is denominated as $\underline{t}_{\Gamma}$. The average amount of the "friction" vector is just the product $\left\langle\underline{\underline{\sigma}}\cdot\underline{m}\right\rangle\cdot\underline{t}_{\Gamma}$, namely the projection of the average traction vector $\left\langle\underline{\underline{\sigma}}\cdot\underline{m}\right\rangle$ onto the tangent vector $\underline{t}_{\Gamma}$. The most important consequence of (45) is that F_{mech}^{i} is formally the same expression as F_{mech} in the case of a frictionless interface. Such an interface does (nearly) exist in the case of high temperature diffusional transformation. Formally one can use the same relations as derived in the section below for the coherent interface. However, the field equations must be solved taking into account a sliding interface leading generally to a different strain and stress state compared to those occurring for a coherent (non-sliding) interface.

At this point we would like to make here a comment on the role of the interface energy σ. Referring e.g. to Simha and Bhattacharya (1998) in the simplest case of a constant value of σ, the mechanical driving force must be reduced by the term $-\kappa\sigma$ yielding $F_{mech} = \left([\![U']\!] - \left\langle\underline{\underline{\sigma}}\right\rangle : [\![\underline{\underline{\varepsilon}}]\!] - \kappa\sigma\right)$. κ is the average curvature $1/R_{ref}$ in the case of a cylinder and $2/R_{ref}$ in case of a sphere, if R_{ref} is the reference radius. Note that for convex interfaces the interface energy term is always negative and acts as a barrier. A very recent paper by Gurtin and Jabbour (2002) should be mentioned here outlining rather subtle interface energy functions which may be valid in the case of thin films etc. and a diffusive flow along the interface. However, one must be careful with respect to the surface energy term if the physical process does not allow for the "classical" term $\kappa\sigma$ leading to the well known Gibbs-Thompson equation.

The situation where a new phase is formed requires some modification, e.g. twins or martensite plates (laths), for details about twins see the example below and for martensitic transformation see e.g. Yan, Reisner and Fischer (1997). In this case the curvature κ of the plates is zero and the term $\kappa\sigma$ becomes 0. This is, however, not correct, since the newly formed phase (e.g. in the special case of a twin layer) is bounded by two planes with a total amount of surface energy being stored which is proportional to the area of both planes bounding the newly formed phase. In this case a curvature term does not exist. The surface energy introduces a length scale since it is proportional to the surface of the newly formed phase and appears as an additional negative term $\int_{\partial\Delta S}\sigma dA$ on the right side of relation (44).

4.4 Twinning as an Example

Twinning is an irreversible deformation mechanism in metals which may be considered as a plastic deformation if observed from the macroscopic point of view. It can be regarded as the sudden rotation of the lattice. The twin itself is separated from the parent material by a distinct twin plane in such a way, that the parent lattice and the twin lattice can be considered as mirror images, if the twin plane were the mirror. Within the twins a shear of the material occurs which may be significant, e.g. $\gamma_T = \sqrt{2}/2$ for Ti-Al. Therefore, a twin can be considered as a mechanically driven transformation of a plate type microregion bounded by planes. A typical picture of a twinned structure is given in Fig. 3.

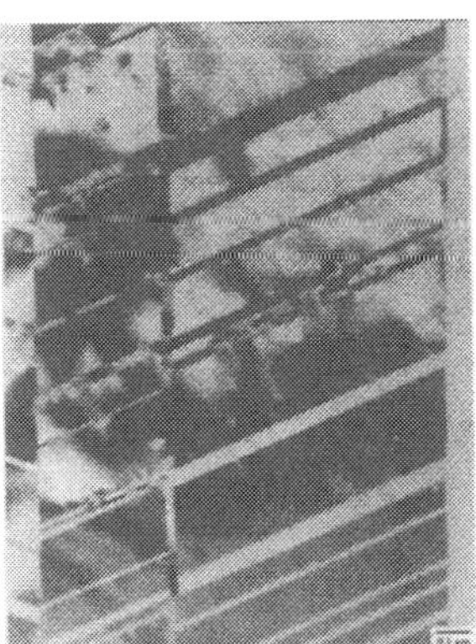

Figure 3. Deformation twins in gamma-based TiAl.

One may ask about the specific situation when twins appear. There is more than one answer. The most important case is the fact that too few slip systems are available to form a "compatible" plastic strain. This is the case in Ti-Al, since only 4 of the 12 slip systems in the f-c-c lattice can be activated – 8 slip systems are related to superdislocations, and sliding would consume too much energy. In this case at least one twin system is activated to allow for, at least, 5 independent slip systems. Another reason for twinning may be that a material shows only a low stacking fault energy which makes twinning occur "easily" as it is the case in some f-c-c metals. It should also be mentioned that twinning is a unidirectional deformation mechanism, which means that twinning can be activated in a distinct direction and not in its reverse direction. This is, of course, not the case for dislocation sliding. With respect to twinning we refer the reader to the prominent overview by Christian and Mahajan (1995), written more from the point of view of materials physics, and to the two papers Petryk, Fischer, Marketz et al. (2003) and Fischer, Schaden, Appel et al. (2003), explaining twinning within the framework of continuum mechanics.

Let us consider an elastic plane (E Young's modulus, μ shear modulus) of unit thickness in a generalized plane strain state. The plane is subjected to a homogeneous stress state $\underline{\underline{\Sigma}}$ with no additional internal stress state, $\underline{\underline{\sigma}}^{int}(t_s) \equiv \underline{\underline{0}}$. Suddenly a twin with the length ℓ and the width h appears. In the case of a twin in Ti-Al the width h is in the order of magnitude of, say, 5-50 nm, ℓ in the order of magnitude of the grain diameter, say 20-100μm. The twin is bounded by nearly exact twin planes, so that it appears in a section (see Fig. 3) as a twin band. This band starts and ends at a grain (or domain) boundary. In a local coordinate system with the x'-axis coinciding with the twin length direction and the y'-axis coinciding with twin thickness direction the eigenstrain tensor reads with the twinning shear γ_T

$$\underline{\underline{\varepsilon}}^* = \gamma_T \begin{bmatrix} 0 & 1/2 \\ 1/2 & 0 \end{bmatrix} = \gamma_T \underline{\underline{p}} .$$

The twin appears after the global stress state has reached a critical resolved shear stress $\tau_{cr,T}$. This means that a barrier of $\tau_{cr,T}\gamma_T$ must be overcome per unit volume to produce a twin. Since no chemical driving force is available, because both the lattice and the chemical composition remain unchanged and only the orientation angle of the lattice is changed, all the energy available for twinning per volume unit is F_{mech}. If, as in the case of Ti-Al, no plastification takes place, only the work term ΔW_{el}^{τ} appears, which is the strain energy occurring if a rectangular region of the thickness h and the length ℓ is suddenly subjected to the eigenstrain $\underline{\underline{\varepsilon}}^*$. Estimations of such a strain energy term are reported in the literature, e.g. by Christian and Mahajan (1995) based on Eshelby's concept of a homogeneous stress state in ellipsoidal inclusion. This concept, however, does not do justice to the situation where the stress state in the twin plate is by no means homogeneous. The strain energy is concentrated in the corners of the rectangular configuration. A detailed finite element study is reported in Fischer, Schaden, Appel et al. (2003). The numerical results fit much better to the specific strain energy according to a dislocation with the Burgers vector $\gamma_T h$ around a circular disc with the radius $\ell/2$;

this represents the strain energy produced, if the body is cut within a plane area, displaced by $\gamma_T h$ and then glued together. The result depends generally on the direction of this displacement. As a conclusion we have

$$\Delta W_{e\ell}^{\tau} = V_T \cdot 4K \cdot \frac{h}{\ell}\gamma_T^2, K = \mu \cdot \frac{\alpha}{4\pi}\ell n\left(\frac{\ell}{h}\right),$$

V_T is the volume of the twin, in our case ℓh multiplied by the unit thickness. α is a factor in the order of 1 and is identified as $1/(1-\nu)$ for the case of generalized plane strain. K/μ is typically in the order of 0.3 to 0.6. Furthermore we have to introduce the surface energy involved due to the existence of the two twin planes, see also sect. 4.3. In our case the surface energy amounts to $2\ell\sigma$ for a twin of unit thickness. The surface energy on the "ends" of the twin is ignored due to $\ell >> h$. Finally we have to balance the total mechanical driving force against the total barrier, which yields with (44)

$$\underline{\underline{\Sigma}} : \underline{\underline{\varepsilon}}^* \cdot V_T = \left(\tau_{cr,T}\gamma_T + 4K \cdot \frac{h}{\ell}\gamma_T^2 + 2\sigma \cdot \frac{1}{h}\right) \cdot V_T .$$

Dividing by V_T and γ_T gives the following relation

$$\underline{\underline{\Sigma}} : \underline{\underline{p}} = \tau_T = \tau_{cr,T} + \left(4K \cdot \frac{h}{\ell}\gamma_T + 2\sigma \cdot \frac{1}{h\gamma_T}\right).$$

τ_T is the resolved shear stress with respect to the twin plan. If we monotonically increase τ_T in a loading process then the equality will be satisfied for the first time for that value of h which minimizes the expression in the brackets on the right hand side. We neglect the dependence of K on h and are immediately able to find a value of h_{min} by minimizing the expression in the bracket with respect to h; ℓ is assumed to be the grain diameter and a given quantity. The minimization yields

$$h_{min} = \frac{1}{\gamma_T}\sqrt{\frac{\sigma\ell}{2K}} .$$

This is a surprisingly simple relation for the estimation of the twin band thickness h_{min}, which can appear earlier in the loading process than any thinner twin at the same value of the resolved shear stress τ_T. Later the twin may grow as long as the resolved shear stress is not reversed. Petryk, Fischer, Marketz et al. (2003) and Fischer, Schaden, Appel et al. (2003), investigated this problem in detail. They found that a twin will not grow if its volume fraction compared to the grain is rather zero. A new twin band just reproduces the existing one but at another place, assuming that any interaction between the twins can be disregarded. The concept developed by Petryk et al. can be considered as a different approach investigating the stability of equilibrium of a disc fixed between two clamps with a twin suddenly appearing. Therefore, in the derivation outlined in the two references above also a strain energy contribution due to the load stress drop, initiated by the twin, is added to right side of the balance relation.

If we use material data for TiAl, $\mu = 70 \cdot 10^9\ \mathrm{N/m^2}, \sigma = 0.06\ \mathrm{J/m^2}$ and consider grain diameters ℓ from 2 to 20μm, we find after some analysis values of h between 1 and 6nm, values which agree well with the observations cited in the two papers quoted above. It should also be mentioned that the resolved shear stress, necessary to produce the twin, can be found by inserting h_{min} in the balance relation yielding

$$\tau_{T,min} = \tau_{cr,T} + 4\sqrt{\frac{2K\sigma}{\ell}} .$$

It is interesting to note that a Hall-Petch-type relation appears for $\tau_{T,min}$ with respect to the grain size ℓ. Obviously, the energy balance allows to find a "twinning" condition as expressed above.

5 Kinetics

5.1 The "Point" Condition

Much of the literature on kinetics of phase-transformations deals with a one-dimensional problem setting (uniaxial or spherically symmetric geometry). In this case the interface degenerates to a point, or only a point at the interface needs to be considered. Therefore, the wording "point" condition is suggested. Ample literature exists on the martensitic transformation with a transformation condition in the form of

$$\begin{aligned} &F_{total} = F_{chem}^{\alpha'} + F_{mech} , \\ &F_{total} \geq F_c , \omega = \omega_0 \text{ or } f(F_{total}), \\ &F_{total} \leq F_c , \omega = 0 . \end{aligned} \tag{46}$$

$F_{chem}^{\alpha'}$ and F_{mech} are taken from (29) and (41) or (42), resp.; ω_0 is a fraction of the velocity of sound or of a translational wave. However, several functional assumptions on ω are used likewise, for references see Fischer and Oberaigner (2001). Of course, the field equations for the strain and stress field and the temperature field must be solved, too.

In the case of a diffusive transformation process a different approach is followed. As F_{total} the expression

$$F_{total} = F_{in,chem} + F_{mech}, \quad F_{in,chem} = [\![\mu_n]\!]/V_{su}, \tag{47}$$

see (30), (31) and (41) or (42), resp., is taken for the total driving force. Since at high temperatures, where usually the diffusive transformation takes place, a much higher atomic mobility is existent, in the simplest case a linear kinetic relation of the type

$$\omega = M\, F_{total}, \tag{48}$$

is used in practice. M is denoted as the interface mobility (which must not be mixed up with atom mobility expressed by the diffusion coefficient). M is a function of the temperature being

$$M = M_0 \exp\left(-Q_M/(R_g\theta)\right). \tag{49}$$

Q_M is the activation energy for interface motion and R_g the gas constant $8.3144\,\mathrm{J/(mol\,K)}$, for pure iron $M_0 = 4.1 \cdot 10^{-7}\,\mathrm{m^2 s/kg}$, $Q_M = 140\,\mathrm{kJ/mol}$. The quantity M_0 may be substituted by a function depending on the velocity itself if a solute drag is considered in the interface, see the chapter by Svoboda in this book and Svoboda, Fischer and Gamsjäger (2002). Then an interactive procedure is necessary to find out the interface velocity. Generally, there are two regimes as depicted in Fig. 4, that of a low velocity fully controlled by the solute drag and that of a high velocity not controlled by the solute drag.

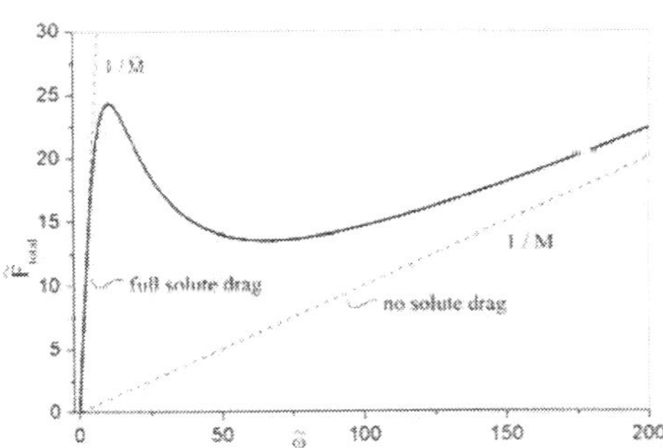

Figure 4. Effect of solute drag on the dimension-free total driving force $\widetilde{F}_{total}$ in dependence on the dimension-free interface velocity $\widetilde{\omega}$.

The transition between both regimes is described by a curve with an inflection point giving rise to a certain instability in the interface motion with respect to the transition; relevant details are described already by Cahn (1962) and are still under investigation.

5.2 The "Transformation" Condition

In many practical cases, it is more convenient, especially for the implementation in numerical algorithms, to establish a transformation condition for a certain microregion, say ΔS, relating the state of this microregion before and after its full transformation. According to the procedure outlined in section 4.3 of this chapter we formulate the expression for the total energy D_{tot} being invested (or, in other words, dissipated) for the transformation of ΔS,

$$D_{tot} = \int_{t_s}^{t_f}\int_{\Gamma}(F_{mech} + F_{chem})dA\,dt\,. \tag{50}$$

The integral with respect to F_{mech} is taken from relation (44). For F_{chem} we take either the expression for the martensitic transformation (29) or for the interface motion (31) or (37) in the case of diffusive transformation. Furthermore, we keep in mind that the volume ΔS is equal to $\int_{t_s}^{t_f}\int_{\Gamma}\omega dA\,dt$.

The transformation condition for the martensitic transformation reads (with the assumption that the chemical energy does not change during the transformation process) as

$$\Delta S\left(\sum_{i=1}^{n}[\![\mu_i]\!]c_i\right) + \int_{\Delta S}\left(\underline{\underline{\Sigma}} + \underline{\underline{\sigma}}^{int}\right):\underline{\underline{\varepsilon}}^{*}dV = \Delta SF_c + \Delta W_{e\ell}^{\tau} + \Delta W_{p\ell}^{\tau} + W_{\Gamma}\,. \tag{51}$$

W_{Γ} stands for the interface energy generated during the transformation process, see the comments at the end of section 4.3. In the case of the diffusional transformation we use relation (49) and eventually find

$$\Delta S\frac{\overline{\omega}}{M} = \Delta S\left([\![\overline{\mu}_n]\!]/V_{su} + \left(\underline{\underline{\Sigma}} + \underline{\underline{\sigma}}^{int}\right):\underline{\underline{\varepsilon}}^{*}\right) - \Delta W_{e\ell}^{\tau} - \Delta W_{p\ell}^{\tau} - W_{\Gamma}\,. \tag{52}$$

$\overline{\omega}$ is the average interface velocity in the time interval $[t_s, t_f]$ and this allows us to directly calculate the transformed volume fraction $\Delta\xi$, e.g. in the case of the generation of transformed spherical regions as

$$\Delta\xi = \frac{4\pi}{3}\frac{\left(\overline{\omega}(t_f - t_s)\right)^3}{\Delta S}\,.$$

$[\![\bar{\mu}_n]\!]$ is the average value of the jump of the chemical potential of the substitutional elements in the time interval $[t_s, t_f]$.

5.3 The Growth of Ferrite as an Example of Diffusional Transformation

With respect to the transformation condition in the case of the diffusional transformation we repeat here an example which is outlined in detail in Fischer, Simha and Svoboda (2003). We consider the growth of a ferritic phase (the product or α-Fe phase) in the austenitic phase (the parent or γ-Fe phase) at 800°C in an iron-carbon-system. The transformation process is controlled by the diffusive motion of carbon from the product phase (the $\alpha-\mathrm{Fe}$ phase) into the parent phase (the $\gamma-\mathrm{Fe}$ phase). We assume an initial molar fraction of carbon of $\overline{X}=0.01$. Since the transformed, ferritic phase is nearly free of carbon, $X^{\alpha}=0$, a significant enrichment of carbon takes place at the interface on the parent phase side defined as $X^{\gamma}\big|_{\mathrm{Interface}}=X_{\mathrm{int}}$. The equilibrium concentration of carbon at the interface, X_{eq}, is assumed to be 0.015. The chemical driving force, expressed by $[\![\bar{\mu}_n]\!]/V_{su}$ follows as

$$[\![\bar{\mu}_n]\!]/V_{su}=1360\left(X_{eq}-X_{\mathrm{int}}\right) \text{ in MPa.}$$

Details on this relation can be taken from the a.m. reference. We represent the volume ΔS by a very thin shell with the thickness ΔR_f surrounding the already formed ferrite with the sphere radius R_f. The phase transformation is accompanied by a transformation volume change $\underline{\underline{\varepsilon}}^{*}=\frac{\delta}{3}\underline{\underline{I}}, \frac{\delta}{3}=0.004, \underline{\underline{I}}$ unity tensor. The stress state and the total work terms in the thin shell can be taken from Fischer and Oberaigner (2000), (2001) with $E=100\mathrm{GPa}$ being the Young's modulus for both phases, $\nu=0.3$ being the Poisson's ratio and $\sigma_y=20\mathrm{MPa}$ being the yield stress of the austenite. Since the transformation volume change $\underline{\varepsilon}^{*}=\frac{\delta}{3}\underline{\underline{I}}$ cannot be accommodated by elastic deformation in any way, a plastic shell with the radius R_p surrounds the transformed sphere with

$$R_p=\kappa_p R,\ \kappa_p=E\frac{\delta}{3}\Big/\left((1-\nu)\sigma_y\right).$$

The elastic energy stored during the generation of a sphere with the final radius R can be found as

$$W_{e\ell}=\frac{(1-\nu)}{E}\sigma_y^2\cdot 4\frac{R_f^3\pi}{3}\cdot\left(2\kappa_p-1\right);$$

the total plastic work spent is

$$W_{p\ell}=\frac{(1-\nu)}{E}\sigma_y^2\cdot 4\frac{R_f^3\pi}{3}\cdot 2\kappa_p\ell n\kappa_p\,.$$

The work increments $\Delta W_{e\ell},\Delta W_{p\ell}$ during the growth of a shell with the thickness ΔR_f can be found by differentiation of the relations above with respect to R_f. If an external stress is active, the work term would be $\Sigma_n\delta\Delta S$ with Σ_n being the hydrostatic stress to $\underline{\underline{\Sigma}}$. We assume in this example $\Sigma_n=0$, and therefore, $\Delta W_{ex}=0$. Finally, the specific mechanical driving force follows as

$$\left(\Delta W_{ex}-\left(\Delta W_{e\ell}+\Delta W_{p\ell}\right)\right)/\Delta S=-\frac{(1-\nu)}{E}\sigma_y^2\cdot\left(2\kappa_p-1+2\kappa_p\ell n\kappa_p\right),$$

yielding with the data above the rather small value of -0.69MPa. If we compare this mechanical driving force, which is now a mechanical barrier, with the chemical driving force at the start of the transformation process with $X_{eq}-X_{int}=0.005$, $[\![\bar{\mu}_n]\!]/V_{su}$ amounts to a value of 6.8MPa, the mechanical barrier is rather small (ca. one tenth of the chemical driving force), and the process is mainly driven by the chemical driving force. A simple relation can now be given for $\bar{\omega}$ as

$$\bar{\omega}=M[20.4-1360X_{int}],\quad \text{in mm/s}\,.$$

M can be of the order of $4\cdot 10^{-4}\,\text{mm}^3/(\text{Ns})$.

It should be noted, that the "reservoir" of chemical energy may be exhausted drastically. Svoboda, Fischer, Fratzl et al. (2001) have derived an analytical expression to estimate the time $\overline{\Delta t}$ necessary to achieve a given ratio $\left(X_{int}-\bar{X}\right)/\left(X_{eq}-\bar{X}\right)$ as

$$\overline{\Delta t}=\left(\frac{X_{int}-\bar{X}}{X_{eq}-\bar{X}}\right)^2\cdot\frac{D\pi}{4\bar{X}^2\beta^2}\cdot\frac{1}{M^2}\,.$$

D is the diffusion coefficient of carbon in an iron lattice in the magnitude of $1.46\cdot 10^{-6}\,\text{mm}^2/\text{s}$. Inserting $\left(X_{int}-\bar{X}\right)/\left(X_{eq}-\bar{X}\right)$ as 1 in this equation yields a time-point $t_1=\overline{\Delta t}$, from which time onwards only the mechanical driving force is responsible for the progress of a mobility-driven diffusive transformation process.

6 Conclusion

Continuum mechanics and continuum thermodynamics provide a rational concept to find the thermodynamic forces at the interface or, in other words, the energy available for the motion of an interface. In the sense of nonequilibrium thermodynamics this is that amount of energy which is dissipated by the interface motion. Of course, to quantify what happens inside and in the neighbourhood of a phase interface makes it necessary to understand the physical processes in the interface region. Here we mention ongoing research with respect to the role of vacancies (their generation and annihilation, both accompanied by some dissipation) or with respect to the solute drag dealing with "extra" atoms carried with the moving interface. One must keep in mind that we have a coupled problem between mechanical and physical quantities. The type of coupling can be a weak one, e.g. between the temperature and the mechanical properties or a rather strong one, e.g. with respect to chemical potentials and the temperature. All the quantities at an interface are also directly related to the corresponding quantities in the bulk. That means, one has to solve the field equations for the mechanical quantities (the displacements, strains and stresses), the physical quantities (the molar fractions or concentrations) and the temperature. At the interface the relations, outlined in detail in this chapter, play the role of transition conditions between the two phases. On the whole the total problem solution is a very extensive and difficult task and can usually not be tackled without the strong support of a computational frame work. Since the experimental investigation of the interface region is generally extremely difficult or even down-right impossible, one usually has to check the bulk quantities with experimental methods. This makes the job even more difficult, since rarely an experimental control of the interface itself is possible. A nice example of a direct control is the growth of a ferrite film (seam) around the austenite grain boundaries. However, this is one of the few examples, where the motion of an interface is observable.

Some words should finally be written on the phase field method, see e.g. the book by Emmerich (2003), compared to a sharp interface concept. There is no doubt that the phase field method, having gradient terms, is able to take on problems with quantities which vary strongly with respect to space. The phase field method may be seen as a bulk-type description of the material with a penalty term with respect to the gradient of a distinct quantity. By penalizing the gradient the phase field method automatically produces "interface type" regions. The concept outlined in this chapter does not introduce extra gradient terms of the physical quantities in addition to the classical terms for the bulk equations. At the interface distinct jumps of the quantities are considered, however, with balance equations exactly prescribing the nature of the jump. What happens in the interface itself is described by the physical phenomena as they can be observed in real interfaces – think of vacancies or the solute drag. However, the authors do not intend to evaluate which method is the better one. Both methods have their individual benefits and shortcomings. Qualitatively, one can argue that the sharp interface concept is "more physical" and the phase field concept is "more mathematical".

References

Böhm, H.J., Fischer, F.D. and Reisner, G. (1997). Evaluation of elastic strain energy of spheroidal inclusions with uniform volumetric and shear eigenstrains. *Scripta Materialia* 36:1053-1059, *Scripta Materiala* 49:107.

Cahn, J.W. (1962). The impurity-drag effect in grain boundary motion. *Acta metall.* 10:789-798.

Cermelli, P., Gurtin, M.E. (1994). The dynamics of solid-solid phase transitions - 2. Incoherent interfaces. *Arch. Rational Mech. Anal.* 127:41-99.

Christian, J.W., Mahajan, S. (1995). Deformation twinning. *Progr. Mater. Sci.* 39:1-157.

Cocks, A.C.F., Gill, S.P.A., Pan, J. (1999). Modeling microstructure evolution in engineering materials. In Giessen, E.v.d. and Wu, T.Y., eds., *Advances in Applied Mechanics*. San Diego: Academic Press, Vol.36. 81-162.

Emmerich, H. (2003). *The Diffusive Interface Approach in Materials Science.* Berlin, Heidelberg, New York; Springer-Verlag.

Fischer, F.D., Oberaigner, E.R. (2001). A micromechanical model of interphase boundary movement during solid-solid phase transformations. *Archive of Applied Mechanics* 71:193-205.

Fischer, F.D., Oberaigner, E.R. (2000). Deformation, stress state and thermodynamic force for a transforming spherical inclusion in an elastic-plastic material. *ASME J. Appl. Mech.* 67:793-796.

Fischer, F.D., Reisner, G. (1998). A criterion for the martensitic transformation of a microregion in an elastic-plastic material. *Acta mater.* 46:2095-2102.

Fischer, F.D., Schaden, T., Appel, F., Clemens, H. (2003). Mechanical twins, their development and growth. *Europ. J. Mechanics A* 22:709-726.

Fischer, F.D., Simha, N.K., Svoboda, J. (2003). Kinetics of diffusional phase transformation in multicomponent elastic-plastic materials. *ASME J. Eng. Mat. & Technology* 125:266-276.

Fried, E., Gurtin, M.E. (1999). Coherent solid-state phase transitions with atomic diffusion: a thermomechanical treatment. *J. Statistical Physics* 95:1361-1427.

Grinfeld, M. (1991). *Thermodynamic Methods in the Theory of Heterogeneous Systems.* Harlow: Longman Scientific & Technical.

Gurtin, M.E., Jabbour, M.E. (2002). Interface evolution in three dimensions with curvature-dependent energy and surface diffusion: Interface-controlled evolution, phase transitions, epitaxial growth of elastic films. *Arch. Rational Mech. Anal.* 163:171-208.

Gurtin, M.E., Voorhees, P. (1993). The continuum mechanics of coherent two-phase elastic solids with mass transport. *Proc. R. Soc. Lond. A* 440:323-343.

Gurtin, M.E. (1993). The dynamics of solid-solid phase transitions - 1. Coherent interfaces. *Arch. Rational Mech. Anal.* 123:305-335.

Hillert, M., Rettenmayr, M. (2003). Deviation from local equilibrium at migrating phase interfaces. *Acta mater.* 51:2803-2809.

Irschik, H. (2003). On the necessity of surface growth terms for the consistency of jump relations at a singular surface. *Acta Mech.* 162:195-211.

Lehner, F.K. (1990). Thermodynamics of rock deformation by pressure solution. In Barber, D.J., Meredith, P.G. eds., *Deformation Processes in Minerals Ceramics and Rocks.* London et al.: Unwin Hyman. 296-333.

Levitas, V.I. (1999). Thermomechanical and kinetic approaches to diffusional - displacive phase transitions in inelastic materials. *Mechanics Research Communications*, 27:217-227.

Levitas, V.I. (2000). Structural changes without stable intermediate state in inelastic material. Part I. General thermomechanical and kinetic approaches; Part II. Applications to displacive and diffusional-displacive phase transformations, strain-induced chemical reactions and ductile fracture. *Int. J. Plasticity*, 16:805-849, 851-892.

Maugin, G.A. (1999). *The Thermomechanics of Nonlinear Irreversible Behaviors.* Singapore et al.: World Scientific. 295-323.

Morland, L.W., Gray, J.M.N.T. (1995). Phase change interactions and singular fronts. *Continuum Mech. Thermodyn.* 7:387-414.

Petryk, H., Fischer, F.D., Marketz, W., Clemens, H., Appel, F. (2003). An energy approach to the formation of discrete twin bands in polycrystalline solids. *Metall. Mater. Trans. A.* 34:2827-2836.

Reisner, G., Fischer, F.D., Wen, Y.H., Werner, E.A. (1999). Interaction energy between martensitic variants. *Metall. Mater. Trans. A* 30A:2583-2590.

Simha, N.K., Bhattacharya, K. (1998). Kinetics of phase boundaries with edges and junctions. *J. Mech. Phys. Solids* 46:2323-2359.

Slattery, J.C. (1990). *Interfacial Transport Phenomena.* New York Berlin Heidelberg et al.: Springer-Verlag.

Svoboda, J., Fischer, F.D., Fratzl, P., Gamsjäger, E., Simha, N.K. (2001). Kinetics of interfaces during diffusional transformations. *Acta mater.* 49:1249-1259.

Svoboda, J., Fischer, F.D., Gamsjäger, E. (2002). Influence of solute segregation and drag on properties of migrating interfaces. *Acta mater.* 50:967-977.

Svoboda, J., Fischer, F.D., Fratzl, P., Kroupa, A. (2002). Diffusion in multi-component systems with no or dense sources and sinks for vacancies. *Acta mater.* 50:1369-1381.

Svoboda, J., Gamsjäger, E., Fischer, F.D., Fratzl, P. (2004). Application of the thermodynamic extremal principle of the diffusional phase transformations., *Acta mater.* 52:959-967.

Thornton, K., Ågren, J., Voorhees, P.W. (2003). Modelling the evolution of phase boundaries in solids at the meso- and nano-scales. *Acta mater.* 51:5675-5710.

Yan, W.Y., Reisner, G., Fischer, F.D. (1997). Micromechanical study on the morphology of martensite in constrained zirconia. *Acta mater.* 45:1969-1976.

Moving Grain Boundaries During Hot Deformation of Metals: Dynamic Recrystallization

Frank Montheillet

Centre SMS, CNRS UMR 5146, Ecole des Mines, Saint-Etienne, France

Abstract

Basic knowledge regarding kinematics and dynamics of grain boundaries in metals is first reviewed. Typical effects of grain boundary migration during hot deformation, *i.e.* flow stress softening and grain coarsening, are illustrated by a simple one-dimensional model. The two mechanisms of continuous and discontinuous dynamic recrystallization are then introduced and compared from recent experimental data. Semi-analytical models involving mainly strain hardening, dynamic recovery, low and high angle boundary generation and migration are proposed. Finally, possible transitions between the two varieties of dynamic recrystallization are discussed.

List of Notations

A, A_1, A_2	constants
a	grain size (Derby) exponent in the σ *vs* $\bar{D}$ relationship
b	Burgers vector modulus
C_1, C_2, C_3	constants
D	grain or crystallite size
$\bar{D}$	average value of D
f_{HAB}	high angle boundary area fraction
f_{LAB}	low angle boundary area fraction
h	microscopic strain hardening parameter
k_N	nucleation parameter in the DDRX model
M	grain boundary mobility
m_h	strain rate sensitivity exponent of h
m_M	strain rate sensitivity exponent of v_M
m_r	strain rate sensitivity exponent of r
p	nucleation exponent in the DDRX model ($p \approx 3$)
Q_B	apparent activation energy of grain boundary diffusion
Q_h	apparent activation energy of strain hardening
Q_M	apparent activation energy of grain boundary migration

Q_r	apparent activation energy of dynamic recovery
Q_S	apparent activation energy of self-diffusion
R	gas constant
r	dynamic recovery parameter
S_V	grain boundary area per unit volume
t_F	lifetime of a dynamically recrystallized grain in the DDRX model
v_{app}	apparent grain boundary migration rate derived from the average grain thickness evolution
v_M	grain boundary migration rate
α	fraction of dynamically recovered dislocations consumed by the creation of new subgrain boundaries in the CDRX model
γ	grain boundary energy per unit surface
$\Delta\rho$	average dislocation density difference between neighbouring grains
Δt_1	time necessary to create a new HAB in the CDRX model
Δt_2	time necessary for a moving boundary to migrate over a distance D
ε	plastic strain
$\dot{\varepsilon}$	plastic strain rate
$\varphi(\theta)$	distribution density function of the subgrain boundary misorientations in the CDRX model
$\phi(t)$	recrystallization probability for a grain in the DDRX model
κ	geometrical constant relating S_V to D
μ	elastic shear modulus
θ	misorientation angle of a subgrain boundary (LAB)
θ_0	minimum misorientation angle of a subgrain boundary (LAB)
θ_c	critical angle for the transformation of a LAB into a HAB
ρ	dislocation density
ρ_i	average dislocation density inside the crystallites in the CDRX model
ρ_{LAB}	dislocation density inside the LABs in the CDRX model
ρ_s	steady state dislocation density
ρ_0	dislocation density of a fully annealed material
$\bar{\rho}$	average dislocation density
σ	flow stress
σ_s	steady state flow stress
τ	line energy of a dislocation

Abbreviations

BMIS	boundary migration induced softening
CDRX	continuous dynamic recrystallization
CSL	coincidence site lattice

DDRX	discontinuous dynamic recrystallization
DRX	dynamic recrystallization
EBSD	electron backscattering diffraction
GB	grain boundary
GBM	grain boundary migration
HAB	high angle boundary
LAB	low angle boundary
SFE	stacking fault energy
TKD	tetrakaidecahedron
YLJ	Yoshie-Laasraoui-Jonas

1 Introduction

Dynamic recrystallization (DRX), which occurs *during* hot working of crystalline materials, was not at first easily recognized, since it was difficult to prove that the "new" grains observed in quenched specimens were not statically (*i.e.*, *after* straining) recrystallized. The issue was addressed in a well-known paper by Stüwe (1968), significantly entitled *Do Metals Recrystallize During Hot Working?* Then, only careful examinations of the stress-strain curves and the development of rapid quenching devices allowed to prove unambiguously the occurrence of DRX. It is not excluded, however, that a number of the "as quenched" investigated microstructures were (and are still) more or less affected by post-dynamic (static) evolutions. More recently, new experimental investigations have given new insights into the high temperature deformation mechanisms of metals: first, large to very large strain torsion data have been obtained; second, the availability of fully automated Electron BackScattering Diffraction (EBSD) devices has allowed detailed and statistical analyses of the microstructures to be collected. Furthermore, models have been developed to better understand the basic mechanisms of DRX.

Dynamic recrystallization is now considered not only as a "scientific curiosity", but also as an "industrial tool" (Jonas, 1994) used for instance to control and optimize the microstructures of rolled steel sheet or forged nickel base superalloys turbine discs. For a long time, it has been considered as restricted to low stacking fault energy (SFE) materials, while in high SFE metals and alloys dynamic recovery was assumed to be the only mechanism taking place during hot deformation. However, it has been shown that grain boundary migration, which may be considered as characteristic of recrystallization processes, plays an important role in *both* cases, as soon as large strains are involved. Therefore, the mechanism involved during hot deformation of high SFE materials, that leads to the creation of "new" grains as well, may also be referred to as a DRX mechanism. This has led to distinguish two types of DRX: *discontinuous* (classical), and *continuous* dynamic recrystallization.

In the present chapter, DRX mechanisms will be described from the main viewpoint of this book, that is as a result of grain boundary movement during hot deformation. Section 2 summarizes the basic knowledge about grain boundary migration in metals, while Section 3 illustrates the effect of grain boundary movements in a simple configuration. Data pertaining to continuous and discontinuous DRX, as well as physical models recently proposed for the two mechanisms are analyzed in Sections 4 and 5, respectively. Finally, Section 6 deals with the possible transitions from one mechanism to the other, with a special emphasis on "model" high purity metals.

2 Recrystallization and Moving Boundaries in Metals

2.1 What Does Recrystallization Mean?

Figure 1 illustrates schematically various microstructural states of a polycrystalline metal. In the "fully annealed" state (Figure 1a), the dislocation density of the material is quite low, about 0.01 μm/μm^3. After straining at room temperature (Figure 1b), it can increase up to 10^4 μm/μm^3 in such a way that, on the basis of early optical observations, it was first believed that "all metals (...) were not crystalline in the rolled state but that they became so on annealing" (Kalisher, 1881, 1882). Annealing after cold working was therefore associated to a *re*-crystallization mechanism. It is now well known, however, that deformed initial grains are still present in the cold-worked state (Figure 1c), while nucleation and growth of new grains, or *static recrystallization*, take place during subsequent annealing (Figure 1e). Whenever hot working is performed on the initial annealed material, nucleation and growth occur during straining, a mechanism that is referred to as *dynamic recrystallization* (Figure 1d). Such dynamically recrystallized microstructure may as well undergo *post-dynamic recrystallization* upon subsequent annealing (Figure 1e).

It is important to note a fundamental difference between the static and dynamic processes: the first case consists in a one way transition from a high energy metastable strain hardened state to the lower energy annealed state. It is therefore analogous to a phase transition. In contrast, dynamic recrystallization leads at large strains to a steady state dynamic equilibrium associated with a self-organized dissipative microstructure with intermediate energy level (Figure 2).

2.2. Grain Structure Geometry

Modeling recrystallization first requires to adopt some geometrical representation of a crystalline aggregate. The choice may depend on the available experimental data, which are generally 2-dimensional (optical metallography, conventional EBSD analysis), although 3-dimensional observations are now possible by X-ray tomography. Models are currently based on three types of representation:

Periodic arrays. One-dimensional "stacks" of grains have been used by some authors for investigating grain growth (Hunderi and Ryum, 1996). A similar approach will be presented in

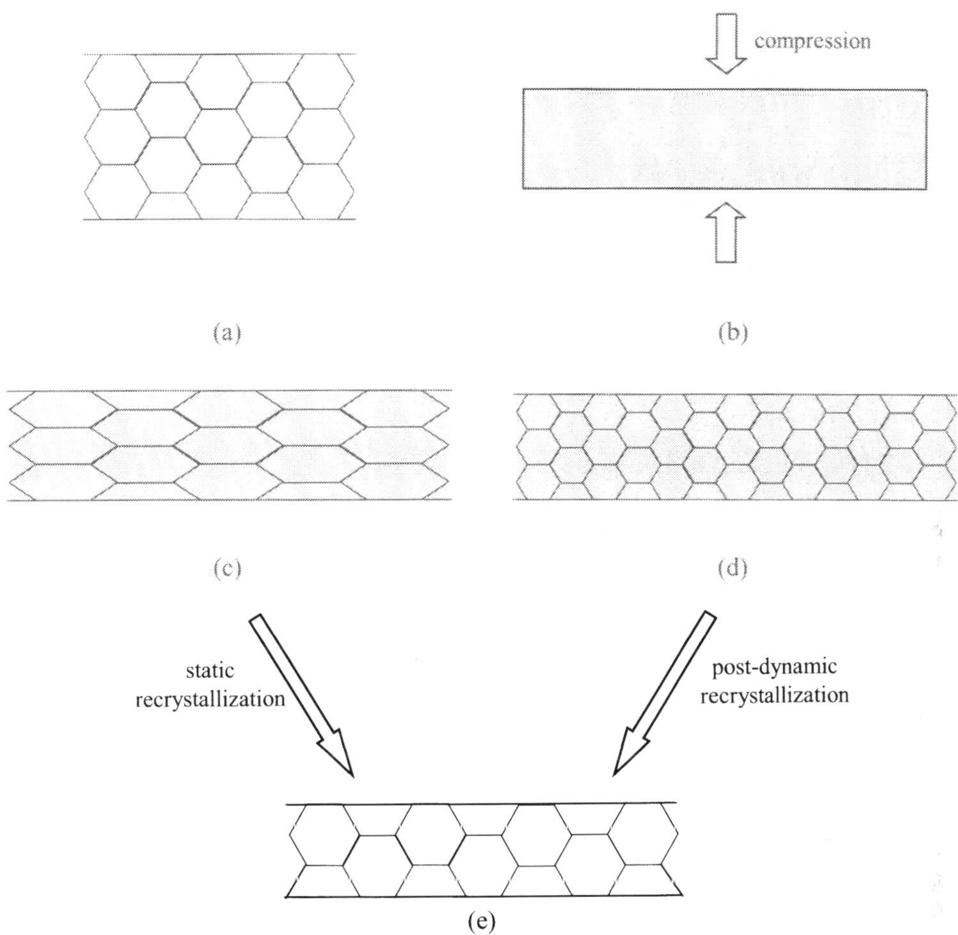

Figure 1. Schematic representations of (a) the "fully annealed" state of a polycrystalline metal, (b) the former interpretation of a cold worked specimen (amorphous state), (c) the modern interpretation of the latter, (d) a hot worked dynamically recrystallized microstructure, and (e) a post-dynamically recrystallized state.

Section 2 of this chapter. Two-dimensional models generally represent grains by regular hexagons, since each grain in a planar section has an average of six first neighbours. A good way to deal with 3-dimensional aggregates of equiaxed grains is to use tetrakaidecahedra, first introduced by Kelvin (1887) for crystalline applications. Such polyhedra can fill space without void nor overlapping (Figure 3). Among the 14 facets, 8 are hexagons and 6 are squares with the same edge length e. The tetrakaidecahedron (TKD) has 24 vertices located on a sphere of

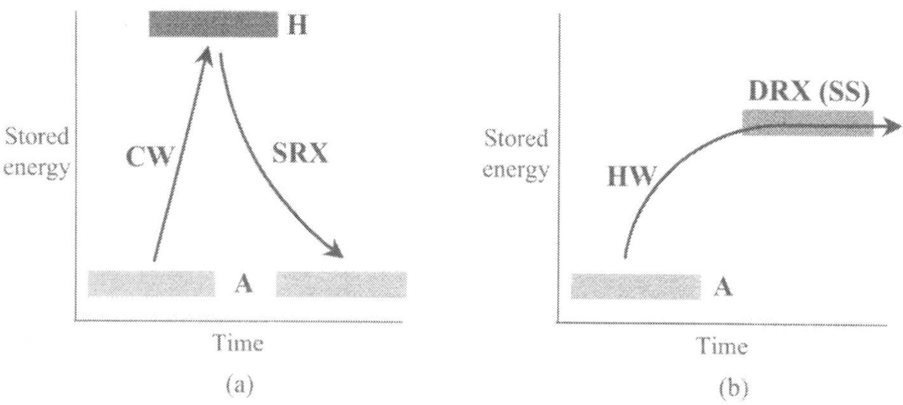

Figure 2. (a) Static recrystallization (SRX) that follows cold working (CW), involves the transition between a high energy work hardened state (H) to the low energy annealed state (A). (b) Dynamic recrystallization taking place during hot working (HW) leads at large strains to a steady state microstructure (SS) with intermediate energy level.

radius $R = \sqrt{5/2}\,e$. The volume is $V = 8\sqrt{2}\,e^3$. Hence, the diameter of the equivalent sphere is $D = \left(6\sqrt{2}/\pi\right)^{1/3} 2e$. Furthermore, the surface of the TKD is given by $S = 6\left(1 + 2\sqrt{3}\right)e^2$, whence the grain boundary area per unit volume is $S_V = (S/2)/V = \kappa/D$, the geometrical constant κ been given by the following expression:

$$\kappa = \left(\frac{81}{\pi}\right)^{1/3} \frac{1 + 2\sqrt{3}}{4} \approx 3.297 \tag{1}$$

This value will be used in the models reported below (Sections 2 and 3). It is worth to note that spherical grains (which, incidentally, do not fill the whole space) and cubic grains lead to κ values of 3.0 and 3.72, respectively. This means that the TKD can be considered as a "good compromise" between the two opposite requirements of filling the space and minimizing the surface energy.

Random aggregates. Various procedures have used Voronoï or Delaunay tessellations, for either two- or three-dimensional numerical simulations. An example is given in Figure 5 below.

Average field models. In this type of model, that will be illustrated in Section 4, each grain is considered as an inclusion embedded in a uniform matrix made of the whole set of grains of

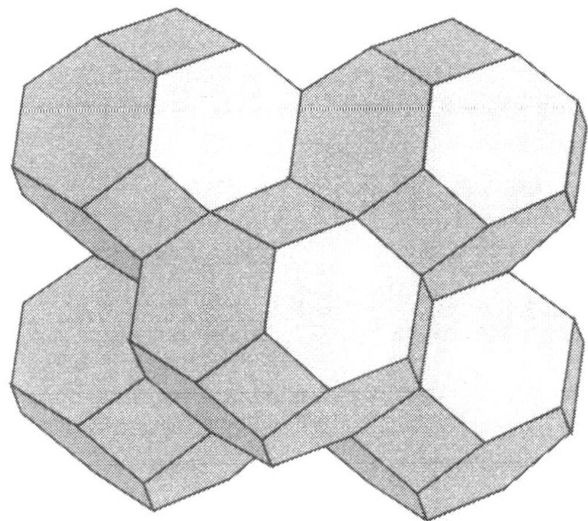

Figure 3. Three-dimensional array of tetrakaidecahedra (Couturier, 2003)

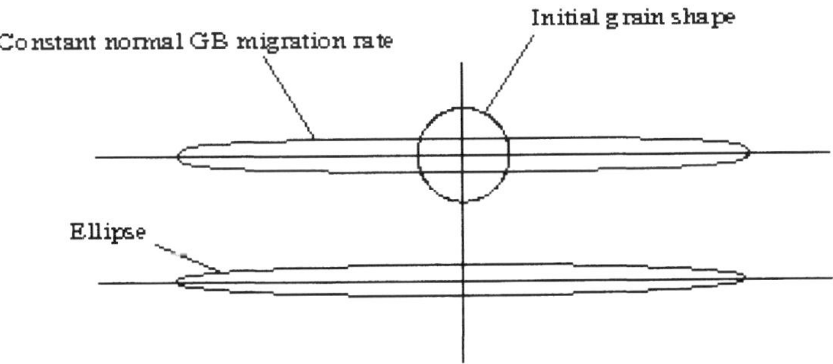

Figure 4. Shape change of an initially circular grain under plane strain compression ($\varepsilon = 1.2$): when GB migration occurs (top), the current shape is not elliptic, by contrast to the case where only material flow is taken into account (bottom).

the aggregate. Each individual grain is generally represented by a circle or an ellipse in 2-dimensions, a sphere or an ellipsoid in 3-dimensions. Such simple quadratic curves or surfaces lead to easy calculations regarding the grain intercepts, surfaces, and volumes. However, it is important to note that ellipsoids (as well as ellipses) do remain ellipsoids under uniform straining of the material, but *not* when grain boundaries are submitted to migration. Figure 4 shows the shape change of an initially circular grain when submitted to plane strain compression, with and without grain boundary migration. In the former case, the elliptical shape is not conserved.

2.3 Grain Boundary Kinematics

During static recrystallization or grain growth, grain boundaries move with respect to matter, which is referred to as grain boundary *migration*. It is basically a diffusion mechanism, that involves only short distance atomic movements. Grain boundary migration is illustrated in Figure 5a, which shows a numerical simulation of static grain growth by a network model. During plastic straining in the cold working range, grain boundaries also move because they are driven by the material flow, which is a *convection* movement (Figure 5b). During hot working, grain boundary migration (GBM) and convection occur simultaneously. Although GBM during straining is very difficult to observe directly by means of *in situ* experiments, and has not generally been included in the DRX models, it is likely to play an important role in the formation of steady state microstructures at large strains.

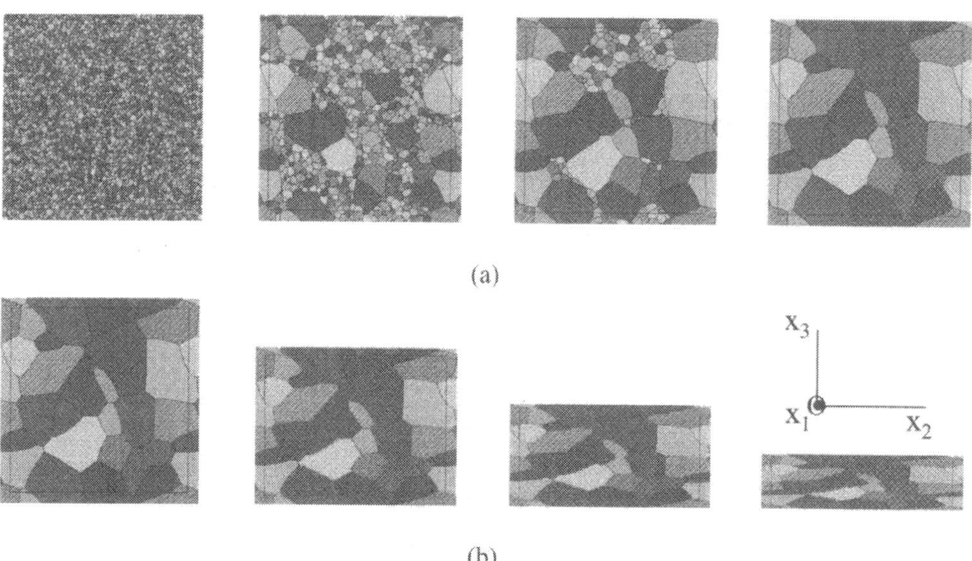

Figure 5. (a) Numerical simulation of static grain growth by a network model illustrating grain boundary migration (Maurice and Humphreys, 1998); (b) Example of grain boundary convection during cold working by plane strain compression, assuming uniform strain (x_1 and x_3 are the elongation and compression directions, respectively).

2.4 Grain Boundary Dynamics

While the convection rate of a grain boundary is obviously determined by the local velocity field: $\mathbf{v_C} = \mathbf{u}(x, y, z)$, the migration rate is assumed to be normal to the grain boundary and proportional to the driving pressure $\mathbf{P} = P\,\mathbf{n}$:

$$\mathbf{v_M} = MP\,\mathbf{n} \tag{2}$$

where M [$\mu m^4/(Js)$] is the grain boundary mobility and P [$N/\mu m^2$ or $J/\mu m^3$] the amplitude of the driving pressure (Figure 6). The latter has two main origins:
- the local dislocation density difference between the two adjacent grains (1) and (2):

$$P = \Delta G = \tau \Delta\rho \tag{3}$$

where $\tau = \mu b^2$ [J/μm] is the line energy of a dislocation, and $\Delta\rho = \rho_2 - \rho_1$ [$\mu m/\mu m^3$]. Note that $M\tau$ [$\mu m^3/s$] has the dimension of a material flux;
- the local grain boundary curvature:

$$P = \gamma\left(\frac{1}{R_1} + \frac{1}{R_2}\right) \tag{4}$$

where R_1 and R_2 are the local principal radii of curvature, and γ [$J/\mu m^2$] the grain boundary energy per unit surface. For a spherical grain of radius R, $\Delta G = 2\gamma / R$, and for a circular grain (two-dimensional analysis), $\Delta G = \gamma / R$.

It should be noted in addition that some authors have mentioned the possible interactions of external (applied) stresses with dislocation stress fields retained in the boundary (Volkov et al., 1979; Senkov et al., 1998; Winning et al., 1999].

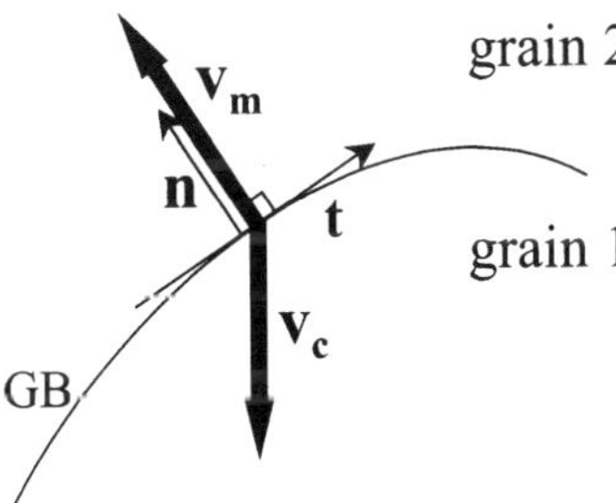

Figure 6. Schematic representation of the two velocity components of a grain boundary (GB).

In the next section, the influence of the main parameters affecting grain boundary mobility, *viz.* grain boundary misorientation, temperature, and metal purity, will be reviewed in turn.

2.5 Grain Boundary Mobility (see Humphreys and Hatherly, 1996)

The geometry of a grain boundary depends on five macroscopic parameters: three Euler angles $(\varphi_1, \phi, \varphi_2)$ (Bunge, 1987) that define the rotation from grain (1) to grain (2), and two of the three components (n_1, n_2, n_3) of the unit vector **n** normal to the boundary (Figure 7). The tranformation matrix from grain (1) to grain (2) can be written in the form:

$$P = \begin{bmatrix} \cos(\varphi_1)\cos(\varphi_2) - \sin(\varphi_1)\sin(\varphi_2)\cos(\phi) & \sin(\varphi_1)\cos(\varphi_2) + \cos(\varphi_1)\sin(\varphi_2)\cos(\phi) & \sin(\varphi_2)\sin(\phi) \\ -\cos(\varphi_1)\sin(\varphi_2) - \sin(\varphi_1)\cos(\varphi_2)\cos(\phi) & -\sin(\varphi_1)\sin(\varphi_2) + \cos(\varphi_1)\cos(\varphi_2)\cos(\phi) & \cos(\varphi_2)\sin(\phi) \\ \sin(\varphi_1)\sin(\phi) & -\cos(\varphi_1)\sin(\phi) & \cos(\phi) \end{bmatrix} \tag{5}$$

where the columns of the matrix consist of the direction cosines of the axes (X_1, X_2, X_3) of grain (2) with respect to the axes (x_1, x_2, x_3) of grain (1).

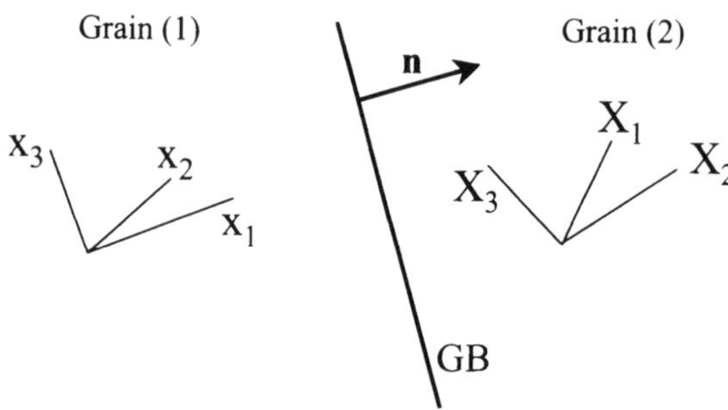

Figure 7. Schematic representation of two adjacent grains (1) and (2) and the grain boundary (GB).

If θ denotes the angle of the single rotation that transforms the set of axes (x_1, x_2, x_3) into (X_1, X_2, X_3), the invariance of the trace of the transformation matrix leads to:

$$2\cos\theta + 1 = \mathrm{tr}(P) \qquad (6)$$

For materials exhibiting cubic symmetry, there are 24 crystallographically equivalent ways to define (X_1, X_2, X_3) once (x_1, x_2, x_3) have been fixed, associated with 24 positive rotation angles θ_i. The smallest one, denoted θ below, is called the *misorientation angle* of the boundary. For cubic crystals, θ varies from 0 to $\theta_M = 2\tan^{-1}\left[\left(\sqrt{2}-1\right)\left(5-2\sqrt{2}\right)^{1/2}\right] \approx$ 62.80 deg (Mackenzie, 1958). When $\theta < \theta_c$, where θ_c ranges from 10 to 15 deg, the boundary is referred to as a *subgrain boundary* or *low-angle boundary (LAB)*. It is then made of one to three arrays of parallel dislocations. When $\theta > \theta_c$, the boundary is a *grain boundary* or *high-angle boundary (HAB)*, generally described in terms of coincidence site lattice (CSL).

Misorientation dependence of grain boundary mobility. M values are mainly associated with climb of individual dislocations for LAB, and to atom jumps across the boundary for HAB. As a general rule, M is much larger for HAB, except for some special orientations (*e.g.*, twin boundaries). However, the temperature dependence of M is larger for LAB, such that the difference vanishes with increasing temperature. This is illustrated in Figure 8, which shows the misorientation dependence of grain boundary velocity in two high purity aluminium grades. Assuming the driving force is constant, the curves are expected to reflect the variations of mobility.

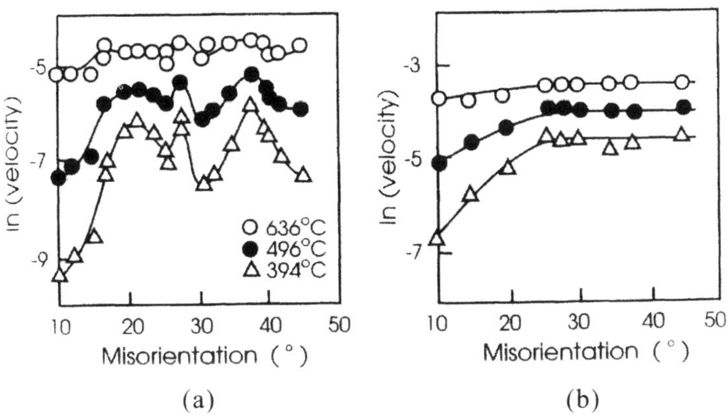

Figure 8. Misorientation dependence of grain boundary migration velocity in high purity aluminium grades: (a) 99.9992 at% Al; (b) 99.99995 at% Al (Fridman et al., 1975).

Temperature dependence of grain boundary mobility. Since M is determined by thermally activated mechanisms, it is assumed to follow an Arrhenius type relationship:

$$M = M_0 \exp\left(-\frac{Q_M}{RT}\right) \tag{7}$$

where M_0 is a constant, R the gas constant, T the temperature, and Q_M the activation energy for grain boundary migration. For LABs, $Q_M \approx Q_S$, the activation energy for bulk self-diffusion, whereas for HABs, $Q_M \approx Q_B$, the activation energy for grain boundary diffusion, which is significantly less than Q_S: $Q_B \approx 0.5$ to $0.7\ Q_S$. Figure 9 shows the misorientation dependence of Q_M measured in high purity copper bicrystals by Viswanathan and Bauer (1973) for pure tilt boundaries (*i.e.*, the rotation axis lies within the boundary plane), and mixed boundaries (*i.e.*, the rotation axis has a nonzero component outside the boundary plane). A sharp drop between the LAB range where Q_M is close to Q_S and the HAB range where Q_M is close to Q_B is clearly visible. Figure 10 compares Q_S, Q_B, and Q for dislocation core diffusion, to Q_M for various metals. It shows that the activation energy for grain boundary migration is generally close to that for boundary and core diffusion, which are significantly less than the activation energy for bulk self-diffusion.

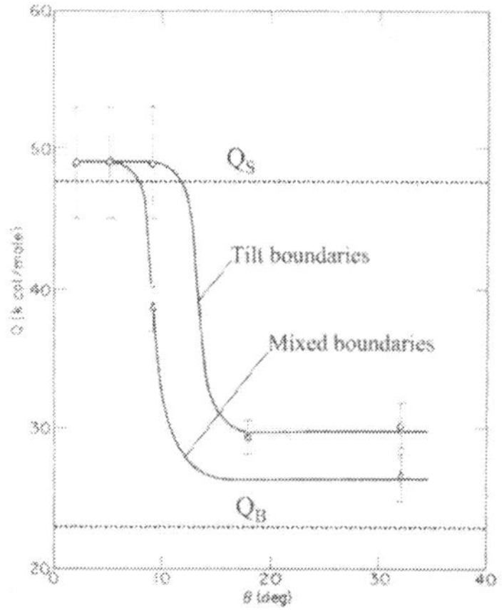

Figure 9. Misorientation dependence of the activation energy of grain boundary migration in high purity copper bicrystals (Viswanathan and Bauer, 1973).

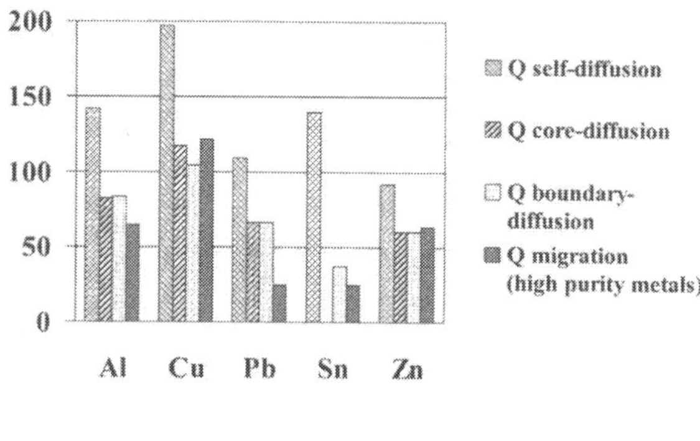

Figure 10. Comparison of the activation energies (kJ/mol) of self-, core- and boundary diffusion with that pertaining to grain boundary migration (after Humphreys and Hatherly, 1996).

Effects of solutes on grain boundary mobility. Figure 11a shows measurements of migration rates, that are considered as above to reflect variations of grain boundary mobility, in Al in the presence of Cu or Mg solutes. At low concentrations, M is large and independent of the solute concentration. At higher concentrations, mobility is much lower and decays with increasing concentration. Furthermore, Figure 11b shows that the activation energy associated with migration is much larger at high solute concentrations. Such increase of the temperature sensitivity may result from the solute drag effect which reduces grain boundary mobility at low temperatures, but vanishes at higher temperatures due to thermal activation.

Finally, The migration rate was assumed till now to be proportional to the driving force P [eq.(2)], which does not hold in the presence of solutes: Lücke and Stüwe (1963) have derived the following relationship relating P to v_M:

$$P = \frac{v_M}{M} + \frac{\alpha C v_M}{1 + \alpha\alpha' v_M^2} \quad (8)$$

where M is the intrinsic mobility of the grain boundary, C the solute concentration, and α and α' are two constants. When v_M (and therefore P) is low, the above equation yields $v_M = M_{app} P$, where the apparent mobility $M_{app} = M/(1+\alpha CM)$ decreases with an increasing solution concentration C (see Figure 11a). By contrast, at large velocities v_M (i.e. large driving forces P), $v_M \approx MP$, which means that the influence of solutes vanishes.

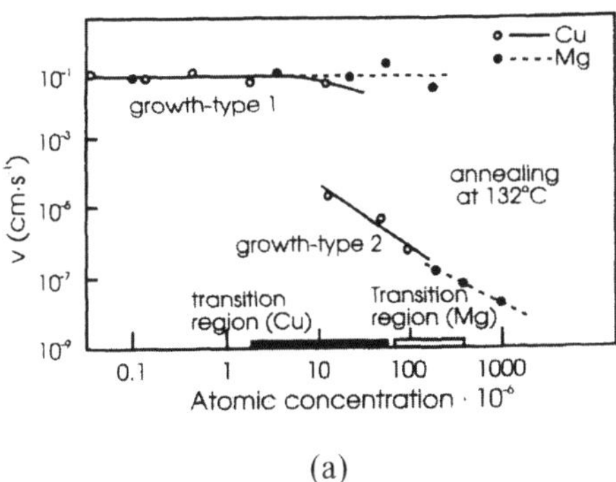

(a)

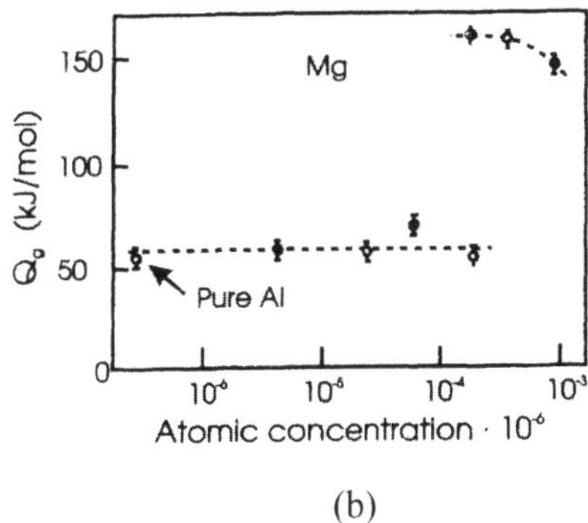

(b)

Figure 11. (a) Influence of the concentration of Cu and Mg in aluminium on the migration rate of grain boundaries during annealing at 132 °C; (b) Influence of the concentration of Mg in aluminium on the activation energy for grain boundary migration (Frois and Dimitrov, 1996).

3 Effects of Dynamic Grain Boundary Migration During Hot Working

Whatever the type of dynamic recrystallization (see Sections 4 and 5), grain boundary migration occurs during hot working of metals. In this Section, the main two effects of GBM, *viz.* flow stress softening and grain coarsening, are first presented and separately analyzed using elementary analytical approaches. A more sophisticated, although still simplified, model is then developed to assess the effects of dynamic GBM (Gourdet and Montheillet, 2002).

3.1 Flow Stress Softening and Grain Coarsening

It is well known that stress-strain curves associated with high temperature deformation most often exhibit a more or less rounded peak at medium to large strains. In low SFE metals, this has been attributed to classical or *discontinuous* dynamic recrystallization (see Section 5), *i.e.* to the nucleation and growth of new grains with low dislocation densities that replace the "old" strain hardened grains. Although such an explanation remains correct, it does not readily explain the flow softening that is observed in high SFE metals as well, since in that case discontinuous DRX does not occur, and it is no longer possible to make any difference between two different populations of grains (see Section 4). However, torsion tests carried out on α-iron or aluminium alloys lead to quite significant flow stress decays at very large strains, as illustrated in Figure 12. It has been shown that texture effects are by far not sufficient to explain such softening amplitudes (Chovet-Sauvage, 2000).

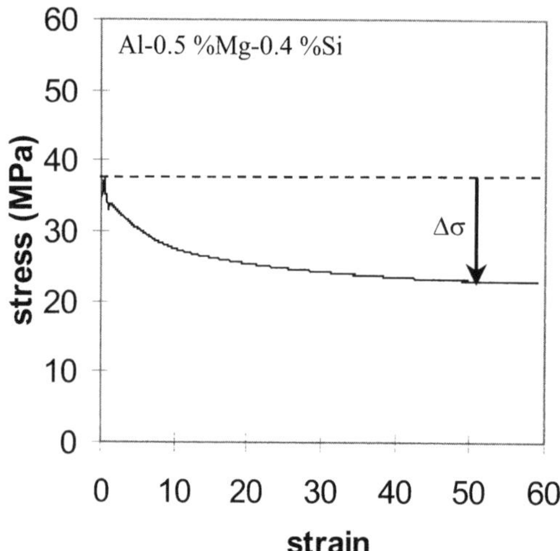

Figure 12. Flow softening of an aluminium alloy during large strain torsion testing at 400 °C and $0.1\ s^{-1}$. The steady state relative softening $\Delta\sigma/\sigma$ is close to 0.37 (Chovet-Sauvage, 2000).

Boundary migration induced softening (BMIS) can be analyzed quite simply in the large strain steady state. The dislocation density evolution in the material can be described by the following equation:

$$\frac{d\rho}{dt} = (h - r\rho)\dot{\varepsilon} - \rho S_V v_M \tag{9}$$

where ρ is the dislocation density and $\dot{\varepsilon}$ the strain rate. The first term on the right hand side corresponds to the Yoshie-Laasraoui-Jonas equation (YLJ), commonly used to account for strain hardening and dynamic recovery through the h and r parameters, respectively (Laasraoui and Jonas, 1991). The additional term is associated with GBM, where v_M is the average rate of migration and S_V denotes the grain boundary area per unit volume. At the steady state, $d\rho / dt = 0$, whence:

$$\rho_S = \frac{1}{1 + \left(\kappa v_M / r\dot{\varepsilon} D\right)} \frac{h}{r} \tag{10}$$

Here the relationship $S_V = \kappa / D$ was used, where D is the average grain diameter and κ a geometrical constant close to 3 (see Section 2.2). Finally, the flow stress is given by the classical equation:

$$\sigma_S = A\mu b\sqrt{\rho_S} \tag{11}$$

where μ is the elastic shear modulus, b the Burgers vector, and A a constant close to unity. For aluminium deformed at 400 °C and 0.1 s^{-1}, using $v_M = 0.5$ µm/s and $D = 8$ µm (Gourdet, 1997), eqs. (10) and (11) lead to $\Delta\rho/\rho \approx 0.1$, which shows that flow softening associated with dynamic GBM can be significant.

The second effect of GBM is grain coarsening. Figure 13 shows the decrease of the average grain thickness H during hot compression of an aluminium alloy. Here H was merely defined as the ratio of the specimen thickness to the number of grains. Using the uniform strain (Taylor) assumption leads to the "geometrical" prediction:

$$H_G = H_0 \exp(-\varepsilon) \tag{12}$$

where H_0 is the initial average grain thickness and ε the compression strain. According to Figure 13, the decrease of H is significantly less than predicted by the above equation. Quite similar observations have been made on a β-titanium alloy (Gourdet et al., 1997). A simple way to account for GBM is to consider the behaviour of an "average grain" (Figure 14). During compression, the grain boundary is submitted to both a convection rate $-H\dot{\varepsilon}$, which tends to decrease H, and an (apparent) migration rate v_{app} which opposes to the latter. This leads to a modified prediction for H:

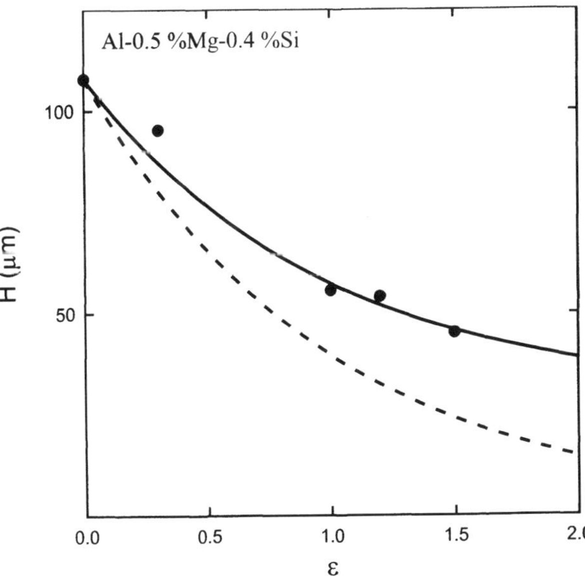

Figure 13. Strain dependence of the average grain thickness during hot compression of an aluminium alloy at 400 °C and 0.1 s^{-1}. Experimental data (dots) are compared to the predictions of eqs.(12) (broken line) and (13) (solid line) with $v_{app} = 1.4$ µm/s (Chovet-Sauvage, 2000).

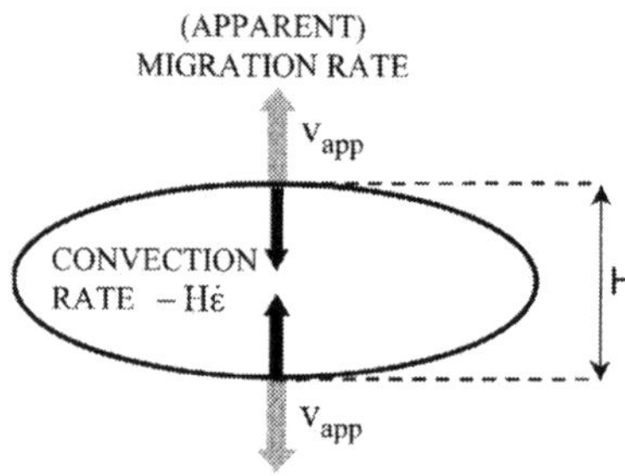

Figure 14. Behaviour of an "average grain" during compression (Gourdet and Montheillet, 2002).

$$H = H_G + H_S\left[1 - \exp(-\varepsilon)\right] \tag{13}$$

where H tends to a steady state value $H_S = 2v_{app}/\dot{\varepsilon}$ at large strains. In Figure 13, the above equation was used to fit the data with v_{app} = 1.4 μm/s. However, such oversimplified approach does not account for the real behaviour of individual grain boundaries, which move upward or downward, according to the local grain to grain dislocation density differences. During straining, some grains grow, while the others shrink. Finally, the average grain thickness increases only when grains locally (*i.e.* along the line used to measure the average H value) disappear. The model presented in Section 3.2 aims at accounting for these various effects and predicting both flow stress softening and grain coarsening.

3.2 A One-Dimensional Model of Hot Compression

Description of the model. The microstructure of a compression specimen is represented schematically by a stack of N grains (Figure 15), where grain (i) is delimited by two boundaries of coordinates x_{i-1} and x_i (i = 1, N). The displacement of the upper boundary is prescribed according to the kinetics of the compression test:

$$x_N = x_{N_0}\exp(-\varepsilon) \tag{14}$$

while each individual grain boundary moves according to the following equation:

$$\frac{dx_i}{dt} = -x_i\dot{\varepsilon} + M\tau(\rho_{i+1} - \rho_i) \qquad \text{for } i = 1, N \tag{15}$$

where ρ_i and ρ_{i+1} are the average dislocation densities in grains (i) and (i+1). The first term on the right hand side is associated with convection (assuming again uniform strain), whereas the second stands for grain boundary migration (see eqs. (2) and (3) above). When two boundaries meet, the grain between them vanishes and N is decreased by one unit. The dislocation density changes during the time increment dt are given in turn by:

$$d\rho_i = d\rho_i^h + d\rho_i^M \qquad \text{for } i = 1, N \tag{16}$$

where $d\rho_i^h = (h_i - r\rho_i)d\varepsilon$ accounts for strain hardening and dynamic recovery (YLJ equation), and $d\rho_i^M$ for BMIS. The derivation of this term requires to consider various possible cases for each grain (Figure 16): ρ_i is not affected by GBM when a boundary moves towards the interior of the grain (case 1, upper boundary in case 2), whereas it decreases when a boundary moves towards the exterior of the grain (lower boundary in case 2, case 3) since in that case a new volume with low dislocation content ρ_0 is added to the grain. The net increment $d\rho_i^M$ is obtained by a rule of mixtures assuming instantaneous homogenization of the dislocation density within each grain.

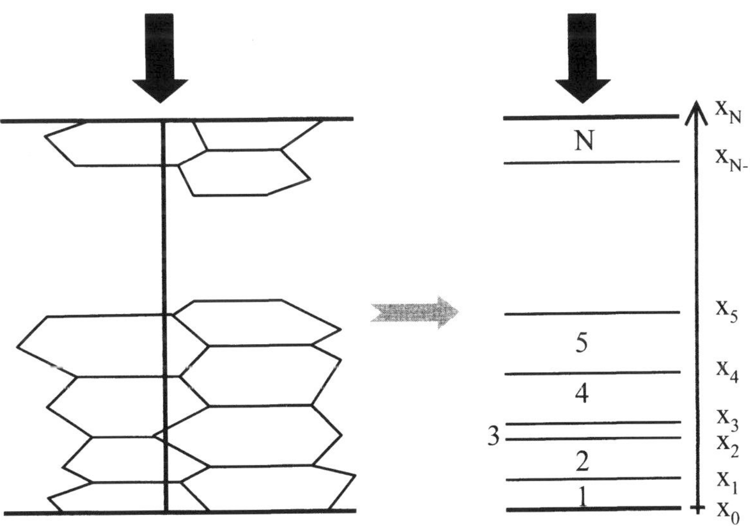

Figure 15. Schematic representation of a grain aggregate by a one-dimensional stack of grains (Gourdet and Montheillet, 2002)

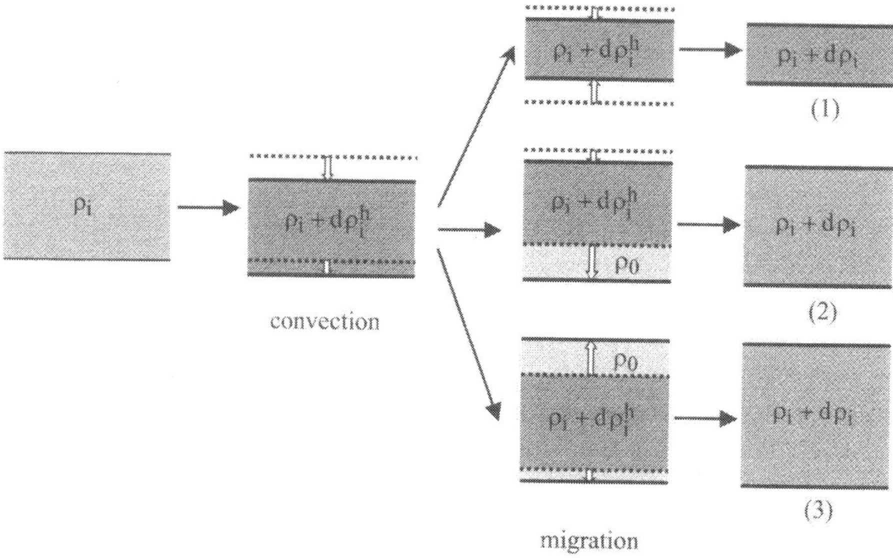

Figure 16. Schematic representation of boundary migration induced softening (BMIS) (Gourdet and Montheillet, 2002).

Materials parameters. The strain hardening h and dynamic recovery r parameters involved in the YLJ equation (9), or any similar equation, are generally considered to be strain rate and temperature dependent in opposite ways:

$$\begin{aligned} h &= A\dot{\varepsilon}^{m_h} \exp(m_h Q_h / RT) \\ r &= B\dot{\varepsilon}^{-m_r} \exp(-m_r Q_r / RT) \end{aligned} \quad (17)$$

where m_h and m_r are the strain rate sensitivities, and Q_h and Q_r the activation energies associated with the above two variables, R the gas constant and A and B two scaling factors. In the absence of grain boundary migration, the YLJ equation leads to the steady state dislocation density $\rho_S = h/r$ [see eq.(10)]. Equation (11) then yields:

$$\sigma_S = C\mu b\, \dot{\varepsilon}^{(m_h + m_r)/2} \exp\left(\frac{m_h Q_h + m_r Q_r}{2RT}\right) \quad (18)$$

where C is a constant. This is a classical power law viscoplastic flow rule with an Arrhenius type temperature dependence. The strain rate sensitivity is therefore $m = (m_h + m_r)/2$, and the apparent activation energy $Q = (m_h Q_h + m_r Q_r)/(m_h + m_r)$.

In the present model, for the sake of simplification, the following assumptions were made for describing the behaviour of commercial purity aluminium deformed within the range 0.6-

$0.8\ T_m$ and 10^{-3}-1 s^{-1} (Gourdet and Montheillet, 2002): $m_h = m_r = m = 0.2$, $Q_h = Q_r = Q_S = 155\ kJ/mol$, *i.e.* the activation energy for self diffusion. For the grain boundary mobility, an Arrhenius dependence was also adopted [see eq.(7)]. Finally, the strain hardening parameters h_i of the various grains in the stack were assumed to be uniformly distributed within the range $h_0 \pm \Delta h$ with $\Delta h = 0.025\ h_0$ to account for the different crystallographic orientations of the grains. The initial grain thicknesses H_i were symmetrically scattered according to a triangular distribution with a maximum deviation $\Delta H = 0.5\ H_0$. The other parameters were chosen according to various experimental data (Gourdet and Montheillet, 2002; Gourdet, 1997; Humphreys and Hatherly, 1996). No significant changes of the output data, except in some cases for the transient stages, were observed when the above distributions were modified.

Migration induced flow stress softening. The current average dislocation density in the stack of grains is given by:

$$\rho = \sum_{i=1}^{N} (x_i - x_{i-1})\rho_i / N \tag{19}$$

from which the flow stress is derived through eq.(11). Figure 17 shows predicted stress-strain curves for commercial purity aluminium at various temperatures. Flow stress softening is negligible at $0.8\ T_m$, but quite significant at lower temperatures.

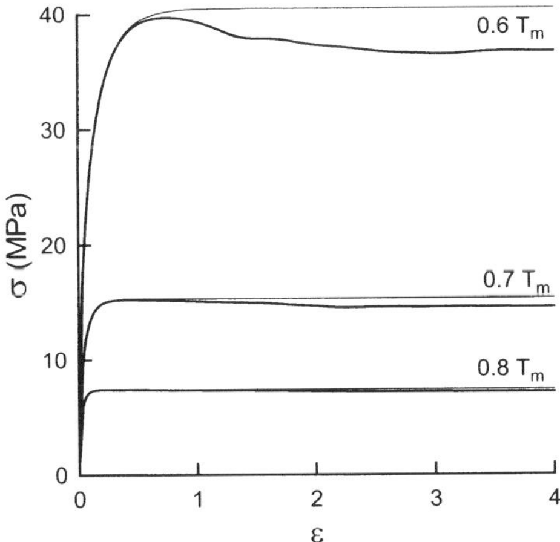

Figure 17. Stress-strain curves predicted by the one-dimensional model of hot compression for commercial purity aluminium at $10^{-2}\ s^{-1}$; thin lines are associated with the absence of GBM (Gourdet and Montheillet, 2002).

Migration induced grain coarsening. The strain dependence of the average grain thickness H is illustrated by Figures 18 a and b. After a transient stage, H tends to a "steady state" value (this holds, of course, as long as the number of grains in the stack is large enough). The latter decreases with increasing strain rate (Figure 18a), but is almost insensitive to temperature (Figure 18b). This is due to the fact that, in aluminium, the increase of grain boundary mobility M with temperature is almost exactly counterbalanced by the decrease of the average driving force $\Delta\rho$. Finally, Figure 18c shows that the steady state grain thickness is independent of the initial average value H_0.

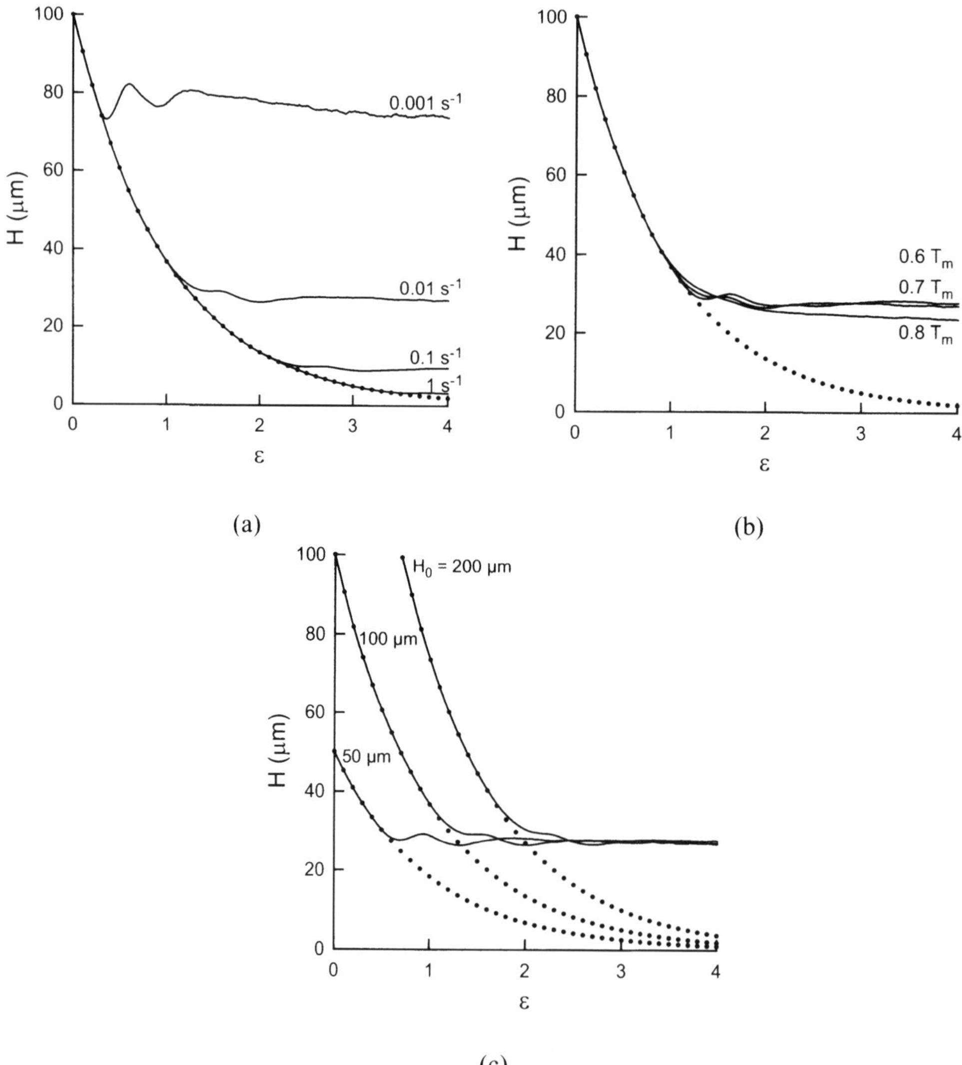

Figure 18. Predicted strain dependence of the grain thickness for various strain rates at 0.7 T_m (a), various temperatures at 10^{-2} s^{-1} (b), and various initial average grain thicknesses at 0.7 T_m and 10^{-2} s^{-1} (c); dotted lines are associated with the absence of GBM (Gourdet and Montheillet, 2002).

It is worth to note that the occurrence of a large strain "steady state" grain thickness was already predicted by the simplified "average grain" approach [see eq.(13)].

Average dislocation density *vs.* average dislocation density difference. The current average dislocation density ρ is defined by the following equation:

$$\rho = \sum_{i=1}^{N} (x_i - x_{i-1})\rho_i / N \tag{20}$$

while the average dislocation density difference $\Delta\rho$ between neighbouring grains is given by:

$$\Delta\rho = \sum_{i=1}^{N} |\rho_i - \rho_{i-1}| / (N-1) \tag{21}$$

where $N - 1$ is the current number of grain boundaries in the stack. Figure 19 shows the relationship between the above two quantities for various temperature and strain rate conditions. In each case, $\Delta\rho$ starts growing slowly and, after a sudden and strong increase, stabilizes around a "steady state" value. In the first stage, corresponding to the strain hardening range, $\Delta\rho \approx 0.017\,\rho$, whereas in the steady state the best fit is given by:

$$\Delta\rho \approx k\rho^p \tag{22}$$

with $p \approx 0.25$ and $k \approx 7.3\,10^{-5}\ \mu m^{-2(1-p)}$. This is an important result, since $\Delta\rho$ is directly related to the average grain boundary migration rate [see eq.(3)] and its estimation is necessary for numerical simulations of DRX.

True *vs.* apparent average migration rate. These two quantities are compared in Figure 20. The former one is defined by:

$$v_M = M\tau\Delta\rho \tag{23}$$

where $\Delta\rho$ is the above estimated average dislocation density difference. The apparent migration rate is derived in turn from the average grain thickness evolution [see eq.(13)]:

$$v_{app} = \frac{(H - H_G)\dot{\varepsilon}}{2[1 - \exp(-\varepsilon)]} \tag{24}$$

The diagram shows that the two variables are related by a linear equation:

$$v_M = \lambda v_{app} \tag{25}$$

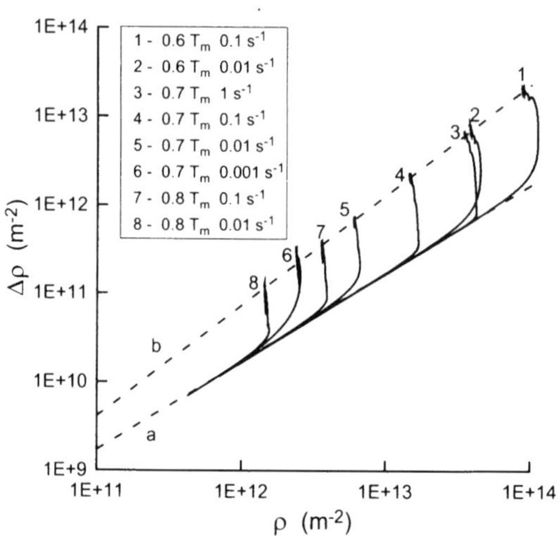

Figure 19. Predicted relationship between the average dislocation density difference Δρ and the average dislocation density calculated over the whole stack of grains, for various straining conditions. Lines a and b are associated with the strain hardening and steady state ranges, respectively (Gourdet and Montheillet, 2002).

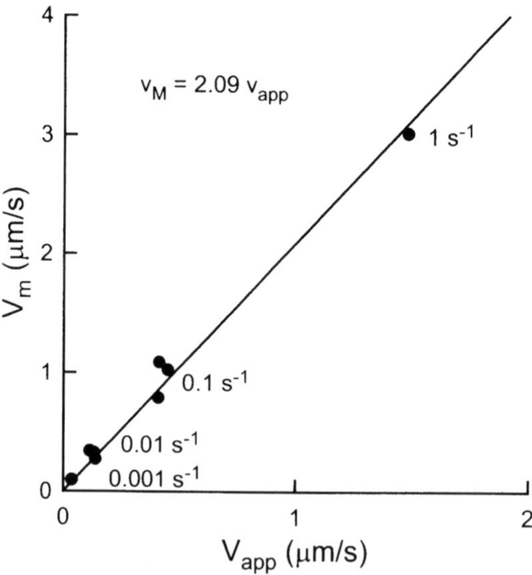

Figure 20. Predicted relationship between the true migration rate v_M and the apparent migration rate v_{app} (Gourdet and Montheillet, 2002).

with $\lambda \approx 2$. Since v_{app} can be directly deduced from experimental measurements of the average grain thicknesses at various strains, the present model thus provides a simple way for estimating v_M. Another important result is that the migration rate is multiplied by a factor 3 when the strain rate is increased by an order of magnitude, whereas temperature changes have only a limited effect, which is consistent with grain thickness evolutions displayed in Figure 18b.

The model described in the above section shows that dynamic grain boundary migration is one of the main mechanisms involved during hot working of metals. However, it does not address dynamic recrystallization itself, since neither the nucleation of new grains, nor the generation of new boundaries are accounted for. Such problems will be dealt with in the next two sections.

4 Continuous Dynamic Recrystallization

4.1 Various Types of Dynamic Recrystallization

It is now well recognized that two types of DRX can take place during hot working of metals, *viz.*, *continuous* DRX (sometimes referred to as "rotation", "apparent", or "*in situ*" DRX, or else "extended dynamic recovery"), and *discontinuous* (or "classical") DRX. The first type (CDRX) is observed in high stacking fault energy materials, such as aluminium an its alloys, α-iron and the ferritic steels, and β-titanium, whereas the second (DDRX) pertains to low SFE alloys, such as copper, γ-iron and the austenitic steels, nickel and most of the nickel base superalloys. In some cases, however, both CDRX and DDRX can be observed according to the material purity and the applied temperature and strain rate (see Section 6). The main characteristics of these two microstructural mechanisms are summarized in Table 1 below.

Table 1. Main features of CDRX and DDRX

CDRX	DDRX
• Occurs by progressive (slow) transformation of subgrain or low-angle boundaries (LABs) into grain or high-angle boundaries (HABs)	• Occurs by local (rapid) cycles involving strain hardening → nucleation → growth of new grains
• Dynamic recovery, *i.e.* dislocation rearrangement and annihilation is strong	• Dynamic recovery is weak
• Dislocation densities are homogeneous ($\Delta\rho$ is weak)	• Dislocation densities are inhomogeneous ($\Delta\rho$ is large)
• The rate of grain boundary migration is low	• The rate of grain boundary migration is high

4.2 Experimental Data Pertaining to CDRX

Stress-strain curves. Figure 21a and b show typical flow stress evolutions for commercial purity aluminium and three ferritic stainless steels, respectively. In both cases, the flow stress goes through a maximum after which it decreases continuously. Large to very large strains are then required before a steady state is established. For the stainless steels, the peak is followed by an intermediate plateau, or even a secondary maximum, which has been mainly attributed to texture changes (Oliveira, 2003). However, such flow curves are basically "smooth", whatever the applied deformation conditions, by contrast to DDRX stress-strain curves which may become "oscillating" at high temperatures and/or low strain rates (see Figure 30 below).

Generation of a crystallite microstructure. EBSD misorientation maps are used in Figure 22a-e to illustrate microstructural changes associated with CDRX during hot torsion of an aluminium alloy (see the associated flow curve in Figure 12). High-angle boundaries ($\theta > 15$ deg) and low-angle boundaries (($\theta < 15$ deg) are represented by black and gray lines, respectively. At low strains (Figure 22a), the deformed shapes of the original grains reflect the prescribed shear strain. A number of LABs have formed, preferentially close to the initial HABs, which exhibit serrations. At medium to large strains (Figures 22b-c), the initial large grain microstructure is progressively replaced by a much finer "recrystallized" one, while accordingly the volume fraction of HABs strongly increases. Such evolution strongly suggests that LABs progressively transform into HABs during straining. Finally, at very large strains (Figures 22d-e), some coarsening occurs before a steady state is established, as shown by the similitude of Figures 22d and e. A quite similar picture is offered by a Fe-11 %Cr ferritic steel strained up to $\varepsilon = 20$ (Figure 22f). For describing such microstructures, it is convenient to introduce the term *crystallite*, which means a part of an aggregate exhibiting uniform crystallographic orientation, and that is delimited partly by HABs (grain boundaries) and partly by LABs (subgrain boundaries). By contrast, *grains* are completely delimited by HABs and *subgrains* by LABs. The above microstructures have therefore been referred to as crystallite microstructures (Gourdet and Montheillet, 2000).

In addition, it is worth to note that some "new grains" are also generated at low strains by *geometric dynamic recrystallization* (McQueen et al., 1985), that involves the increase of the initial HAB area per unit volume due to the flattening or elongation of the original grains. This is associated with the development of serrations, and the subsequent "pinching-off" of some parts of that grains, when their thickness has been sufficiently reduced by grain boundary movements (see for instance area marked G in Figure 22a). This mechanism is distinct from CDRX, since it requires the presence of pre-existing HABs. It is likely, however, that the two mechanisms may operate simultaneously, their respective contributions depending on the initial grain size, the strain path, and the influence of grain boundary migration which opposes the reduction of grain thicknesses (see Section 3 above).

The occurrence of CDRX can be brought into evidence in a more quantitative way by the evolution of misorientation distributions derived from EBSD data: Figure 23a shows that before straining the distribution is close to that derived by Mackenzie (1958) for cubic

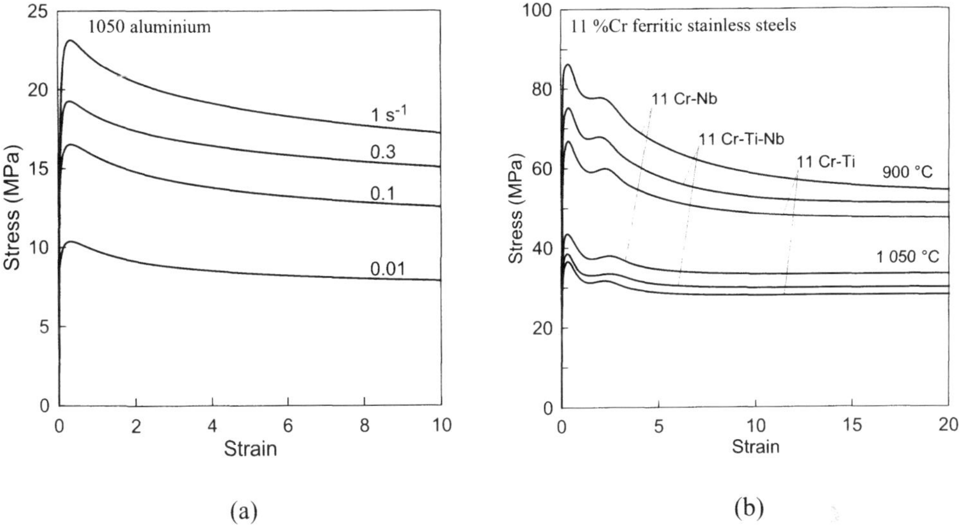

Figure 21. Stress-strain curves obtained from torsion tests of materials exhibiting CDRX: (a) a commercial purity (1050 grade) aluminium at 450 °C and various strain rates; (b) three grades of a 11 %Cr ferritic stainless steel in which the ferrite phase is stabilized by additions of Ti, Ti + Nb, and Nb, respectively, at 1 s^{-1} and two temperatures (Oliveira and Montheillet, 2002).

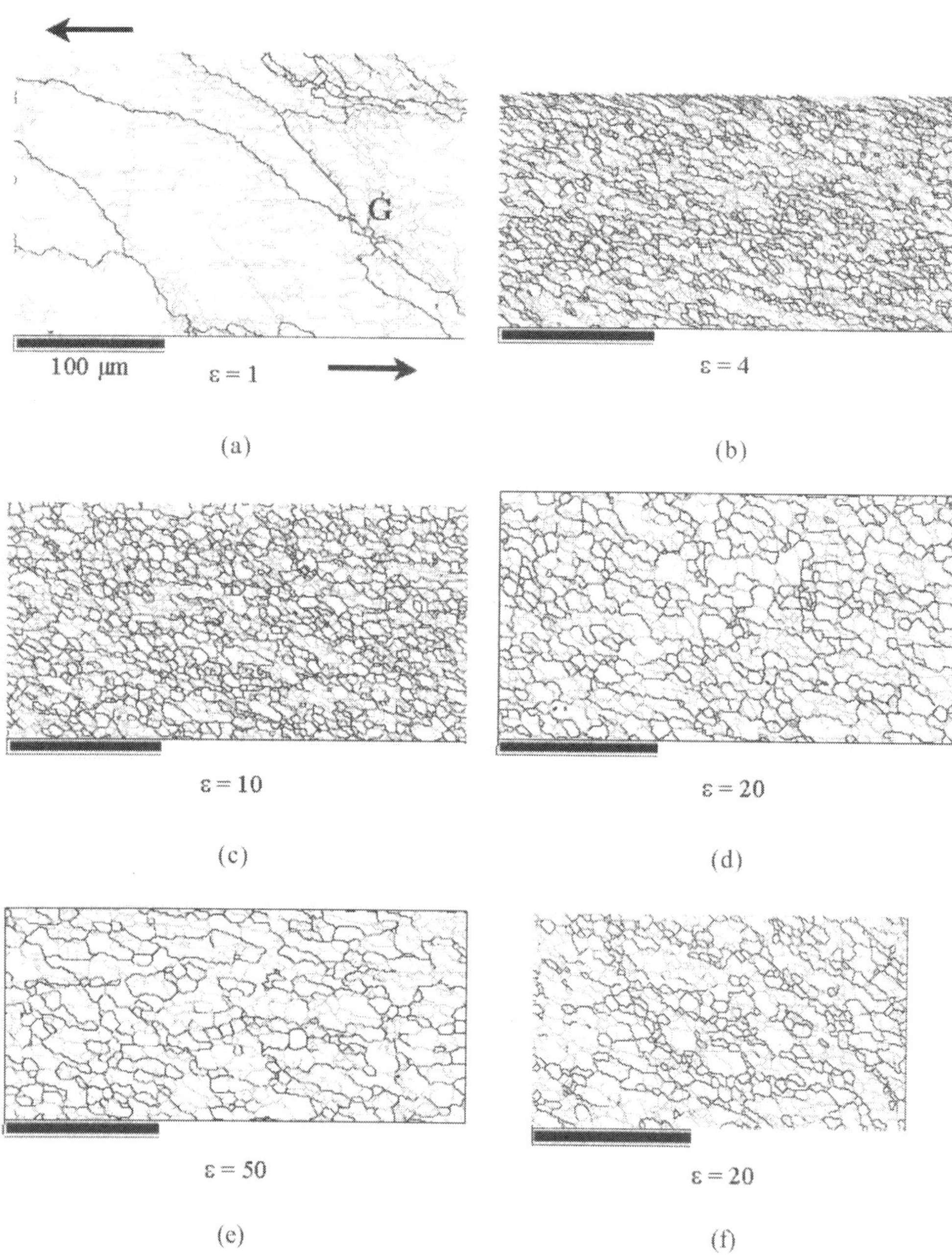

Figure 22. Microstructures generated during torsional straining of materials exhibiting CDRX: (a)-(e) Al-0.5 %Mg-0.4 %Si alloy at 400 °C, 0.1 s^{-1} and various strains (Chovet-Sauvage, 2000); (f) Nb stabilized Fe-11 %Cr ferritic steel at 800 °C, 1 s^{-1} and $\varepsilon = 20$ (Oliveira, 2003). Arrows indicate the shear direction. G denotes the occurrence of geometric dynamic recrystallization in Figure 22a.

crystals uniformly distributed in the orientation space, with a few LABs and a maximum of HABs close to 45 deg. At a strain of 20 (Figure 23b), associated with the microstructure of Figure 22d, the histogram is quite different: the area fraction of LABs is much larger and decreases from 0 to 15 deg, while HABs are almost uniformly distributed between θ_c and θ_M, with a limited increase at large values close to 60 deg.

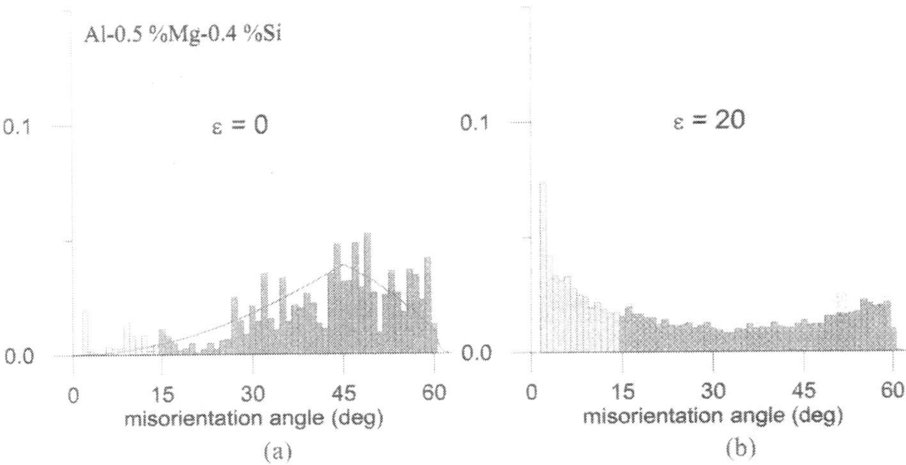

Figure 23. Misorientation distribution functions derived from EBSD measurements for an aluminium alloy before (a) and after (b) torsional straining to $\varepsilon = 20$ at 400 °C, 0.1 s^{-1}. The solid line in (a) depicts the theoretical distribution for a set of uniformly oriented grains (Chovet-Sauvage, 2000).

Large strain CDRX torsion textures. To close this brief account of CDRX characteristics, it is necessary to deal with hot deformation textures: as a general rule, the intensities of the latter are quite strong at large strains (steady state), as illustrated in Figure 24 in the case of torsion. In aluminium and its alloys (Figure 24a), the main component is $B/\bar{B}$ $\{112\}<1\bar{1}0>$, where $\{hkl\}$ and <uvw> denote here the shear plane perpendicular to the torsion axis z, and the shear direction θ, respectively. This component is termed "twin-symmetric" since it is made of two ideal orientations B and $\bar{B}$, symmetric with respect to the shear plane. Simultaneous development of B and $\bar{B}$ with identical intensities is obviously necessary to fulfill the symmetry requirements of simple shear. In the ferritic stainless steel illustrated in Figure 24b (as well as in any ferritic alloy), the steady state component is D_2 $\{112\}<\bar{1}\bar{1}1>$, which is "self-symmetric", *i.e.* made of one single ideal orientation that fulfills the symmetry conditions by itself. The component D_2 is clearly the counterpart of $B/\bar{B}$ in the transformation $<110> \leftrightarrow <111>$ which relates the slip systems of the FCC to that of the BBC crystallographic structures. It is remarkable, however, that D_1, the orientation symmetric of D_2 with respect to the shear plane, is never observed, unless the shear direction is reversed. It has been recently proposed that this could be attributed to the ability of D_2, rather than D_1, grains to grow at the expense of their neighborhood during deformation (Oliveira, 2003). This underlines the fact that, although CDRX textures are basically deformation textures, they are also influenced by grain growth, *i.e.* dynamic grain boundary migration.

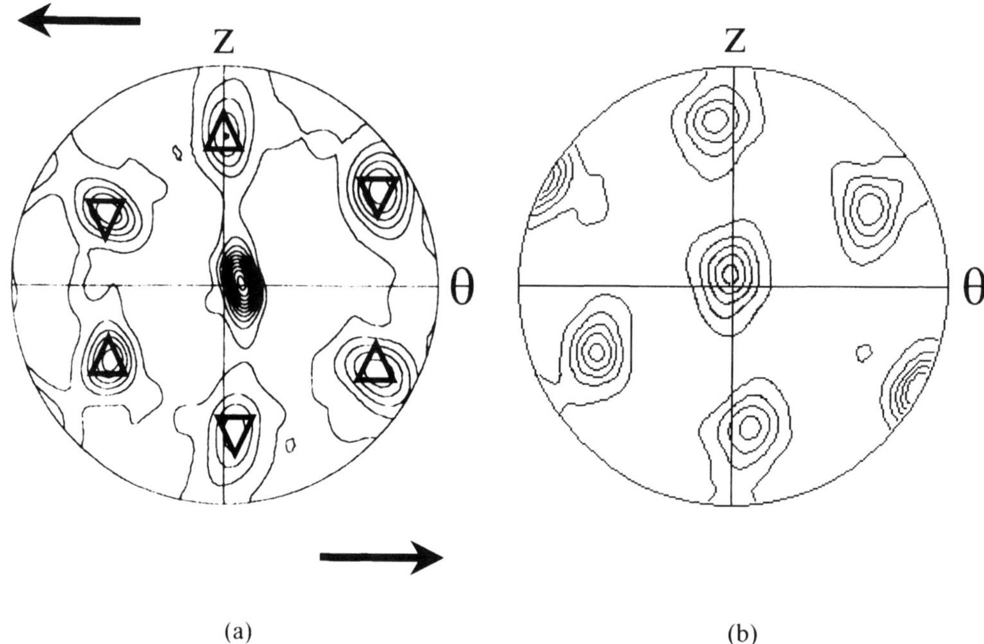

Figure 24. Large strain (steady state) torsion textures represented by {111} (a) and {110} (b) pole figures derived from EBSD measurements: (a) Al-0.5 %Mg-0.4 %Si alloy at 400 °C, 0.1 s^{-1}, and $\varepsilon = 20$ (Chovet-Sauvage, 2000); (b) Nb stabilized Fe-11 %Cr ferritic steel at 1050 °C, 1 s^{-1}, and $\varepsilon = 20$ (Oliveira, 2003). In (a), upward and downward triangles stand for ideal orientations B and $\overline{B}$, respectively, the maximum at the center of the pole figure is common to the two orientations. Arrows indicate the shear direction.

4.3 A Model of Continuous Dynamic Recrystallization

Only a very few number of authors have tried to analyze the physical mechanisms associated with CDRX to give them a mathematical representation. Interesting ideas have nevertheless been proposed by Lyttle and Wert (1994a and b), who formulated three simulative models of microstructure and microtexture evolutions. However, their approach involved grain boundary sliding and crystallite switching, two mechanisms that are more relevant for the superplastic range, and did not lead to really quantitative predictions.

The model presented in the present section basically assumes that plastic deformation occurs by dislocation slip and climb processes, which is relevant for classical high temperature thermomechanical processing at strain rates above $\approx 10^{-2}$ s^{-1}. The model involves a minimum number of physical mechanisms and parameters, and can be dealt with almost analytically.

Although texture formation and topological effects are not addressed, it aims at providing a comprehensive insight into the basic mechanisms of CDRX (Gourdet and Montheillet, 2003).

Description of the model. Figure 25 shows a schematic representation of a CDRX microstructure, *i.e.* an aggregate of crystallites, as defined in Section 4.2: each of them is delimited partly by grain boundaries (HABs), and partly by subgrain boundaries (LABs). The latter are considered to be made of one to three arrays of parallel dislocations.

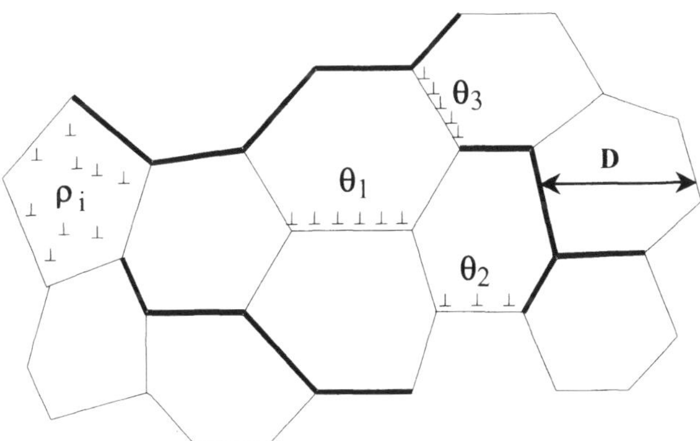

Figure 25. Schematic drawing of a crystallite aggregate. HABs and LABs are represented by thick and thin lines, respectively (Gourdet and Montheillet, 2003).

The model involves three internal variables: the average dislocation density ρ_i inside the crystallites, the average size D of the crystallites, and the distribution density function $\varphi(\theta)$ of the subgrain boundary misorientations, such that $\varphi(\theta)d\theta$ is the fraction of LABs with misorientations ranging between θ and $\theta + d\theta$. The diagram in Figure 26 summarizes the behaviour of dislocations during hot deformation: the main part $d\rho_i^{-(1)}$ of the dislocation density $d\rho_i^+$ generated during a strain increment $d\varepsilon$ is subject to dynamic recovery. While a fraction α is consumed by the creation of new subgrain boundaries (with a misorientation angle θ_0), the remaining fraction $1 - \alpha$ is merely absorbed by LABs or HABs, in direct ratio to their respective surface fractions f_{LAB} and f_{HAB}. The absorption of dislocations by subgrain boundaries induces in turn a progressive increase of their misorientation angles, which may range from θ_0 to θ_c, as indicated by the gray arrows in the diagram. Whenever θ reaches the critical value θ_c, the subgrain boundary transforms into an ordinary grain boundary, the misorientation of which is not accounted for in the model. Finally, dynamic migration of LABs and HABs during a strain increment $d\varepsilon$ leads to the annihilation of a part $d\rho_i^{-(2)}$ of the dislocations generated as well as a boundary area dS^- (black arrows).

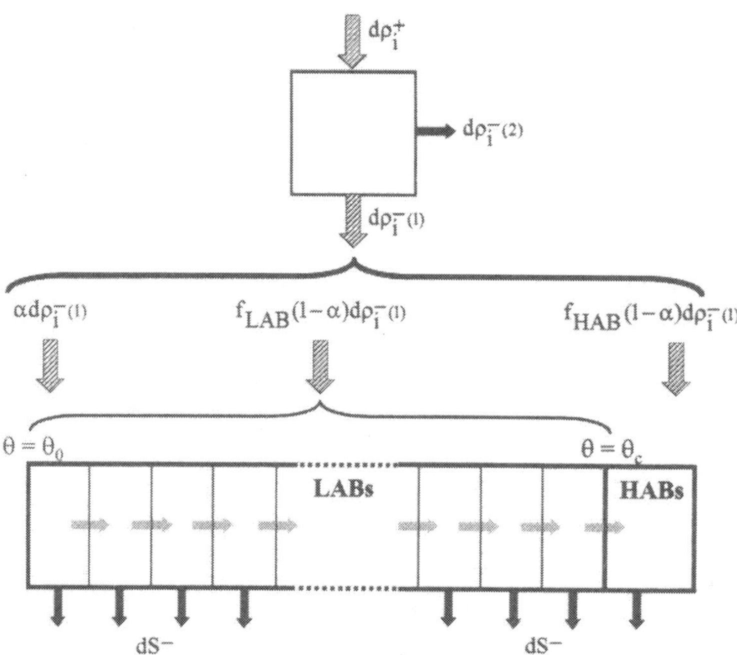

Figure 26. Diagram illustrating the way dislocations generated during straining distribute among the various microstructural elements (hatched arrows). Absorption of dislocations and boundaries by migrating LABs and HABs is symbolized by the black arrows. Gray arrows show the continuous increase of LAB misorientations from $\theta = \theta_0$ to $\theta = \theta_c$ (Gourdet and Montheillet, 2003).

Like in Section 3.1, eq.(9), a modified YLJ equation is used for the evolution of ρ_i:

$$d\rho_i = (h - r\rho_i)\,d\varepsilon - \rho_i dV \tag{26}$$

with $h\,d\varepsilon = d\rho_i^+$, $r\rho_i d\varepsilon = d\rho_i^{-(1)}$, and $\rho_i dV = d\rho_i^{-(2)}$, where h and r are the above-mentioned microscopic strain hardening and dynamic recovery parameters, respectively, and dV is the volume swept by mobile boundaries during a strain increment. The misorientation increase of LABs is given by:

$$d\theta = (b/2n)(1-\alpha)\,r\rho_i D\,d\varepsilon \tag{27}$$

where b is the Burgers vector modulus, n the average number of dislocation sets in a LAB, and D the average crystallite size. It should be noted that the model assumes here that the misorientation rate $d\theta/d\varepsilon$ does not depend on θ, which means that it is identical for all LABs. This seems reasonable ands leads to correct predictions for the behaviour of materials deforming homogeneously like aluminium, as will be seen below. However, recent results pertaining to a 718 grade nickel base superalloy are better reproduced when $d\theta/d\varepsilon$ increases with the misorientation angle θ, which is thought to reflect heterogeneous straining of the material (Thomas et al., 2003).

Calculations were carried out using h and r parameters depending on strain rate and temperature according to equation (17). Furthermore, in view of the predictions reported in Section 3.2, the average HAB migration rate was assumed to depend only on strain rate, according to a power law relationship:

$$v_M = C\dot{\varepsilon}^{m_M} \tag{28}$$

where the exponent m_M was estimated to 0.5 from both theoretical and experimental results (Gourdet and Montheillet, 2002; Gourdet et al., 1997). The average number n of dislocation arrays in a LAB was taken equal to its average value 2. The only parameter of the model that cannot be directly measured is α, which determines the rate of generation of new LABs. According to prior work carried out on aluminium, it was set equal to 0.1 in the following (see Section 6). Finally, the flow stress was derived from the dislocations densities inside the crystallites ρ_i and the subgrain boundaries ρ_{LAB} according to:

$$\sigma = \mu b\left(A_1\sqrt{\rho_i} + A_2\sqrt{\rho_{LAB}}\right) \tag{29}$$

where A_1 is a constant close to unity, and A_2 a constant at least an order of magnitude smaller.

Some predictions of the CDRX model. Stress-strain curves are shown in Figure 27a for commercial purity aluminium deformed at 0.1 s^{-1} and two different temperatures. Such predictions are very sensitive to the values of constants A_1 and A_2 of eq.(29) and the case $A_2 = 0$ appears to be closer to the classical experimental results. Even with this assumption, however, only limited amount of softening is visible, by contrast to the data as well as the predictions of the one-dimensional model proposed in Section 3.2. Although self-heating induced softening is likely to affect part of the experimental data, it has been suggested that such difference between the two above models could be attributed to topological effects, which are not accounted for in the above CDRX approach (Gourdet and Montheillet, 2003).

Figure 27b displays the evolution of the average crystallite size D with strain for various initial grain sizes. It is remarkable that the steady state value is independent of the latter, which is consistent with the predictions of the one-dimensional model (see Figure 18c). By contrast, whenever the HAB migration rate v_M is set equal to zero, D decreases continuously at large strains, which illustrates the prominent role of grain boundary migration during hot working.

The strain dependences of the HAB and LAB areas per unit volume are shown in Figure 27c. The latter increases continuously from zero to its steady state value, while the HAB area first undergoes decay associated with dynamic grain boundary migration (that is tantamount to grain growth), before it increases due the generation of new HABs up to the steady state.

Finally, the evolution of the grain boundary misorientation distributions with strain is illustrated in Figure 28: up to a strain of 0.6, no LAB has yet transformed into a HAB; at

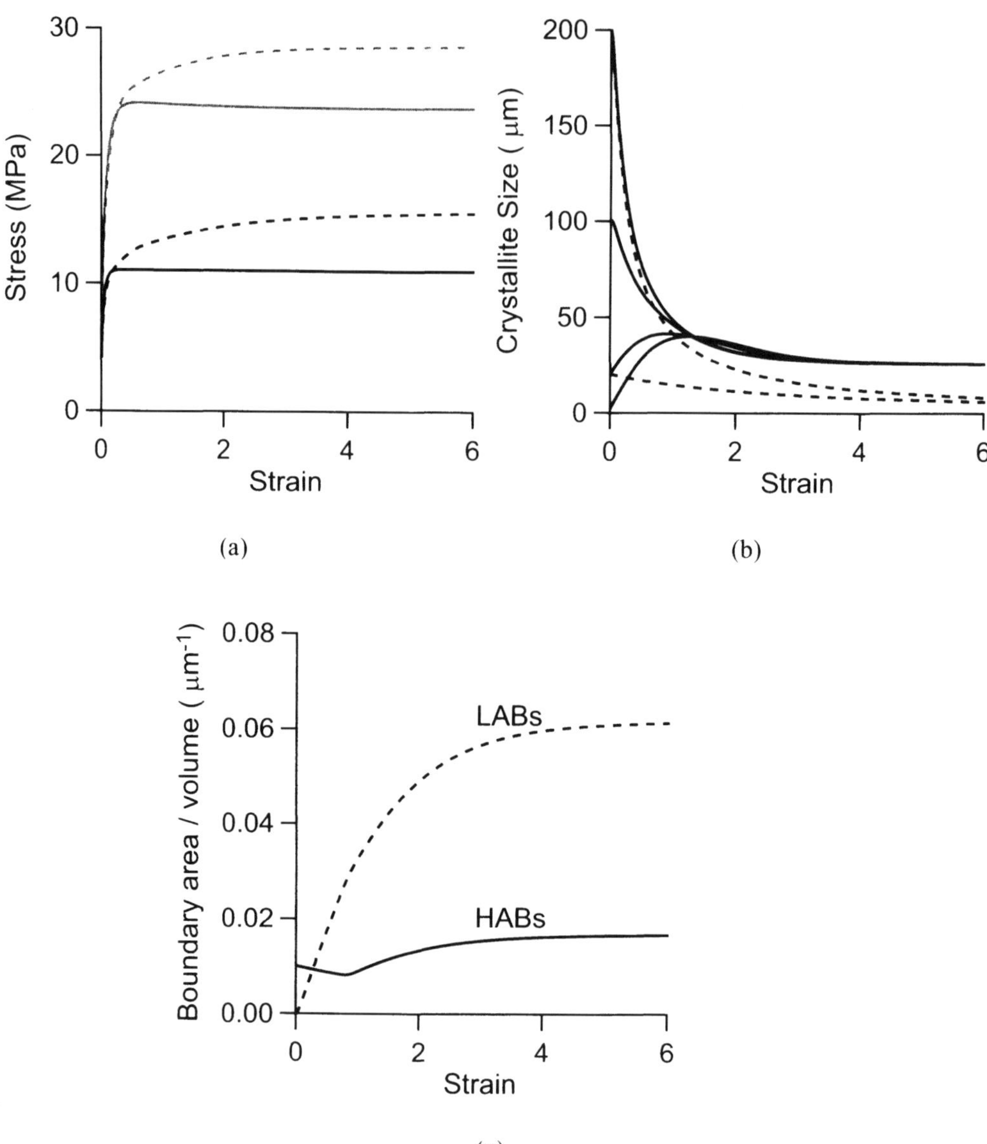

Figure 27. Strain dependence of the flow stress (a), the average crystallite size (b), and the HAB and LAB areas predicted by the CDRX model. In (a), flow stresses were calculated at 0.7 T_m, 0.1 s^{-1} (gray lines), and 0.8 T_m, 0.1 s^{-1} (black lines), for two variants of eq.(29), *viz.* $A_1 = 1$, $A_2 = 0$ (solid lines), and $A_1 = 0.9$, $A_2 = 0.1$ (broken lines). In (b), the crystallite size evolutions were calculated at 0.8 T_m, 0.01 s^{-1} starting from four different initial grain sizes; broken lines show evolutions without HAB migration. The same deformation conditions were used in (c) (Gourdet and Montheillet, 2003).

$\varepsilon = 0.9$, however, subgrain misorientations extend over the whole range between $\theta_0 = 1$ deg and $\theta_c = 15$ deg, and some of the LABs have already transformed. Nevertheless, the relative surface fraction of HABs remains lower than at low strains, which results from the continuous generation of new LABs. No significant change in the distribution is observed between $\varepsilon = 0.9$ and 1.2, which means that a steady state is achieved. The distribution then takes the shape of a decreasing exponential function, as will be shown below. It is interesting to note that the various microstructural parameters reach their respective steady state values with different kinetics.

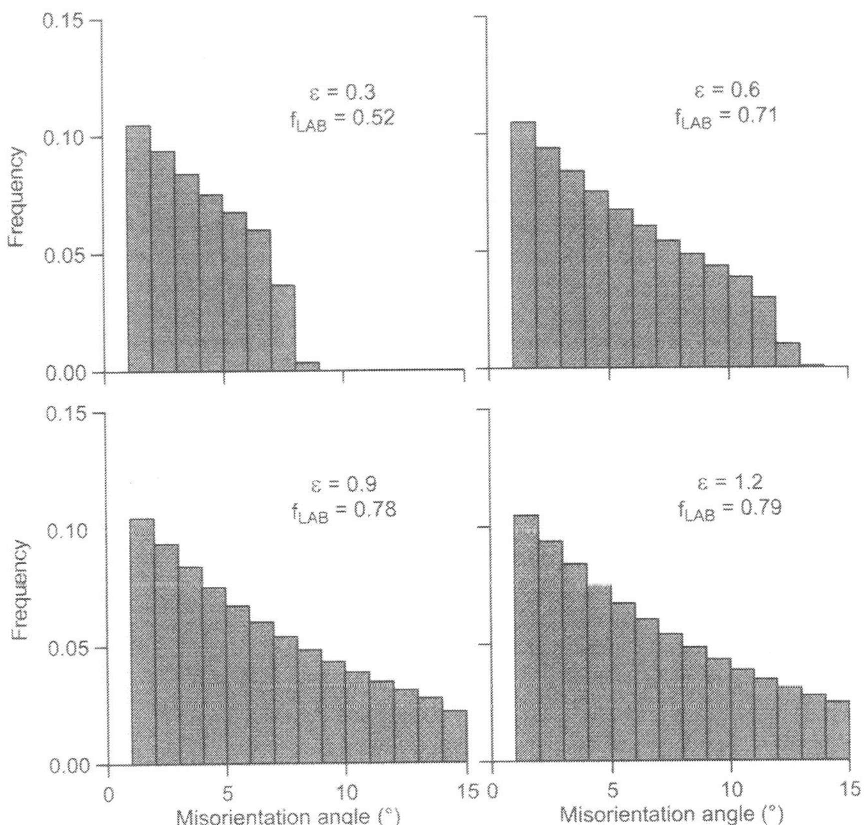

Figure 28. Strain dependence of the grain boundary misorientation distribution functions at 0.8 T_m, 0.01 s^{-1}. The surface fractions f_{LAB} of low-angle boundaries pertaining to each strain are indicated (Gourdet and Montheillet, 2003).

Analysis of the steady state regime. With the two conditions that the total area of boundaries per unit volume and the dislocation density inside the crystallites remain constant, the above model leads to analytical expressions for the steady state. The misorientation distribution density function is then given by:

$$\varphi(\theta) = \varphi(\theta_0)\exp\left[-\varphi(\theta_0)(\theta - \theta_0)\right] \tag{30}$$

where $\varphi(\theta_0) = \alpha/\left[(1-\alpha)\theta_0\right]$. The HAB and LAB area fractions derived from this equation are:

$$f_{HAB} = 1 - f_{LAB} = \exp\left[-\varphi(\theta_0)(\theta_c - \theta_0)\right] \tag{31}$$

whence the average LAB misorientation is:

$$\bar{\theta} = \frac{\theta_0}{\alpha} - (\theta_c - \theta_0)\frac{f_{HAB}}{f_{LAB}} \tag{32}$$

The expression for ρ_{LAB} is then:

$$\rho_{LAB} = C_1 / D \tag{33}$$

with $C_1 = 2n\, f_{LAB}\, \bar{\theta} / b$.

It is remarkable that the four above expressions do not depend on temperature and strain rate, although such dependence may admittedly be introduced through the parameters θ_0, θ_c, and α. By contrast, the deformation conditions are explicitly involved in the expression of the steady state dislocation density inside the crystallites:

$$\rho_i = C_2 / D^2 \tag{34}$$

with $C_2 = 4n\,\theta_0\, f_{HAB}\, v_M / (\alpha\, b\, r\, \dot{\varepsilon})$. The flow stress can now be derived from eqs.(29), (33), and (34). Finally, the steady state average crystallite size is given by:

$$D = 2\sqrt{C_2 r / 3h}\cos\left[(1/3)\cos^{-1}\left(C_3\sqrt{27h}\big/2\sqrt{C_2 r^3}\right)\right] \tag{35}$$

where $C_3 = 2 f_{HAB}\, v_M / \dot{\varepsilon}$.

In conclusion, CDRX can be interpreted as the combination of three elementary mechanisms, *viz.* strain hardening, dynamic recovery, and high-angle boundary migration, associated with parameters h, r, and v_M, respectively. The first two can be readily derived from experimental stress-strain curves. With respect to the last one, since direct measurements are quite difficult, it has been suggested that it can be estimated from microstructural investigations, using the one-dimensional approach of Section 3.2. A specific feature of CDRX is that it does not involve any nucleation mechanism, by contrast to DDRX which will be dealt with in the next section.

5 Discontinuous Dynamic Recrystallization

The main characteristics of DDRX summarized in Table 1 are first reviewed in the next subsection.

5.1 Experimental Data Pertaining to DDRX

Stress-strain curves. The flow stress curves in Figure 29 may be considered as one of the first (indirect) proofs of recrystallization taking place during straining. They were obtained from compression tests carried out on lead specimens. Room temperature belongs to the hot working range of lead, since $T/T_m \approx 0.5$, where T is the deformation temperature (K) and T_m the absolute melting temperature. It is noticeable that the flow stress exhibits strong oscillations during straining. Such typical behaviour reflects the occurrence of repeated cycles of strain hardening and recrystallization, which in this case are synchronized within the specimen. It is never observed in high SFE materials in which CDRX takes place, as described in the previous section.

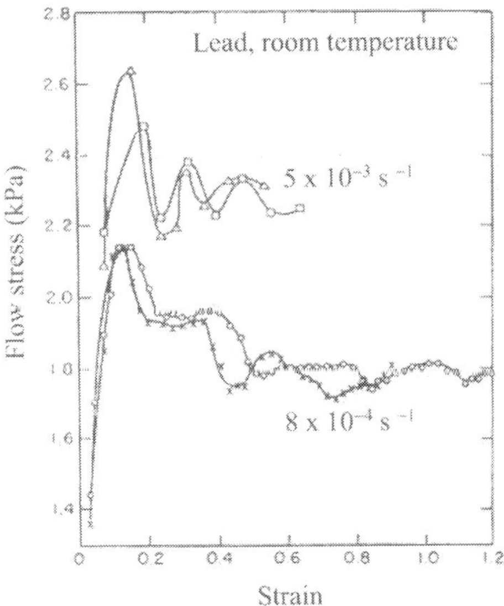

Figure 29. Oscillating compression stress-strain curves for lead at room temperature (Thomsen et al., 1954)

Multiple peak stress-strain curves are not observed, however, for all strain rates and temperatures, as illustrated in Figure 30a for 0.25 %C steels deformed at 1100 °C and various strain rates, and Figure 30b for OFHC (Oxygen Free High Conductivity) copper at 2×10^{-3} s^{-1} and various temperatures. It is convenient to introduce here the Zener-Hollomon parameter $Z = \dot{\varepsilon}\exp(Q/RT)$, where $\dot{\varepsilon}$ and T are the applied deformation conditions, R the gas constant, and Q the apparent activation energy for hot deformation. Whenever Q is known, it is generally considered that the above definition states an "equivalence" between strain rate and temperature. The diagrams in Figure 30 show that a transition occurs from multiple peak flow curves at low Z values (*i.e.*, low strain rates and/or high temperatures, hence low flow stresses) to single peak curves at large Z values. Such behaviour is quite different from that associated with CDRX (see Figure 21). Furthermore, the two types of curve have been associated with grain coarsening and grain refinement during straining, respectively (Sakai and Jonas, 1984). Finally, the steady state flow stress is reached at strains of a few units, which shows that the kinetics of DDRX is much faster than the kinetics of CDRX.

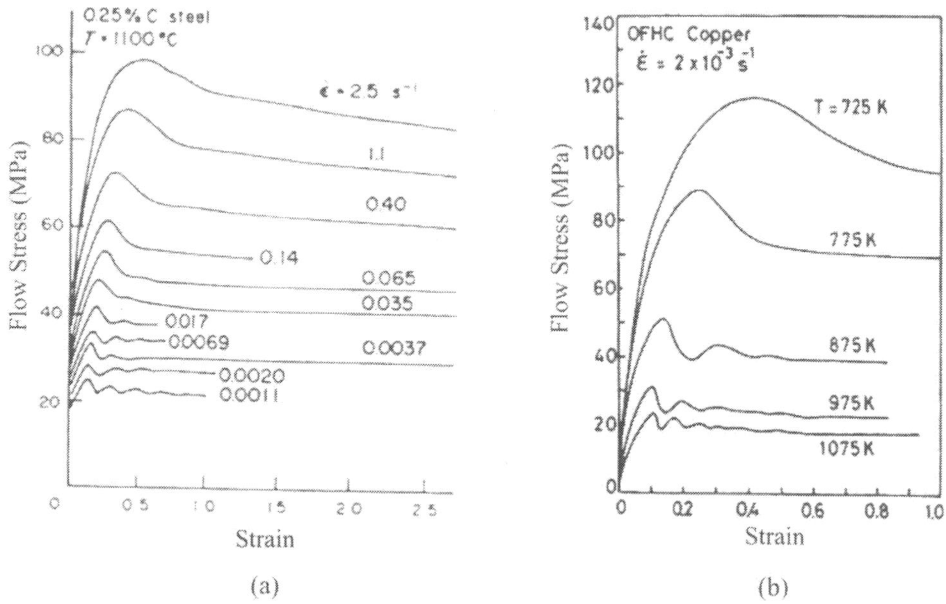

Figure 30. Stress-strain curves associated with DDRX, illustrating the transition between multiple peak behaviour at low Z values and single peak behaviour at large Z values: (a) torsion (Rossard and Blain, 1959); (b) compression (Blaz et al., 1983).

Microstructural evolutions. A typical example is given by the EBSD maps of Figure 31 for a nickel base grade 718 superalloy deformed in compression within the hot forging range. At $\varepsilon = 0.4$, recrystallized grains have nucleated, mostly along the original grain or twin boundaries. (The surface fraction of the latter, which are represented by gray lines in the initial state, is quite large in such low stacking fault energy alloy.) At larger strains ($\varepsilon = 0.7$ and 1.0), the number of new grains is much larger, and they form "necklaces" around the initial deformed and partially consumed grains. Such mechanism is clearly visible in the present case, since the dynamically recrystallized grains are much smaller than the initial ones. At $\varepsilon = 1.0$, remnants of the original structure are still visible, which means that the steady state has not been achieved yet.

A steady state regime microstructure in shown in Figure 32 for a high purity base 304 grade austenitic stainless steel. A large number of twin boundaries are identified by white lines (they are defined by a rotation of 60 degrees around a <111> type axis). Two types of grain can be readily distinguished: some of them, that contain almost no substructure (*i.e.*, subgrain boundaries, represented by light gray lines) and one or several twin boundaries, can be considered as currently growing "young" grains. In contrast, the other grains, generally of small size, display strong substructures but no twin boundaries and can therefore be viewed as "old" grains. Such grains are currently shrinking and will next disappear for the benefit of younger ones.

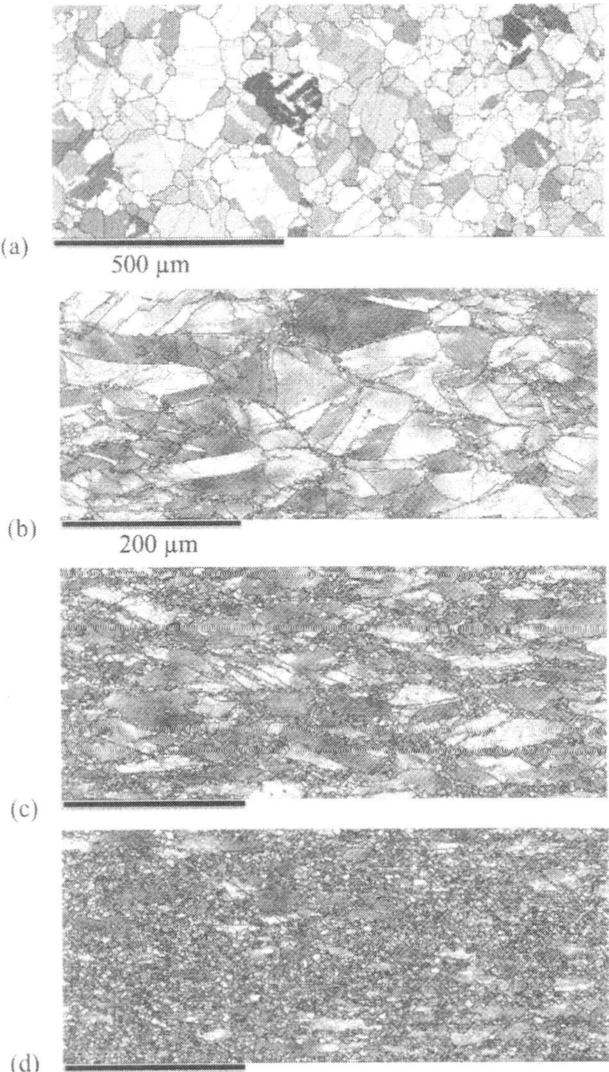

Figure 31. "Necklace" DDRX in a nickel base superalloy during compression at 980 °C, 0.01 s^{-1}; (a) $\varepsilon = 0$; (b) $\varepsilon = 0.4$; (c) $\varepsilon = 0.7$; (d) $\varepsilon = 1.0$; the compression axis is vertical (Thomas, 2003).

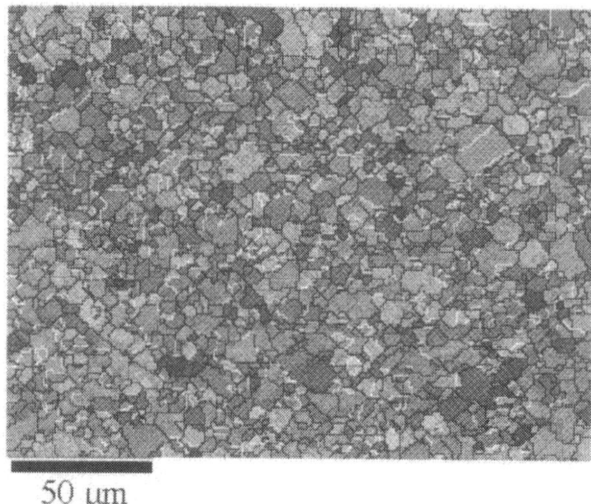

Figure 32. DDRX steady state in a high purity base 304 grade austenitic stainless steel (Gavard, 2001).

Twin boundaries thus appear as linked with the generation and growth of new grains. The diagrams in Figure 33 show schematically the way a new grain can be "nucleated" by growth twinning associated with grain boundary migration. In Figure 33a, grain 1 grows at the expense of grain 2. In Figure 33b, which corresponds to a larger strain, a twin boundary has formed for some reason possibly linked with the local stress and strain conditions, and the twin keeps growing into grain 2, while the low energy twin boundary does not move. Finally, in Figure 33c, at still larger strain, the crystallographic orientations of the three grains have changed, so that the twin boundary has lost its special character and has started to migrate in turn; a new grain has been generated. Such mechanism seems to be responsible for most of the nucleation events in low SFE metals, especially during steady state deformation (Ponge and Gottstein, 1998; Bocher et al., 1997; Kaibyshev and Sitdikov, 2000; Gavard, 2001).

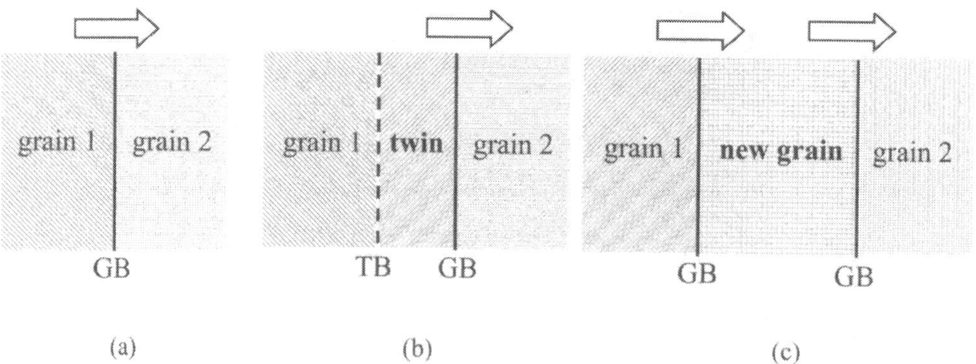

Figure 33. Schematic representation of nucleation by growth twinning during hot deformation. Arrows indicate grain boundary migration (GB: ordinary grain boundary; TB: twin boundary).

Large strain DDRX textures. As illustrated in Figure 34 by the pole figures relative to the above-mentioned 304 type steel, textures associated with DDRX are much weaker than that produced by CDRX (see Figure 24). In uniaxial compression (Figure 34a), a slight <110> fiber texture is nevertheless observed, while in torsion (Figure 34b), the main classical deformation texture components, *viz.* $A/\bar{A}$ $\{111\}<1\bar{1}0>$, $B/\bar{B}$ $\{112\}<1\bar{1}0>$, and C $\{001\}<110>$ (Montheillet et al., 1984), are present, although with limited intensities compared with $B/\bar{B}$ in Figure 24. This indicates that DDRX tends to randomize grain orientations, which is likely to result from repeated twinning events. Such assumption is supported by Figure 35 which displays EBSD misorientation maps and the associated boundary misorientation distributions for a Cu-9 %Sn alloy deformed by uniaxial compression: after a strain of 0.2 (Figures 35a and c), the distribution is close to that for a set of randomly oriented grains (Mackenzie distribution), while a strong peak close to 60 deg indicates a large fraction of twin boundaries. Still after a strain of 1.2, where DDRX is fully operating (Figures 35 b and d), the misorientation distribution follows that of Mackenzie. The peak at 60 deg is strongly reduced, since twin boundaries loose rapidly their special character due to strain induced grain rotations.

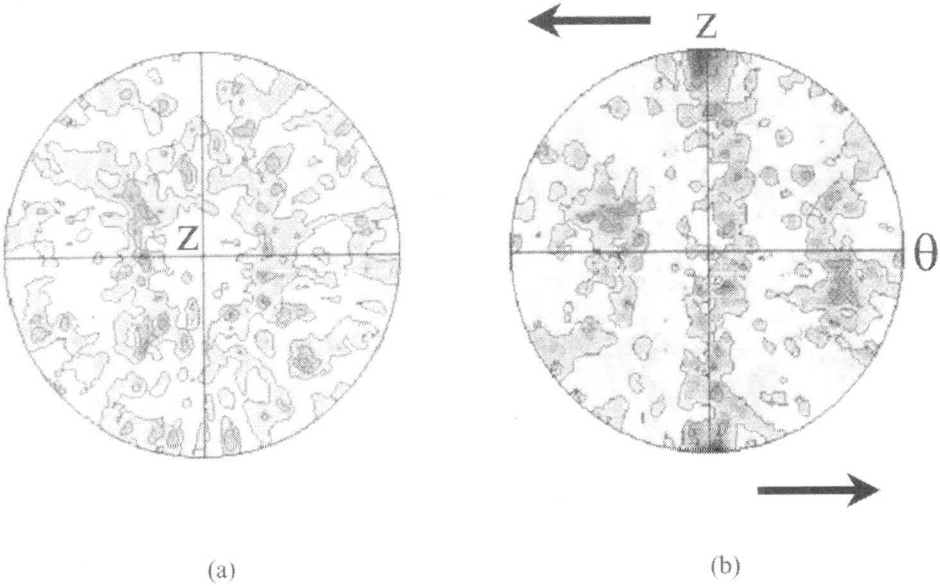

Figure 34. Large strain (steady state) textures represented by {111} pole figures derived from EBSD measurements: high purity base 304 grade austenitic stainless steel deformed at 1050 °C, 10^{-2} s^{-1}; (a) uniaxial compression along the z-axis, $\varepsilon = 1$; (b) torsion, $\varepsilon = 3.5$; arrows indicate the shear direction (Gavard, 2001).

Average steady state grain size. It has long been observed that in the steady state, the average grain size $\bar{D}$ is related to the flow stress σ by the following empirical relationship (Derby, 1992):

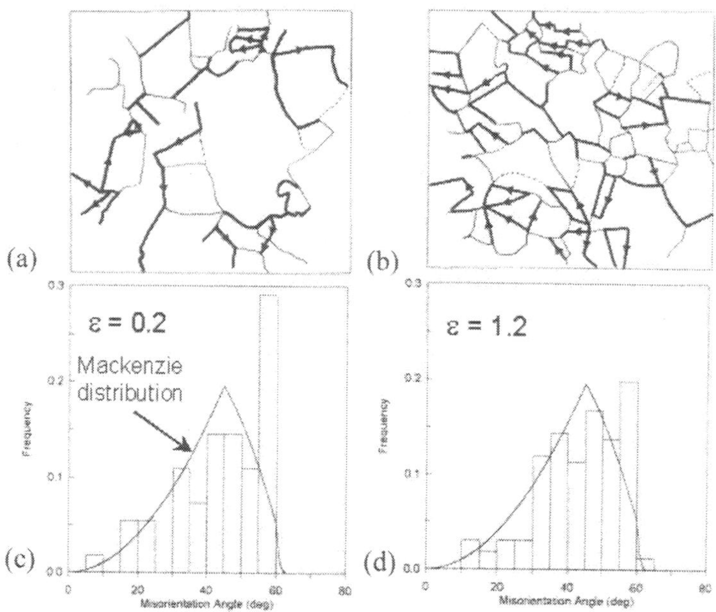

Figure 35. EBSD misorientation maps (a,b) and boundary misorientation distributions (c,d) for a Cu-9 %Sn bronze alloy deformed in compression at 750 °C, 10^{-1} s^{-1}. In (a) and (b), HABs are represented by thick lines, while triangles indicate twin boundaries; LABs are represented by thin lines. The Mackenzie distribution is shown in (c) and (d) for comparison (Bayle et al., 1999).

$$\sigma = \frac{A}{\bar{D}^a} \tag{36}$$

where the exponent a ranges between 0.6 and 0.8 (Figure 36).

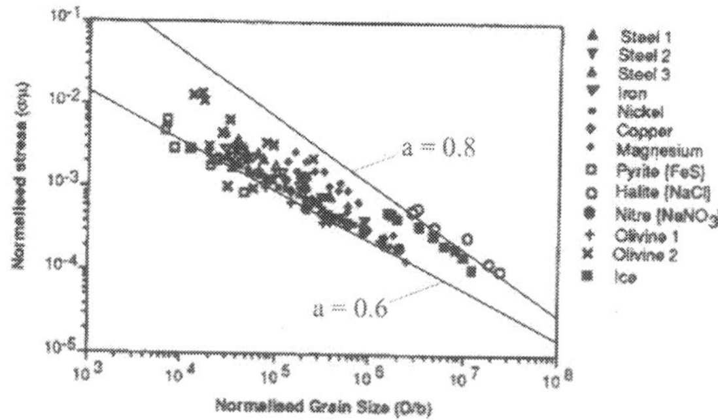

Figure 36. Collected data from various metals and minerals illustrating the empirical relationship between $\bar{D}$ and σ in the DDRX steady state (after Derby, 1992).

5.2 Modeling Discontinuous Dynamic Recrystallization

Unlike the case of continuous dynamic recrystallization, various models have already been proposed for DDRX. Analytical or semi-analytical approaches coupling dynamic grain growth and the evolutions of dislocation densities in a more or less complex way have been developed by Stüwe and Ortner (1974), Sandström and Lagneborg (1975), Kaptsan et al. (1992), and more recently by Gao et al. (1999), and Busso (1997) who implemented the model into a finite element code. An alternative is to use (stochastic) Monte-Carlo methods, like for instance Rollet et al. (1992), Luton and Peczak (1992), Peczak (1995), or (deterministic) cellular automata, like Goetz and Seetharaman (1998). Such network approaches have allowed topological effects, such as necklace formation, to be taken into account. However, they are well suited to describe grain boundary migration, but not straining neither grain boundary convection (see Section 2.3). As a general rule, all the above models are able to reproduce quite well the shape of the stress-strain curves and the transition from multiple peak to single peak behaviour with increasing Z. Much less attention has been paid, however, to the "fraction recrystallized" and the grain size evolutions. A simple model is proposed below, that deals with a set of interacting grains. In a similar way to that of CDRX, it involves a limited number of parameters; furthermore, in the steady state, the equations can be simplified to lead to closed form results (Montheillet, 1999).

Basic equations of the DDRX model. The material is represented by a set of spherical grains, each of them being characterized at any time by two internal variables: the size (diameter) D_i, and the dislocation density ρ_i. To a first approximation, the spherical shape is consistent with the limited lifetime of a grain, and agrees with experimental observations (see Figure 32). Instantaneous homogenization of ρ_i within the grain is assumed. Furthermore, each grain is considered as an inclusion imbedded in a homogeneous matrix, whose properties are obtained by averaging that of all the grains in the aggregate. Such an approach is close to the self-consistent models commonly used in continuum mechanics of inhomogeneous materials. It precludes topological properties, such as necklacing, which are expected, however, to be second order effects, especially during the steady state.

The dislocation density difference between a grain and the surrounding matrix is assumed to be the predominant driving force for the dynamic migration of grain boundaries (see Section 2.4); hence,

$$\frac{dD_i}{dt} = 2M\tau(\bar{\rho} - \rho_i) \tag{37}$$

for grain (i), with $\tau = \mu b^2$, where M is the grain boundary mobility (the factor 2 means that two opposite parts of a spherical grain boundary move into opposite directions), μ the shear modulus, b the Burgers vector, and $\bar{\rho}$ is an average dislocation density within the matrix. Therefore, whenever $\rho \leq \bar{\rho}$, the considered grain is presently growing (Figure 37a), whereas it shrinks when $\rho > \bar{\rho}$ (Figure 37b). Since the overall volume is proportional to $\sum D_i^3$, its time derivative is proportional to $\sum D_i^2 \dot{D}_i = M\tau \sum D_i^2 (\bar{\rho} - \rho_i) = M\tau\left[\bar{\rho} \sum D_i^2 - \sum \rho_i D_i^2\right]$, which must be zero for insuring incompressibility. It is therefore relevant to define $\bar{\rho}$ by the following equation:

$$\bar{\rho} = \sum \rho_i \, D_i^2 / \sum D_i^2 \tag{38}$$

where the summation is extended to all the grains of the aggregate. When the above summation is restricted to the first neighbours of the grain considered, eq.(38) merely means that a grain interacts with them in proportion to their surface areas rather than their volumes.

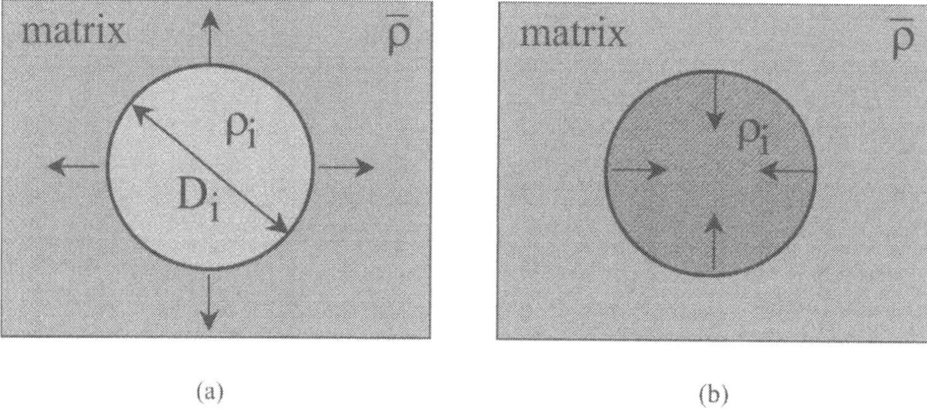

Figure 37. Schematic representation of a grain in the aggregate, which grows when $\rho_i \le \bar{\rho}$ (a) and shrinks when $\rho_i > \bar{\rho}$ (b).

The YLJ equation (Laasraoui and Jonas, 1991) is again used here for describing strain hardening and dynamic recovery, since it has been shown to apply at large strains in the hot working range, in particular for austenitic steels. However, any other formulation of the same type could be substituted to the latter. Nevertheless, in the same way as in eqs.(9) and (26), it is necessary to modify such formulation to account for the effect of dynamic grain boundary migration: when a grain is growing, the volume increment during time dt is almost dislocation free (density ρ_0), since the dislocations present in front of the moving grain boundary (density $\bar{\rho}$) are absorbed by the latter (Figure 38). With the assumption of instantaneous homogenization of the dislocation density within the grain, this leads to an additional (softening) term in the rate equation for ρ_i. This mechanism has been referred to as boundary migration induced softening (BMIS) in Section 3. By contrast, when a grain is shrinking, its dislocation density is not affected by grain boundary migration, since the volume swept by the boundary is removed from the grain. The strain hardening-dynamic recovery equation can thus be written:

$$\frac{d\rho_i}{dt} = \left(h_i - r_i\,\rho_i\right)\dot{\varepsilon} - 3\frac{\rho_i - \rho_0}{D_i}\frac{dD}{dt} \qquad \text{if} \quad \rho_i \le \bar{\rho} \qquad \text{(growth)} \qquad (39a)$$

$$\frac{d\rho_i}{dt} = \left(h_i - r_i\,\rho_i\right)\dot{\varepsilon} \qquad \text{if} \quad \rho_i \ge \bar{\rho} \qquad \text{(shrinkage)} \qquad (39b)$$

where the strain hardening h_i and dynamic recovery r_i parameters may vary from one grain to the other, for instance with respect to their crystallographic orientations. In the following, r will be considered as identical for all the grains. In the same way, a specific strain rate $\dot{\varepsilon}_i$ could be applied to each grain according to its current flow stress or dislocation density and current shape. For the sake of simplicity, however, the classical Taylor assumption (uniform $\dot{\varepsilon}$) will be adopted here.

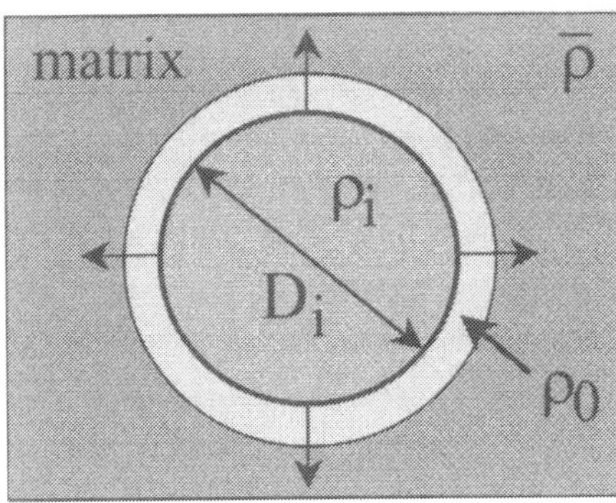

Figure 38. Schematic representation of the boundary migration induced softening (BMIS) mechanism associated with grain growth during DDRX.

A third equation is necessary to describe the nucleation of new grains. The physical mechanisms of nucleation during hot deformation are not yet well established to date, since it is very difficult to carry out *in situ* representative experiments and observations. For static recrystallization, nuclei have been proposed to form either by bulging of grain boundaries or progressive misorientation of special subgrains (Humphreys and Hatherly, 1996). More recently, it has been suggested that new grains could be generated by twinning, as described in Section 5.1. In the present model, a nucleation probability ϕ is associated with each grain, whose general form is likely to account for any physical type of nucleation. This third internal variable is defined by:

$$\frac{d\phi_i}{dt} = f\left(\rho_i, \bar{\rho}, D_i, \dot{\varepsilon}, T\right) \qquad (40)$$

where T is the deformation temperature. A simple form was first chosen for the function f, *viz.*:

$$\frac{d\phi}{dt} = k_N \bar{\rho}^p D^2 \qquad (41)$$

for any grain, where k_N is a temperature dependent nucleation parameter and p a positive exponent that will be determined below. During the time increment dt, the increase of ϕ is assumed to be proportional to the square of D, since dynamic nucleation is known to operate principally at grain boundaries. Alternatively, nucleation within the volume of the grains could be accounted for by a cubic dependence in D. With this definition, $\dot{n} = k_N \bar{\rho}^p$ represents the classical nucleation rate per unit surface. When ϕ reaches the critical value 1, a nucleus of size D_0 and dislocation density ρ_0 is added to the set of grains, and ϕ is reset to zero (in the results presented below, $D_0 = \rho_0 = 0$).

It is worth to note that, according to the above formulation, three distinct softening mechanisms are involved in DDRX, *viz.* (i) dynamic recovery; (ii) growth of new grains at the expense of the harder matrix (*geometrical softening*); and (iii) dislocation annihilation by grain boundary migration or BMIS (*physical softening*). Although they are intimately connected in the real material, it is possible to assess their respective contributions with the help of the present model.

Some predictions of the DDRX model. The following results were obtained using material parameters for a high purity base 304 grade austenitic stainless steel deformed at 1050 °C and 10^{-2} s^{-1}. The strain hardening parameters h_i in eqs.(39a) and (39b), and the initial grain sizes where uniformly distributed within an interval $h_0 \pm \Delta h$ and $D_0 \pm \Delta D$, with $\Delta h = 0.1\, h_0$ and $\Delta D = 0.3\, D_0$, respectively. Figure 39a shows stress-strain curves for various initial average grain sizes D_0 while the corresponding grain size evolutions are displayed in Figure 39b. Grain growth (D_0 = 5, 10, and 20 μm) or moderate decrease (D_0 = 50 μm) is clearly associated with multiple peak flow stress curves. By contrast, when grain size is strongly reduced by DDRX (D_0 = 100 and 150 μm), the stress-strain curves exhibit one single maximum. Furthermore, steady state behaviour independent of the initial grain size occurs after a strain of about 0.4, that is much smaller than in the case of CDRX (compare with Figure 27). Figure 39c shows in turn the strain dependence of the "fraction recrystallized" f_{RX}, which must be understood as the volume fraction of grains that were not present in the initial state of the material. It should be noted, however, that part of these grains may contain large dislocation densities because of continued straining, by contrast to the case of static recrystallization.

From the curves of Figure 39c, it is possible to specify the influence of the initial grain size on DDRX kinetics (Figure 40): in the range $D_0 < 20\,\mu m$, DDRX starts at low strains, since the large grain boundary area per unit volume favours the nucleation of new grains. However, "complete" DDRX (*i.e.*, $f_{RX} = 1$) is delayed since BMIS lowers the driving force for grain boundary migration. In the opposite range $D_0 > 100\,\mu m$, DDRX is retarded as well, but in that case because of the reduced nucleation rate. The fastest DDRX kinetics is therefore observed for intermediate initial grain size values ($D_0 \approx 50\,\mu m$).

Steady state behaviour. Consider now the steady state limit of the system. Assuming that all grains of the aggregate have the same behaviour, i.e. the associated h, r, and k_N parameters are the same, their histories described by the three internal variables D(t), ρ(t), and ϕ(t) are identical. The ergodic assumption can therefore be applied to the steady state system. This

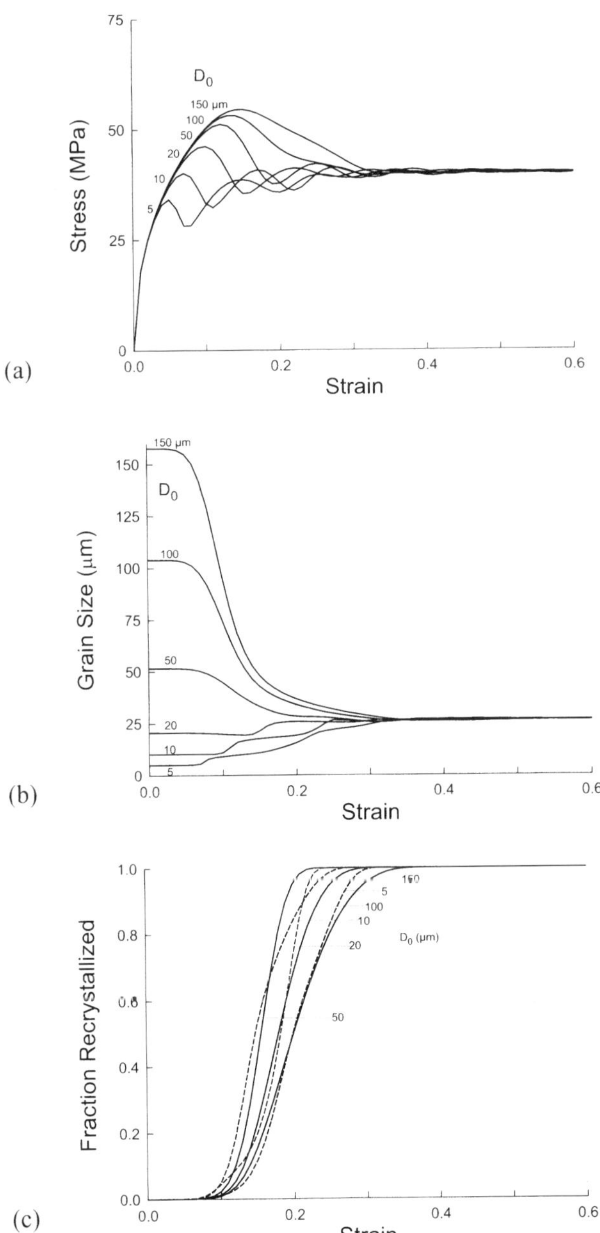

Figure 39. Strain dependence of the flow stress (a), the average grain size (b), and the fraction recrystallized (c) predicted by the DDRX model for a high purity base 304 grade austenitic stainless steel deformed at 1050 °C and 10^{-2} s^{-1}, with various initial grain sizes D_0.

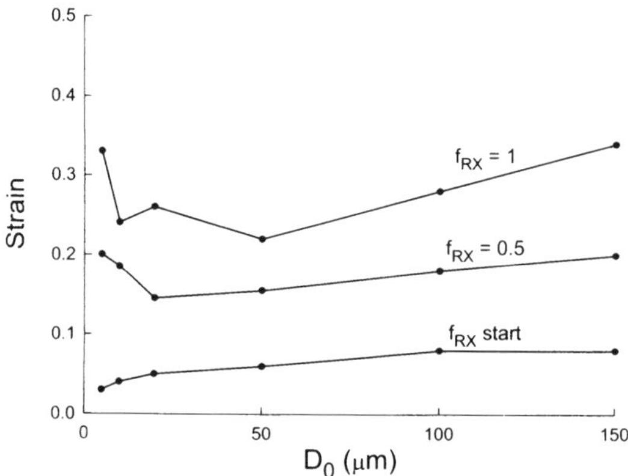

Figure 40. Diagram showing the predicted dependence on the initial grain size of the strains for initiation of DDRX (f_{RX} start), "half recrystallization" ($f_{RX} = 0.5$), and "complete recrystallization" ($f_{RX} = 1$). High purity base 304 grade austenitic stainless steel deformed at 1050 °C and 10^{-2} s^{-1}.

means that the investigation of the whole set of grains at any given time is tantamount to the observation of a single grain, from its nucleation (time 0) to its disappearance (time t_F). In particular, any summation over the grains can be replaced by a time integral, *e.g.* the average grain size $\bar{D} = \sum D_i / N$ is equivalent to $\bar{D} = \int_0^{t_F} D(t)\,dt / t_F$. Furthermore, in order to achieve a steady state, each grain must give rise, on average, to one and only one nucleus during its life. Therefore, the steady state nucleation condition can be written:

$$\phi(t_F) = k_N\, \bar{\rho}^p \int_0^{t_F} D^2\, dt = 1 \tag{42}$$

With the above condition, equations (37) and (39) can be dealt with numerically to find the grain size and dislocation density histories of a grain in the steady state regime. Such results are illustrated in Figure 41a and b for two strain rates, where the material parameters of a high purity base 304 grade austenitic stainless steel were used. At low strain rate ($\dot{\varepsilon} = 10^{-2}$ s^{-1}, Figure 41a), the grain disappears much before the dislocation density has reached its steady state regime. The average dislocation $\bar{\rho}$ in the material, *i.e.* the average value of the function $\rho(\varepsilon)$ is much lower than the latter, which means that geometrical and physical softening typical of DDRX are predominant. By contrast, at larger strain rate ($\dot{\varepsilon} = 1$ s^{-1}, Figure 41b), most of the lifetime of the grain takes place under constant saturation dislocation density ρ, and $\bar{\rho}$ is close to that value, which means in turn that the main contribution to the overall softening is dynamic recovery.

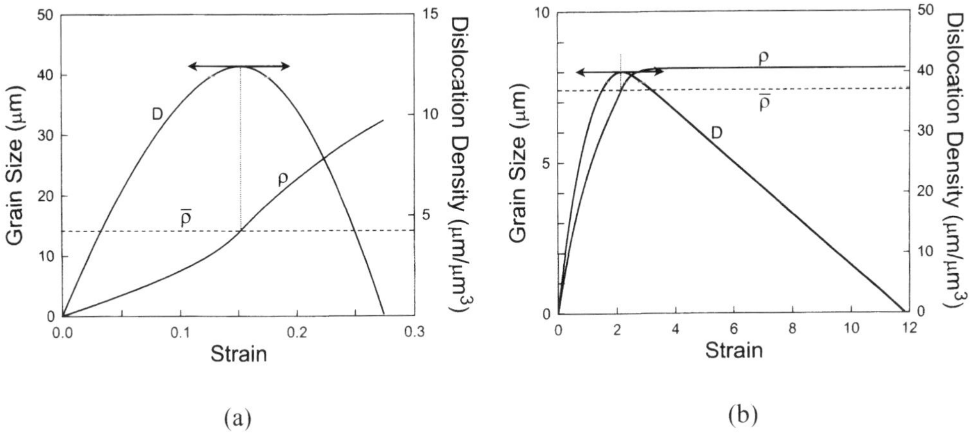

Figure 41. Predicted grain size D and dislocation density ρ histories for any grain in the steady state regime of DDRX for two applied strain rates: (a) 10^{-2} s^{-1}; (b) 1 s^{-1}. High purity base 304 grade austenitic stainless steel deformed at 1050 °C (Montheillet, 1999).

5.3 A Simplified Analytical Model of Steady State DDRX

Two further simplifications are now introduced, that allow the above set of equations (37), (39), and (42) to be solved analytically (Montheillet, 1999):
(i) dynamic recovery is neglected, *i.e.* r = 0;
(ii) BMIS, *i.e.* the second term on the right hand side of equation (39a), is neglected.
Among the three above-mentioned mechanisms, geometrical softening is therefore the only one to operate. In this simplified case, the time dependence of the grain size D and dislocation density ρ are given by (Figure 42):

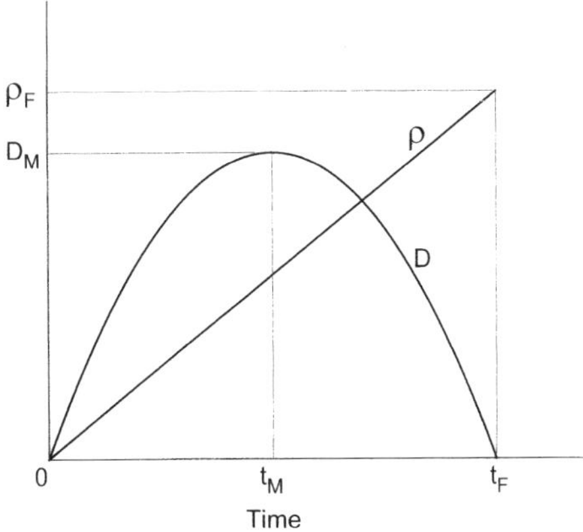

Figure 42. Schematic representation of the simplified grain size D and dislocation density ρ histories of any grain in the steady state regime of DDRX, under the above assumptions (i) and (ii).

$$D = M\tau h(\varepsilon_F t - \dot{\varepsilon} t^2)/2 \tag{43}$$

and

$$\rho = h\dot{\varepsilon} t \tag{44}$$

where $\varepsilon_F = \dot{\varepsilon} t_F$ denotes the strain undergone by the grain when it disappears. Substitution of $\bar{\rho} = h\dot{\varepsilon} t_F / 2$ into equation (42) allows t_F to be determined, whence:

$$\bar{\rho} = \left[\frac{15}{4} \frac{(h\dot{\varepsilon})^3}{k_N (M\tau)^2} \right]^{1/(p+5)} \tag{45}$$

The steady state flow stress is then derived from the classical relationship $\sigma = A\mu b\sqrt{\bar{\rho}}$ (where A is a constant close to unity), which yields:

$$\sigma = A\mu b \left[\frac{15}{4} \frac{(h\dot{\varepsilon})^3}{k_N (M\tau)^2} \right]^{1/[2(p+5)]} \tag{46}$$

Hence, the strain rate sensitivity exponent is $m = 1/[2(p + 5)]$.

In the same way, the average grain size deduced from equation (43), $\bar{D} = M\tau h\dot{\varepsilon} t_F^2 / 12$, can be written in the form:

$$\bar{D} = \frac{1}{3} \left(\frac{15}{4} \right)^{2/(p+5)} \left[\frac{(M\tau)^{p+1}}{k_N^2 (h\dot{\varepsilon})^{p-1}} \right]^{1/(p+5)} \tag{47}$$

Finally, solving equation (47) for $h\dot{\varepsilon}$ and substituting this quantity into equation (46) yields:

$$\sigma = A\mu b \left(\frac{5}{36} \right)^{1/[2(p-1)]} \left(\frac{M\tau}{k_N} \right)^{1/[2(p-1)]} \frac{1}{\bar{D}^{3/[2(p-1)]}} \tag{48}$$

Therefore, the grain size exponent in eq.(36) is $a = 3/[2(p - 1)]$.

Since a unique relationship (*i.e.*, independent of the strain rate-temperature couple) between flow stress and the average grain size is generally observed, the above equation suggests that the ratio $M\tau/k_N$ should be independent or weakly dependent of temperature. The apparent activation energies for the grain boundary mobility M and the nucleation parameter k_N should therefore be similar. From eq.(48), it is easy to show that the exponent p of the nucleation equation (41) must be chosen between 2 ($a = 3/2$) and 4 ($a = 1/2$). The intermediate value $p = 3$ leads to $a = 3/4 = 0.75$, which is very close to the data commonly reported (see Figure 36). The associated value of the strain rate sensitivity, $m = 1/16 \approx 0.19$ is also quite realistic. Note, however, that m is much less sensitive to p than a. Other relevant quantities can be derived from the DDRX model as well, using either the more general numerical version, or the above simplified analytical approach, for instance the steady state grain size distributions (Thomas and Montheillet, 2002).

In conclusion of this section devoted to DDRX, it is worth to note that the involved elementary mechanisms are *the same* as for CDRX, *viz.* strain hardening, dynamic recovery, and high angle grain boundary migration. In the latter case (Section 4), the proposed model dealt mainly with grain (HAB) and subgrain (LAB) *boundaries*, to account for the evolutions of boundary misorientation distributions, and more specifically the transformation of LABs into HABs. In the former case of DDRX, however, although such mechanism is also likely to take place inside the grains (*e.g.*, in medium stacking fault energy metals), the emphasis was put instead on the *grain* populations, since the main mechanisms are the nucleation and growth of new grains. Such dissymmetry between the two approaches should note hide the physical analogy between the two types of dynamic recrystallization. The above remark even suggests that transitions between CDRX and DDRX are possible under specific circumstances. This point will be briefly addressed in the last section of the chapter.

6 Transitions Between CDRX and DDRX

In the 1970's, Guillopé and Poirier (1979) brought into evidence a transition from discontinuous (referred to as "rotation" DRX) to continuous dynamic recrystallization in sodium chloride with decreasing purity. More recently, similar observations were made on ice (De la Chapelle et al., 1998). Such transition has been first established in a metallic material by Yamagata (1992, 1995), who showed that DDRX occurs in high purity aluminium whereas aluminium undergoes CDRX at lower purities. This is illustrated very clearly in Figures 43a and b that show stress-strain curves from hot compression tests on both 99.99 %Al and 99.999 %Al single crystals along the <100>, <110>, and <111> crystallographic directions. Although the value of the Zener-Hollomon parameter Z is larger for the higher purity material, the latter exhibits flow oscillations typical of DDRX, by contrast to the lower purity grade.

Figure 44 displays wavy flow curves pertaining to 99.999 %Al (with a 2 ppm silicon content) single crystals submitted to compression along a <111> axis. Comparison with the stress-strain curves of the same grade added with 23 ppm silicon shows that the occurrence of DDRX is prevented by the presence of the solutes. A large amount of work remains to be carried out to understand the role of solutes in the DRX mechanisms. It is likely, however, that they mainly affect grain boundary mobility as illustrated in Section 2.5, although they might as well influence dynamic recovery through the mobility of dislocations, and the nucleation kinetics.

To a first approximation, the transition between CDRX and DDRX can be predicted as follows, starting from a material that currently undergoes CDRX: let Δt_1 be the time necessary to create a new HAB by the progressive misorientation of a LAB during straining. If the misorientation rate $\dot{\theta}$ is independent of θ, as assumed in Section 4, an estimation of that time is given by:

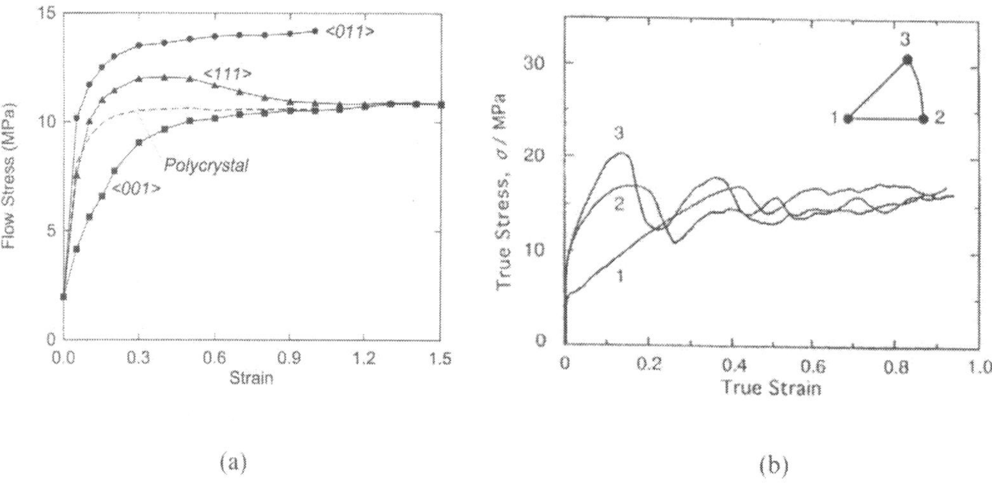

(a) (b)

Figure 43. Influence of purity on the dynamic recrystallization mode of aluminium single crystals (uniaxial compression along the <100>, <110>, and <111> axes): (a) 99.99 %Al, 380 °C, 10^{-2} s^{-1}, $Z \approx 10^{10}$ s^{-1}, flow curves are typical of CDRX (Gourdet and Montheillet, 2000); (b) 99.999 %Al, 260 °C, 1.67 x 10^{-3} s^{-1}, $Z \approx 8 \times 10^{11}$ s^{-1}, flow curves are typical of DDRX (Tanaka et al., 1999).

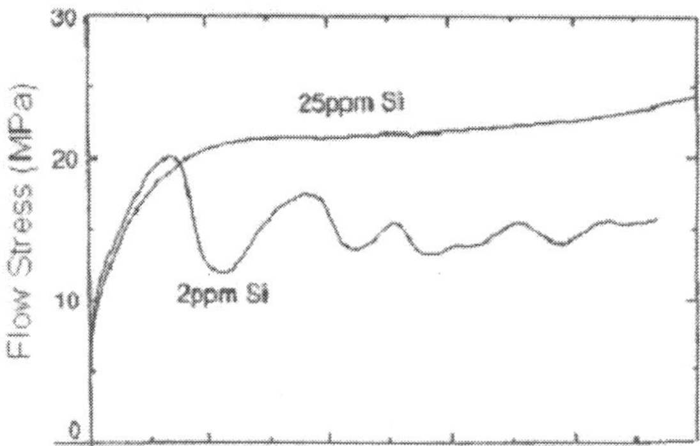

Figure 44. Influence of the silicon content on the dynamic recrystallization mode of aluminium single crystals (uniaxial compression along the <111> axis), 260 °C, 1.67 x 10^{-3} s^{-1} (Tanaka et al., 1999).

$$\Delta t_l = \theta_c / \dot{\theta} \tag{49}$$

where θ_c is the critical angle defined in Section 2.5.

Consider now the time Δt_2 required for a grain boundary to sweep the old microstructure, *i.e.* to migrate over the average crystallite size D. The latter can be written in the following form:

$$\Delta t_2 = D / f_{HAB} v_M \tag{50}$$

where f_{HAB} is the surface fraction of high angle boundaries, and v_M the rate of migration. A condition for CDRX to occur is then merely $\Delta t_1 < \Delta t_2$, since otherwise new HABs could not be generated by the progressive misorientation of LABs. Such condition clearly accounts for the effect of solutes, if Δt_1 remains unchanged while v_M is decreased. Furthermore, by combination of the above condition with the steady state equations of the CDRX model of Section 4.3, the following simple inequality can be derived (Gourdet, 1997):

$$\alpha < \frac{2\theta_0}{\theta_c + 2\theta_0} \tag{51}$$

i.e., with $\theta_0 = 1$ deg and $\theta_c = 15$ deg, $\alpha < 0.12$. Although this result is nothing more as an order of magnitude, it means nevertheless that if the α parameter is too large, CDRX is unable to occur. Grain boundary migration is then predominant, and DDRX takes place. Special attention should be paid now to the influence of the solute content on the α parameter.

References

Bayle, B., Bocher, Ph., Jonas, J.J., and Montheillet, F. (1999). Flow stress and recrystallization during the hot deformation of Cu–9%Sn alloys. *Materials Science and Technology* 15:803–811.

Blaz, L., Sakai, T., and Jonas, J.J. (1983). Effect of initial grain size on the dynamic recrystallization of copper. *Metal Science* 17:609–616.

Bocher, Ph., Montheillet, F., and Jonas, J.J. (1997). Microstructural evolution during the dynamic recrystallization of a 304 stainless steel. In McNelley, R., ed., *The third International Conference on Recrystallization and Related Phenomena*. Monterey (CA). 355–362.

Bunge, H.J. (1987). Three-dimensional texture analysis. *International Materials Reviews* 32:265–291.

Busso, E.P. (1998). A continuum theory for dynamic recrystallization with microstructure-related length scales. *International Journal of Plasticity* 14:319–353.

Chovet-Sauvage, C. (2000). *Evolution des microstructures et des textures en grande déformation à chaud d'un alliage Al-Mg-Si*. Ph.D. Dissertation, Ecole Nationale Supérieure des Mines, Saint-Etienne, France.

Couturier, G. (2003). Contribution à l'étude de la dynamique du Zener pinning: simulations numériques par éléments finis. Ecole Nationale Supérieure des Mines, Saint-Etienne, France.

De La Chapelle, S., Castelnau, O., Lipenkov, V., and Duval, P. (1998). Dynamic recrystallization and texture development in ice as revealed by the study of deep ice cores in Antarctica and Greenland. *Journal of Geophysical Research* 103:5091–5105.

Derby, B. (1992). Dynamic recrystallization: the steady state grain size. *Scripta Metallurgica et Materialia* 27:1581–1586.

Fridman, E.M., Kopezky, C.V., and Shvindlerman, L.S (1975). Effects of orientation and concentration factors on migration of individual grain boundaries in aluminium. *Zeitschrift für Metallkunde* 66:533–539.

Frois, C., and Dimitrov, M.O. (1966). Influence de quelques éléments d'addition sur la recristallisation de l'aluminium très pur. *Annales de Chimie* 1:113–128.

Gao, W., Sakai, T., and Miura, H. (1999). Modeling the new grain development under dynamic recrystallization. In Sakai, T., and Suzuki, H.G., eds., *The Fourth International Conference on Recrystallization and Related Phenomena*. Tsukuba: The Japan Institute of Metals. 659–664.

Gavard, L. (2001). *Recristallisation dynamique d'aciers inoxydables austénitiques de haute pureté*. Ph.D. Dissertation, Ecole Nationale Supérieure des Mines, Saint-Etienne, France.

Goetz, R.L., and Seetharaman, V. (1998). Modeling dynamic recrystallization using cellular automata. *Scripta Materialia* 38:405–413.

Gourdet, S. (1997). *Etude des mécanismes de recristallisation au cours de la déformation à chaud de l'aluminium*. Ph.D. Dissertation, Ecole Nationale Supérieure des Mines, Saint-Etienne, France.

Gourdet, S., Girinon, A., and Montheillet, F. (1997). Discussion and modelling of continuous dynamic recrystallization. In Chandra, T. and Sakai, T., eds., *Thermec '97*, Wollongong (NSW), Australia. 2117–2124.

Gourdet, S., and Montheillet, F. (2000). An experimental study of the recrystallization mechanism during hot deformation of aluminium. *Materials Science and Engineering A* 283:274–288.

Gourdet, S., and Montheillet, F. (2002). Effects of dynamic grain boundary migration during the hot compression of high stacking fault energy metals. *Acta Materialia* 50:2801–2812.

Gourdet, S., and Montheillet, F. (2003). A model of continuous dynamic recrystallization. *Acta Materialia* 51:2685–2699.

Guillopé, M., and Poirier, J.-P. (1979). Dynamic recrystallization during creep of single crystalline halite; an experimental study. *Journal of Geophysical Research* 84:5557–5567.

Humphreys, F.J., and Hatherly, M. (1995). *Recrystallization and related annealing phenomena*. Oxford: Pergamon.

Hunderi, O., and Ryum, N. (1996). The influence of spatial grain size correlation on normal grain growth in one dimension. *Acta Materialia* 44:1673–1680.

Jonas, J.J. (1994). Dynamic recrystallization–Scientific curiosity or industrial tool? *Materials Science and Engineering A* 184:155–165.

Kaibyshev, R.O., and Sitdikov, O.Sh. (2000). On the role of twinning in dynamic recrystallization. *The Physics of Metals and Metallography* 89:384–390.

Kalisher, S. (1881). Über der Einfluss der Wärme auf die Molekularstruktur des Zinks. *Berichtungen der deutschen chemischen Gesellschaft* 14:2747–2753.

Kalisher, S. (1882). Über der Molekularstruktur der Metalle. *Berichtungen der deutschen chemischen Gesellschaft* 15:702–712.

Kaptsan, Y.V., Gomostyrev, Yu.N., Urtsev, V.N., Levit, V.I., and Maslennikov, V.A. (1993). Mathematical model of dynamic recrystallization. *Materials Science Forum* 113–115:341–348.

Laasraoui, A., and Jonas, J.J. (1991). Prediction of steel flow stresses at high temperatures and strain rates. *Metallurgical Transactions* A22:1545–15558.

Lücke, K., and Stüwe H.P. (1963). On the theory of grain boundary migration. In Himmel, L., ed., *Recovery and Recrystallization in Metals*. Interscience Publications. 171–209.

Luton, M.J., and Peczak, P. (1993). Monte Carlo modeling of dynamic recrystallization: recent developments. *Materials Science Forum* 113–115:67–80.

Lyttle, M.T., and Wert, J.A. (1994a). Modelling of continuous recrystallization in aluminium alloys. *Journal of Materials Science* 29:3342–3350.

Lyttle, M.T., and Wert, J.A. (1994b). Simulative modeling of continuous recrystallization of aluminum alloys. In Jonas, J.J., Bieler, T.R., and Bowman, K.J., eds., *Advances in Hot Deformation Textures and Microstructures*. Warrendale (PA):The Minerals, Metals and Materials Society. 373–383.

Mackenzie, J.K. (1958). Second paper on statistics associated with the random disorientation of cubes. *Biometrika* 45:229–240.

Maurice, C., and Humphreys, F.J. (1998). 2- and 3-D curvature driven vertex simulations of grain growth. In Weiland, H., ed., *Grain Growth in Polycrystalline Materials III*, Warrendale (PA): The Minerals, Metals and Materials Society. 81–90.

McQueen, H.J., Knustad, O., Ryum, N., and Solberg, J.K. (1985). Microstructural evolution in Al deformed to strains of 60 at 400 °C. *Scripta Metallurgica* 19:73–78.

Montheillet, F. (1999). Modeling the steady state regime of discontinuous dynamic recrystallization. In Sakai, T., and Suzuki, H.G., eds., *The Fourth International Conference on Recrystallization and Related Phenomena*. Tsukuba: The Japan Institute of Metals. 651–658.

Montheillet, F., Cohen, M., and Jonas, J.J. (1984). Axial stresses and texture development during the torsion testing of Al, Cu and α–Fe. *Acta Metallurgica* 32:2077–2089.

Montheillet, F., Thomas, J.-Ph., and Damamme, G. (2002). Distribution de la taille des grains recristallisés dynamiquement dans les matériaux métalliques. In Congrès Matériaux 2002, Tours, France. CD-ROM publication.

Oliveira, T. (2003). *Effet du niobium et du titane sur la déformation à chaud d'aciers inoxydables ferritiques stabilisés*. Ph.D. Dissertation, Ecole Nationale Supérieure des Mines, Saint-Etienne, France.

Oliveira, T., and Montheillet, F. (2002). Evolution de la microstructure et de la texture d'aciers inoxydables ferritiques stabilisés pendant la torsion à chaud. In Congrès Matériaux 2002, Tours, France. CD-ROM publication.

Peczak, P. (1995). A Monte Carlo study of influence of deformation temperature on dynamic recrystallization. *Acta Metallurgica et Materialia* 43:1279–1291.

Ponge, D., and Gottstein, G. (1998). Necklace formation during dynamic recrystallization: mechanisms and impact on flow behavior. *Acta Materialia* 46:69–80.

Rollett, A.D., Luton, M.J., and Srolovitz, D.J. (1992). Microstructural simulation of dynamic recrystallization. *Acta Metallurgica et Materialia* 40:43–55.

Rossard, C., and Blain, P. (1959). Evolution de la structure de l'acier sous l'effet de la déformation plastique à chaud. *Mémoires Scientifiques de la Revue de Métallurgie* 56:285–300.

Sakai, T., and Jonas, J.J. (1984). Dynamic recrystallization: mechanical and microstructural considerations. *Acta Metallurgica* 32:189–209.

Sandström, R., and Lagneborg, R. (1975). A model for hot working occurring by recrystallization. *Acta Metallurgica* 23:387–398.

Senkov, O.N., Jonas, J.J., and Froes, F.H. (1998). Steady-state flow controlled by the velocity of grain-boundary migration. *Materials Science and Engineering* A255:49–53.

Stüwe, H.P. (1968). Do metals recrystallize during hot working? In Tegart, W.J.McG., and Sellars, C.M., eds., *Deformation under Hot Working Conditions*, ISI Special Report 108, Iron and Steel Institute, London, 1–6.

Stüwe, H.P., and Ortner, B. (1974). Recrystallization in hot working and creep. *Metal Science* 8:161–167.

Tanaka, K., Otsuka, M., and Yamagata, H. (1999). Effect of orientation and purity on the dynamic recrystallization of aluminium single crystals with multi glide systems. *Materials Transactions JIM* 40:242–247.

Thomas, J.-Ph. (2003). Ph.D. Dissertation, Ecole Nationale Supérieure des Mines, Saint-Etienne, France. To be published.

Thomas, J.-Ph., Montheillet, F., and Dumont, Ch. (2003). Microstructural evolutions of superalloy 718 during dynamic and metadynamic recrystallizations. In Chandra, T., Torralba, J.M., and Sakai, T., eds., *THERMEC'2003*, Leganés, Madrid, Spain, *Materials Science Forum* 426–432:791–796.

Thomsen, E.G., Yang, C.T., and Bierbower J.B. (1954). In: *An Experimental Investigation of the Mechanics of Plastic Deformation of Metals*. Berkeley: University of California Press.

Thomson, W. (1887). On the division of space with minimum partitional area. *Philosophical Magazine* 24:503–514.

Viswanathan, R., and Bauer, C.L. (1973). Kinetics of grain boundary migration in copper bicrystals with [001] rotation axes. *Acta Metallurgica* 21:1099–1109.

Volkov, A.Ye, Likhachev, V.A., and Shikhobalov, L.S. (1980). Theory of grain boundaries as autonomous imperfections of a crystal. *Physics of Metals and Metallurgy* 47:1–12.

Winning, M., Gottstein, G., and Shvindlerman, L.S. (1999). Influence of external shear stresses on grain boundary migration. In Sakai, T., and Suzuki, H.G., eds., *The fourth International Conference on Recrystallization and Related Phenomena*. Tsukuba: The Japan Institute of Metals. 451–456.

Yamagata, H. (1992). Multipeak stress oscillations of five-nine-purity aluminum during a hot compression test. *Scripta Metallurgica et Materialia* 27:201–203.

Yamagata, H. (1995). Dynamic recrystallization and dynamic recovery in pure aluminum at 583 K. *Acta Metallurgica et Materialia* 43:723–729.

Zeitfracht Medien GmbH
Ferdinand-Jühlke-Straße 7
99095 Erfurt, Deutschland
produktsicherheit@kolibri360.de